Analysis for Power Quality Monitoring

Analysis for Power Quality Monitoring

Special Issue Editors

Juan-José González de la Rosa
Manuel Pérez Donsión

MDPI • Basel • Beijing • Wuhan • Barcelona • Belgrade

Special Issue Editors

Juan-José González de la Rosa
University of Cádiz
Spain

Manuel Pérez Donsión
University of Vigo
Spain

Editorial Office
MDPI
St. Alban-Anlage 66
4052 Basel, Switzerland

This is a reprint of articles from the Special Issue published online in the open access journal *Energies* (ISSN 1996-1073) from 2018 to 2019 (available at: https://www.mdpi.com/journal/energies/special_issues/power_quality_monitoring).

For citation purposes, cite each article independently as indicated on the article page online and as indicated below:

LastName, A.A.; LastName, B.B.; LastName, C.C. Article Title. *Journal Name* **Year**, *Article Number*, Page Range.

ISBN 978-3-03928-110-7 (Pbk)
ISBN 978-3-03928-111-4 (PDF)

Contents

About the Special Issue Editors

Juan-José González de la Rosa is a Full Professor at the Department of Automation Engineering, Electronics, Architecture, and Computers Networks; Electronics Area; University of Cádiz. He received an M.Sc. in Physics-Electronics in 1992, from the University of Granada, Spain, and his Ph.D. in Industrial Engineering in 1999 (summa cum laude), from the University of Cádiz, Spain. With three awards-recognitions over a six-year research period in the field of Engineering and Communication Engineering, Computation, and Electronics, by the National Assessment Commission of the Research Activity of the Spanish Ministry of Education and Science (CNEAI), he is the author of nearly 100 JCR research articles, and is an attendant of multiple conferences and international research meetings (e.g., chairman and speaker in the main track), which are the result of his participation and leadership in a number of nearly 20 research projects and 40 research contracts with international companies. During his career, he has also visited and attended universities (invited professor), research institutes, and companies in France, Switzerland, Italy, and Spain. He is currently the assistant director for Strategic Planning and Postgraduate Studies in the Higher Polytechnic Engineering School of Algeciras, where he has served as the academic secretary and is responsible for students' training in companies. He is the main researcher and the founder of the Research Group in the Computational Instrumentation and Industrial Electronics (PAIDI-TIC-168), one of the most important in its disciplines among Andalusian, and is responsible for the creation of the Research Institute of Energy Engineering and Sustainability at the University of Cádiz, and the for the accreditation of the doctorate program with the same name. His research line is summarized as "Computational Intelligence and Statistical Signal Processing for Smart Measurement Systems". More precisely, his research interests include instrumentation for Industry 4.0, power quality (PQ), statistical signal processing, measurement methods, and higher-order statistics (HOS). Regarding membership, he belongs to the following associations (among others): IEEE Senior Member (Instrumentation and Measurement Society, TC-39-Measurement in Power Systems, and Technical Society of Smart Grids and Spanish Chapter of Sensors); Latin-American Group in Acoustic Emission (GLEA); International Frequency Sensor Association (IFSA); European Signal Processing Association (EURASIP); Spanish Professional College of Physicists (COFIS), in which he obtained the Excellency in Physics Technologies in the year 2010; and the Spanish Centre of Authors and Editors (CEDRO). He belongs to the professional network of Professional Excellency in Physics. He is an assessor of the Spanish Ministry of Education and Professional Formation (MINECO) within the National Agency of Education and Prospective (ANEP); the Spanish-American Program of Science and Technology for Development (CYTED); and Austrian Government, Buenos Aires University, and Argentinian Government. Also, he belongs to the staff of experts at the European Quality Assistance (EQA), DNV-GL, and SGS companies, for which he has assessed myriads of projects of companies and different organizations for certification purposes. Among other activities, acting as guest editor, he has run Special Issues (e.g., for the MDPI Energies journal, in which he also acts as the topic editor). He belongs to the reviewers' boards of numerous journals from Elsevier, IEEE, Springer, etc., and is a books reviewer for ELSEVIER, MARCOMBO, PARANINFO, IGI GLOBAL, IET, and SPRINGER. Also, he has edited and written books and book chapters for research and educational purposes. Additionally, he has produced patents, other international properties, and worked for international companies with the goal of transferring technology.

Manuel Pérez Donsión is a full professor at the Department of Electrical Engineering, University of Vigo. He received Ph.D. in Industrial Engineering from the same University. Donsión is an important asset for research and scientific work, which is reflected in his numerous publications in technical journals. He has also participated in a total of 18 national and international projects. Throughout his deep career, he has completed more than 30 engineering projects, and led the construction management of many of them. He has conducted five doctoral theses, as well as 70 final degree projects. Throughout his career, and with the constant desire to spread knowledge, Pérez Donsión participated as a speaker and/or as a guest speaker in a total of 37 conferences nationwide and 86 internationally. The transmission of knowledge and contact with other colleagues of the Spanish University made him conceive the idea of holding congresses in the field of Electrical Engineering, as the saga of the International conferences on Renewable Energies and Power Quality (ICREPQ), where his been intensely focused. Regarding his memberships and recognitions, he was a member of the Governing Council of the Technological Institute of Galicia (Spain), and he is a member of the Steering Committee of the International Conference on Electrical Machines (ICEM). He is chairman of the Steering Committee of the Spanish–Portuguese Congress of Electrical Engineering, and a member of the Spanish Subcommittee on Electromagnetic Compatibility. He is the president of the Spanish Association for the Development of Electrical Engineering (AEDIE); the European Association for the Development of Electrical Engineering (EADEE); and the European Association for the Development of Renewable Energy, Environment, and Power Quality (EA4EPQ).

Preface to "Analysis for Power Quality Monitoring"

Power quality (PQ) analysis is evolving continuously, mainly as a result of the incessant growth and development of the Smart Grid (SG) and the incipient Industry 4.0, which demands the quick and accurate tracking of the electrical power dynamics. Much effort is put into two main issues. First, numerous distributed energy resources and loads provoke highly fluctuating demands that alter the ideal power delivery conditions, introducing, at the same time, new types of electrical disturbances. For this reason, permanent monitoring is needed in order to track this a priori unpredictable behavior. Second, and consequently, the huge amount of data (Big Data) generated by the measurement equipment during a measurement campaign are usually difficult to manage because of different causes, such as complex structures and communication protocols that hinder accessibility to storage units, and the limited possibilities of monitoring equipment, based on regulations that do not reflect the current network operation.

This book gathers new advances in techniques and procedures to describe, measure, and visualize the behavior of the electrical supply, from physical instruments to statistical signal processing (SSP) techniques, and new indexes for PQ that try to go beyond traditional norms and standards. The authors are recognized experts in the field, committed to the following main goal: to provide new instrumental and analytical tools to help mitigate the serious consequences of a bad PQ in our digitized society, enhancing energy efficiency for more sustainable development.

Juan-José González de la Rosa, Manuel Pérez Donsión
Special Issue Editors

Editorial

Special Issue "Analysis for Power Quality Monitoring"

Juan-José González de-la-Rosa [1],* and Manuel Pérez-Donsión [2]

[1] Research Group PAIDI-TIC-168 in Computational Instrumentation and Industrial Electronics (ICEI), Higher Polytechnic School, University of Cádiz, Ramón Puyol Av., E-11202 Algeciras, Spain
[2] Department of Electrical Engineering, ETSII, Campus de Lagoas-Marcosende, University of Vigo, E-36310 Vigo, Spain; donsion@uvigo.es
* Correspondence: juanjose.delarosa@uca.es; Tel.: +34-956028069

Received: 20 January 2020; Accepted: 21 January 2020; Published: 21 January 2020

Abstract: We are immersed in the so-called digital energy network, continuously introducing new technological advances for a better way of life. As a consequence, numerous emerging words are relevant to this point: Internet of Things (IoT), big data, smart cities, smart grid, industry 4.0, etc. To achieve this formidable goal, systems should work more efficiently, a fact that inevitably leads to power quality (PQ) assurance. Apart from its economic losses, a bad PQ implies serious risks for machines and, consequently, for people. Many researchers are endeavouring to develop new analysis techniques, instruments, measurement methods, and new indices and norms that match and fulfil the requirements regarding the current operation of the electrical network. This book, and its associated Special Issue, offer a compilation of some of the recent advances in this field. The chapters range from computing to technological implementation, going through event detection strategies and new indices and measurement methods that contribute significantly to the advance of PQ analysis and regulation. Experiments have been developed within the frameworks of research units and projects and deal with real data from industry practice and public buildings. Human beings have an unavoidable commitment to sustainability, which implies adapting PQ monitoring techniques to our dynamic world, defining a digital and smart concept of quality for electricity.

Keywords: power quality (PQ); PQ indices and thresholds; reliability; sensors and instruments for PQ; big data; machine learning; soft computing; statistical signal processing; data scalability; data compression

---

1. Introduction

Power quality (PQ) consists of a group of electrical limits as defined by several norms and standards thought to allow electrical equipment to operate as designed without a significant loss of performance or life expectancy; this definition implies the constant and stable supply of electricity supply through the electrical network. To achieve this objective, it is necessary to permanently monitor the power conditions within the grid. However, there are a large number of parameters involved in this goal, and many of these still have not been completely defined. This has forced the manufacturers of electronic instruments to develop their own methodologies, which has resulted in incomparable values and rules between instruments such that, despite the fact that they have been conceived for the same purpose, the implemented measurement methods are different. This point is where IEC 6100-4-30 comes in, as it standardizes the measurement methodologies and creates the ability to make a direct comparison of the results of different analyzers. However, there is still a need for new parameters that reflect the current situation in the smart grid (SG), even when a noncomparable measurement is required or when the customer is interested in detecting some specific types of electrical disturbance. Furthermore, current measurement campaigns are demanding new Class S methods and their associated instruments. Despite the fact that their accuracy levels are less demanding, this emerging family of power analysers is conceived to develop a new approach to specific campaigns with "à la carte" measures, with the goal of performing specific energy efficiency analysis.

Indeed, the need goes even further as the increasing complexity of the current electrical network makes it necessary to introduce new methods, parameters, and measurement indices that allow for a characterization that is not only more reliable, but also more energy efficient, and that includes aspects related not only to producers, but also to consumers. Current electrical networks are immersed in a transformation process in order to adapt to new technologies in the framework of a future SG. This conception of the so-called "smart" grid is mainly based on the reliable and permanent capacity of the systems to provide energy behavior information in real time.

Thus, instrumentation solutions tackle big data issues as a consequence of permanent PQ monitoring. For this reason, several fields and disciplines are converging in the analysis for PQ monitoring (e.g., machine learning, data compression, advanced signal processing, communications and network connectivity). Of special interest, Internet of Things (IoT) is a trending topic with serious technical, social, and economic implications. Industrial systems, utility components, and sensors are being combined with Internet connectivity and powerful data analytic capabilities that promise to transform the way we work, live, and play.

Looking again at the norms, the UNE-EN 50160 standard approved by CENELEC in 1994 and entitled "characteristics of the voltage supplied by the general distribution networks" defines the main characteristics that the voltage supplied by a general low and medium voltage distribution network must have, under normal operating conditions, at the point of delivery. However, this standard is far from being updated, in accordance with the reality of the SG. Numerous initiatives are being carried out in this direction, as promoted by organizations such as CIGRE and congresses such as CIRED. The works developed within this organization have provided an overview of PQ monitoring that is supported by the results of the surveys derived from the industry practice. With all of this, CIGRE defines the six major objectives of PQ monitoring, identified (not in order of importance) as compliance verification, performance analysis/benchmarking, site characterization, troubleshooting, advanced applications and studies, and active PQ management. CIGRE also provides a set of recommendations and guidelines for efficient and cost-effective PQ monitoring in existing and future electric power systems.

As Guest Editors of this Special Issue, we were able to open the contributions to several technical areas that address PQ, being well concerned about the need for converging synergies with the goal of developing new instruments that assess the quality of the electrical supply.

Consequently, the goal of this Special Issue is to offer a multidisciplinary approach to PQ, bringing together fields and disciplines that converge in techniques and procedures for enhancing PQ. Energy policy will soon be reflected in PQ more directly, as the need to manage energy systems more efficiently is triggering the development of new standards and norms that will match the current network infrastructure.

All in all, it is of high interest to classify research literature on PQ into the following flourishing topics or branches that are directly and transversely addressed in multidisciplinary work teams:

- Statistical signal processing (SSP) and intelligent methods for PQ analysis

    - Statistical planning and characterization in PQ campaigns;
    - Higher-order statistics (HOS) for PQ characterization;
    - Intelligent methods for PQ analysis;
    - New estimators for PQ monitoring.

- Power quality and reliability characterization

    - PQ indices and thresholds;
    - Customized PQ for utilities, customers and specific geographical areas;
    - Industry research benchmark reports on PQ metrics;

- New types of electrical perturbations.
- Management of PQ big data in the smart grid
  - Spatial and temporal compression of measurements;
  - Spatial and temporal scalability of measurements;
  - Modelling and forecasting of PQ time-series;
  - Graphical visualization of PQ: plots, diagrams, and trajectories.
- PQ monitoring systems: architectures and communications
  - New tendencies in smart instruments for PQ;
  - Uncertainty in PQ instruments;
  - Sensors networks for PQ monitoring;
  - Nonintrusive load monitoring;
  - PQ for renewable energy systems;
  - Low-cost measurement equipment.
- PQ losses and mitigation assessment
  - Energy efficiency and PQ;
  - Economic impact and losses due to poor PQ;
  - PQ maintenance strategies in networks;
  - PQ mitigation.
- New PQ monitoring norms and standards
  - PQ indices;
  - PQ norms;
  - PQ standardized measurements for phasor measurement units (PMUs);
  - PQ monitoring in the industry 4.0.

In the next section, a brief review of the papers published, which transversally address the above-classified topics, is provided. They constitute the very positive response to this Special Issue and gather a broad range of thematic areas issued by recognized researchers. The works range from emerging signal processing techniques applied to PQ monitoring to low-cost instruments for specific PQ events detection, going through the postulation of prospective indices and statistical indicators within measurement campaigns. They all have a common axis: a new or evolved conception of power quality for a more efficient use of energies.

2. A Short Review of the Contributions in this Issue

Phasor measurement units (PMUs) constitute examples of how manufactures need to converge for a better measurement interpretation. The paper by P. Castelo et al., "PMU's Behaviour with Flicker-Generating Voltage Fluctuations: An Experimental Analysis" [1], presents and discusses the results of experimental tests carried out on commercial PMUs in the presence of voltage fluctuations that give rise to the flicker. The performed characterization of commercial devices remarks on how different possible interpretations could be given to the PMU outputs considering the same signal and depending on the quantity of interest, which is a key issue of instrument development. It reports that the first step in solving the possible misinterpretations of the measurement results is to clearly define the objective of the measurement (e.g., remove all nonfundamental frequency components or tracking the dynamics of the signal being tested), which in turn depends on the requirements of the specific measurement context. Only in this sense, new and more suitable test conditions and performance evaluation criteria could be defined for PMUs targeted to distribution networks. This paper can be allocated within the topic "PQ monitoring systems: architectures and communications: new tendencies in smart instruments for PQ" and in "new PQ monitoring norms and standards: PQ standardized measurements for PMUs", as it addresses the need for homologate measurements in the modern SG.

The second work, "Power Quality in DC Distribution Networks", by J. Barros et al. [2], goes into the emerging topic of PQ in low-voltage DC networks in low-scale network design. The specific types of disturbances dealt with include voltage supply interruptions, voltage ripple, and rapid voltage changes. Different types of sources were tested using measures from different indices over different waveforms. Their conclusions suggest that each type of disturbance has its associated preferable index with which performance is optimum. The paper falls within the topic "power quality and reliability characterization: PQ indices and thresholds" as well as inside the topic "New PQ monitoring norms and standards: PQ indices and PQ monitoring in the industry 4.0."

The work by Sierra-Fernández et al. "Application of Spectral Kurtosis to Characterize Amplitude Variability in Power Systems' Harmonics" [3] shows how higher-order statistics (HOS) in the frequency domain enhance the detection of electrical disturbances, more precisely, low-level harmonics not detected with the traditional power spectrum. An estimator of the spectral kurtosis (SK) was used to assess the amplitude trends of each spectral component. The results confirmed that SK is capable of tracking constant-amplitude harmonics, performing high-resolution frequency analysis for higher-order harmonics, even with low-level amplitudes. Two signals were chosen to validate the method with adequate results: an electric current from an arc furnace and a voltage signal from the power grid of a public building. The paper falls within the topic "statistical signal processing (SSP) and intelligent methods for PQ analysis: HOS for PQ characterization" as well as inside "power quality and reliability characterization: PQ indices and thresholds".

Inside the same set of former topics, the fourth paper, "Reliability Monitoring Based on Higher-Order Statistics: A Scalable Proposal for the Smart Grid" [4] by O. Florencias et al., proposed a new index for both PQ and reliability assessment (depending on the considered analysis window's length) thought to be used in measurement campaigns that require deep statistical characterization. The index consists of a summation of three differential terms: variance, skewness, and kurtosis, each with respect to the ideal value of the statistic. Skewness and kurtosis account for the waveform features, which are really interesting elements to introduce in new standards and norms. Furthermore, 2D graphs were used as complementary tools to track the energy behavior. A long-term monitoring analysis was shown over a power signal in a public building, and the conclusions show that the power supply adopts different patterns in the time domain and in the 2D graphs, depending on the day period. Additionally, the 2D graphs compressed information in the time domain and can also be used for compression issues in the space.

The contribution by Flores-Arias et al., "A Memory-Efficient True-RMS Estimator in a Limited-Resources Hardware" [5], presents an RMS voltage estimator that eludes the inherent uncertainty of complex arithmetic operations related to the discretized RMS algorithm on an ATmega328p microcontroller. Its capability as a sag/swell detector was exhibited on substation signals. The proposal constitutes a step to implement PQ algorithms in low-cost platforms and may be implemented in simple FPGA systems. The results were compared with a TRMS voltmeter (Fluke ScopeMeter^{TM} series 120) in order to check its accuracy. This work is an example of how signal processing functions can be integrated into simple physical platforms in order to produce cheap PQ indicators. Thus, it falls inside the topic "PQ monitoring systems: architectures and communications: low-cost measurement equipment".

M. Ptacek et al. conducted an "Analysis of Dense-Mesh Distribution Network Operation Using Long-Term Monitoring Data" [6] from a municipal distribution network (E.ON) based on long-term data from PQ monitors. The paper showed the lack of usability of data recorded by these instruments that come from transformers and exhibited new results from processing these big data. One of the main results is that it is not necessary to assess the voltage magnitudes while using the measuring over the unified phase data. This constitutes another method for efficiently evaluating large amounts of PQ measurement data within the topic "management of PQ big data in the smart grid", addressing all the subtopics within this issue.

The seventh work, "An Extended Kalman Filter Approach for Accurate Instantaneous Dynamic Phasor Estimation" [7] was conducted by De Apráiz et al. The literature showed that variations of Kalman filters have been proposed in phasor estimation to improve the dynamics of emerging measurement systems conceived to be integrated in the SG. This paper proposed a nonlinear adaptive extended Kalman filter (EKF) to improve the adaptability to the dynamic requirements of power system signals, in the sense that, thanks to the use of the Kalman filters' residuals, it manages to track online the fundamental component, harmonics, and the subsynchronous interharmonic phasors, along with the detection of transient conditions in the waveform under test. The work is to be placed inside the category of "PQ monitoring systems: architectures and communications: new tendencies in smart instruments for PQ and also in "statistical signal processing (SSP) and intelligent methods for PQ analysis: new estimators for PQ monitoring".

The following work [8] by Cifredo-Chacón et al. compares the performance of two different autonomous units that implement the same measurement algorithm: an estimator of the spectral kurtosis (SK). In the "Implementation of Processing Functions for Autonomous Power Quality Measurement Equipment: A Performance Evaluation of CPU and FPGA-Based Embedded System", the authors managed to implement an estimator of the fourth-order spectrum in an FPGA-based system and showed that FPGAs improved the processing capability of the best processor using an operating frequency 33 times lower. One of the main a priori differences, put into practice, between FPGA and processor-based implementations is that the processing time is constant for FPGAs, but not for processor-based implementations. This work showed interesting results related to different performance parameters (e.g., average time per iteration in the main algorithm for the SK). Consequently, the work is considered within the topic "PQ monitoring systems: architectures and communications: new tendencies in smart instruments for PQ and low-cost measurement equipment". Additionally, the paper addresses "statistical signal processing (SSP) and intelligent methods for PQ analysis: higher-order statistics (HOS) for PQ characterization".

The study [9] by Yue Shen et al., "Power Quality Disturbance Monitoring and Classification Based on Improved PCA and Convolution Neural Network for Wind-Grid Distribution Systems", falls within the set "statistical signal processing (SSP) and intelligent methods for PQ analysis". The authors made a deep revision on signal processing and the multivariate techniques applied to feature extraction in PQ contexts. Figures of the intelligent systems architecture are very illustrative and help the reader to understand the operation of the nucleus of the system. The work compares specific PCA-based algorithms to existing ones, assessing the overall performance over a set of different electrical disturbances.

The work [10] "Power Quality Disturbances Assessment during Unintentional Islanding Scenarios" by A. Serrano et al. presents a novel voltage sag topology that occurs during an unintentional islanding operation. This is precisely the value of the paper, because it analyzed and documented a particular and very common case in the industry practice: the effects of large induction motors in islanding cases. The authors proposed an analytical expression for this new type of sag, which was confirmed by simulations and terrain measurements. The work belongs to the topic "power quality and reliability characterization: new types of electrical perturbations".

Finally, the paper [11] by Guerrero-Rodríguez et al., entitled "An Embedded Sensor Node for the Surveillance of Power Quality", investigated a small and compact PQ detector using a low-cost microcontroller and a very simple conditioning circuit and analyzed different methods to implement various surveillance algorithms. The paper belongs to the group "PQ monitoring systems: architectures and communications: low-cost measurement equipment".

Author Contributions: The authors contributed equally to this work. All authors have read and agreed to the published version of the manuscript.

Acknowledgments: The editors Juan José González de la Rosa and Manuel Pérez Donsión would like to express their gratitude to the MDPI Publisher for the invitation to act as Guest Editors of this Special Issue. Our special thanks go to the editorial staff of *Energies* for the fruitful cooperation based on a previous yearly committed engagement. The final acknowledgement should be perhaps the first and goes to the indebted work of the worldwide recognized reviewers, whose positive critiques have enhanced the accepted papers, polishing every detail, and helped researchers to improve their careers, making the reviews profitable.

Conflicts of Interest: The authors declare no conflicts of interest.

References

1. Castelo, P.; Muscas, C.; Pegoraro, P.A.; Sulis, S. PMU's Behaviour with Flicker-Generating Voltage Fluctuations: A Experimental Analysis. *Energies* **2019**, *12*, 3355. [CrossRef]
2. Barros, J.; De Apráiz, M.; Diego, R.I. Power Quality in DC Distribution Networks. *Energies* **2019**, *12*, 848. [CrossRef]
3. Sierra-Fernández, J.-M.; Rönnberg, S.; González de la Rosa, J.-J.; Bollen, M.H.J.; Palomares-Salas, J.-C. Application of Spectral Kurtosis to Characterize Amplitude Variability in Power Systems' Harmonics. *Energies* **2019**, *12*, 194. [CrossRef]
4. Florencias-Oliveros, O.; González-de-la-Rosa, J.-J.; Agüera-Pérez, A.; Palomares-Salas, J.-C. Reliability Monitoring Based on Higher-Order Statistics: A Scalable Proposal for the Smart Grid. *Energies* **2019**, *12*, 55. [CrossRef]
5. Flores-Arias, J.-M.; Ortiz-López, M.; Quiles Latorre, F.J.; Bellido-Outeiriño, F.J.; Moreno-Muñoz, A. A Memory-Efficient True-RMS Estimator in a Limited-Resources Hardware. *Energies* **2019**, *12*, 1699. [CrossRef]
6. Ptacek, M.; Vycital, V.; Toman, P.; Vaculik, J. Analysis of Dense-Mesh Distribution Network Operation Using Long-Term Monitoring Data. *Energies* **2019**, *12*, 4342. [CrossRef]
7. De Apráiz, M.; Diego, R.I.; Barros, J. An Extended Kalman Filter Approach for Accurate Instantaneous Dynamic Phasor Estimation. *Energies* **2018**, *11*, 2918. [CrossRef]
8. Cifredo-Chacón, M.-Á.; Perez-Peña, F.; Quirós-Olozábal, A.; González-de-la-Rosa, J.-J. Implementation of Processing Functions for Autonomous Power Quality Measurement Equipment: A Performance Evaluation of CPU and FPGA-Based Embedded System. *Energies* **2019**, *12*, 914. [CrossRef]
9. Shen, Y.; Abubakar, M.; Liu, H.; Hussain, F. Power Quality Disturbance Monitoring and Classification Based on Improved PCA and Convolution Neural Network for Wind-Grid Distribution Systems. *Energies* **2019**, *12*, 1280. [CrossRef]
10. Serrano-Fontova, A.; Casals Torrens, P.; Bosch, R. Power Quality Disturbances Assessment during Unintentional Islanding Scenarios. A Contribution to Voltage Sag Studies. *Energies* **2019**, *12*, 3198. [CrossRef]
11. Guerrero-Rodríguez, J.-M.; Cobos-Sánchez, C.; González-de-la-Rosa, J.-J.; Sales-Lérida, D. An Embedded Sensor Node for the Surveillance of Power Quality. *Energies* **2019**, *12*, 1561. [CrossRef]

Article

PMU's Behavior with Flicker-Generating Voltage Fluctuations: An Experimental Analysis

Paolo Castello, Carlo Muscas *, Paolo Attilio Pegoraro and Sara Sulis

Department of Electrical and Electronic Engineering, University of Cagliari, 09123 Cagliari, Italy
* Correspondence: carlo.muscas@unica.it

Received: 9 July 2019; Accepted: 25 August 2019; Published: 30 August 2019

Abstract: Phasor measurement units (PMUs), which are the key components of a synchrophasor-based wide area monitoring system (WAMS), were historically conceived for transmission networks. The current trend to extend the benefits of the synchrophasor technology to distribution networks requires the PMU to also provide trustworthy information in the presence of signals that can occur in a typical distribution grid, including the presence of severe power quality (PQ) issues. In this framework, this paper experimentally investigates the performance of PMUs in the presence of one of the most important PQ phenomena, namely the presence of voltage fluctuations that generate the disturbance commonly known as flicker. The experimental tests are based on an ad-hoc high-accuracy measurement setup, where the devices under test are considered as "black boxes" to be characterized in the presence of the relevant signals. Two simple indices are introduced for the comparison among the different tested PMUs. The results of the investigation highlight possible critical situations in the interpretation of the measured values and provide a support for both the design of a new generation of PMUs and the possible development of an updated synchrophasor standard targeted to distribution systems.

Keywords: power quality; phasor measurement units; voltage fluctuations; flicker; modulation; power distribution systems; smart grids

---

1. Introduction

Whatever new management/business models can be envisaged for modern power systems, they are based on the availability of suitable information and, consequently, new measurement solutions are required for their practical implementation. In particular, the increasing complexity of the electric distribution grids, with, for example, the growing penetration of distributed generation plants fed by renewable energy sources, as well as the increasing relevance of PQ disturbances, calls for critical changes in network monitoring. The smart grid (SG) paradigm, in its several different declinations, emphasizes the power system as a cyberphysical system, where information quality is critically dependent on coordination among elements composing a distributed system. In this context, the primary involved factors are accuracy, cost-effectiveness, synchronization, communication quality, reliability, and timeliness. The transition toward a smarter network management approach thus implies the need for a new and better performing measurement infrastructure.

To this purpose, the possibility of exploiting at a distribution level the benefits of high-performance measurement devices and systems, currently deployed in the transmission grids, and can be explored. This refers in particular to the synchrophasor technology, which is the key element of modern WAMSs. In WAMSs, synchronized phasors, frequency, and rate of change of frequency (ROCOF) are measured by the PMUs which are sent to the corresponding phasor data concentrator (PDC), where these data are collected, stored, and correlated, using the absolute time reference associated with every measured value [1–3]. The main features of PMUs and WAMSs allow them to outperform the classical architectures based on supervisory control and data acquisition (SCADA) in terms of accuracy, synchronization, reporting rate, etc. For this reason, the number of PMUs and PDCs installed is growing quickly in different countries.

Even though the overall performance of a WAMS depends on the behavior of all its components, including the communication infrastructure, the first uncertainty source to be considered arises from its "sensing" element, namely the PMU. Thus, it is of utmost importance to define the metrological performance expected from PMUs under different possible operating conditions.

This goal is currently accomplished mainly by applying the latest standard IEEE/IEC 60255-118-1-2018 [1], as well as the guides [2,3] and standards [4–6], representing reference documents for synchronization, calibration, testing, and installation of PMUs. The test conditions and the relevant limits defining the performance classes in these standards clearly reflect the fact that PMUs were originally conceived for the monitoring requirements of transmission systems [7–10].

Several projects have been already funded worldwide on the possible applications of synchrophasor systems to distribution grids [11]. To assess the practical feasibility of such applications, the fact must be stressed that the characteristics of the electrical quantities in these systems may differ substantially from those considered as reference test conditions in [1] and, consequently, the metrological performance of the PMUs may substantially differ as well. Therefore, to effectively extend the synchrophasor technology to the distribution level, it is necessary to test the PMUs with more realistic signals. In particular, several power quality (PQ) disturbances may affect the voltage in distribution networks, see for example [12–14]. They are defined, for example, in the standard describing the main characteristics of the AC voltage in public low, medium, and high voltage networks, under normal operating conditions [15]. It is possible to mention several papers discussing PMU characterization results, obtained considering all or most of the standardized tests (see for instance [16–20]). On the contrary, PMU measurements obtained considering signals that better represent the realistic conditions of distribution systems, and thus are affected by possible PQ issues, are still not appropriately analyzed. Among them, voltage fluctuations, causing the PQ disturbance known as flicker, are assuming an ever-increasing role [21,22]. This is witnessed, among others, by the recent document released by the CIGRE WG C4.111 [23].

Flicker-generating voltage fluctuations can be challenging for PMU characterization, since the frequency ranges considered for the quantification of these signals are spread across quite a wide frequency band. Thus, depending on its specific frequency, a fluctuation can be considered as either a dynamic signal of interest, which PMUs are expected to follow appropriately, or a disturbance to be rejected.

By highlighting the lack of a clear interpretation, in the international standards on synchrophasors of the PMUs behavior during the voltage fluctuations outside the modulation range, this work aims at emphasizing the possible different behaviors of current commercial PMUs in the presence of these signals and, thus, the risk of an incorrect/ambiguous analysis of the electrical phenomena under investigation. In particular, the relationship of the PMU's behavior with different reporting rates and maximum modulation bandwidth will be discussed.

Unfortunately, most commercial PMUs do not provide clear and comprehensive information on the implemented algorithm, making it necessary, for the user, to have a proper characterization process tailored for the unconventional conditions of interest. A first study about the possible misinterpretations of the data measured by a PMU in the presence of voltage fluctuations was presented in [24]. Starting from the outcomes of [24], this paper presents a systematic and detailed study of the performance of commercial PMUs in the presence of different types of voltage fluctuations. To this purpose, the methodological approach has been changed with respect to [24], moving from a "controlled" situation, where the algorithms used by the measurement devices are known, to a more realistic and practical condition, where all the measurement devices are "black boxes" to be experimentally characterized. Furthermore, the analysis has been extended also by considering, besides the synchrophasors, the frequency and ROCOF, and by introducing and discussing two simple indices, for the sake of a more quantitative and meaningful comparison among the different tested PMUs.

The study is based on an ad-hoc experimental setup and an appropriately designed test suite following the technique recently suggested in the Annex I of [1] to determine the actual measurement bandwidth of the PMUs. The characterization tests have been inspired by [15] and by the latest release of the international standard about flickermeters [25]. The dependence of the PMU behavior on the specific configuration of the device (reporting rate, parameters of the measurement algorithm, etc.) is discussed. The potential practical issues, or even the conceptual incongruences, that can emerge in the interpretations of PMU measurement results obtained in these operating conditions are highlighted.

The structure of the paper is organized as follows: Section 2 discusses the relationship between the models of voltage fluctuations and the modulation and out-of-band signals described in the PMU standards; Section 3 presents the architecture of the test setup and the evaluation process considering two possible interpretation of the results; Section 4 presents the results obtained from three different commercial PMUs, focusing on both total vector error (TVE) performance and frequency measurement results, and Section 5 highlights the conclusions.

2. Voltage Fluctuations and PMU Requirements

2.1. Models of Voltage Fluctuations

In the international electrotechnical vocabulary, the phenomenon known as flicker is defined as "impression of unsteadiness of visual sensation induced by a light stimulus whose luminance or spectral distribution fluctuates with time" and is generated by fluctuations of the voltage root mean square (rms), caused by the varying operating conditions of industrial loads and/or generators.

According to [15,25], the flicker severity, i.e., the intensity of flicker annoyance, is evaluated on both a short term ten-minute period and a long term two-hour interval, estimating the indices "short-term flicker", P_{st}, and "long-term flicker", P_{lt}, respectively, both expressed *per unit* (see also [26] for further details on P_{st}). The standard [25] defines the parameters of the reference signals to test the flickermeter.

Nevertheless, the report [23] focused on the effects that voltage fluctuations may have on the behavior of non-lighting equipment, causing possible malfunctions, heating effects, loss of life, and even equipment failure. Thus, this study opened the door to the definition of compatibility level specifications not based solely on lighting products.

Voltage fluctuations can be modeled using an amplitude modulated signal. As is well known, modulating a sinusoidal signal having fundamental frequency f_0 and amplitude A_0 by a sinusoidal signal with modulating frequency f_m and amplitude A_m gives rise to two additional spectral components ($f_0 \pm f_m$) symmetrically placed around f_0 (Figure 1).

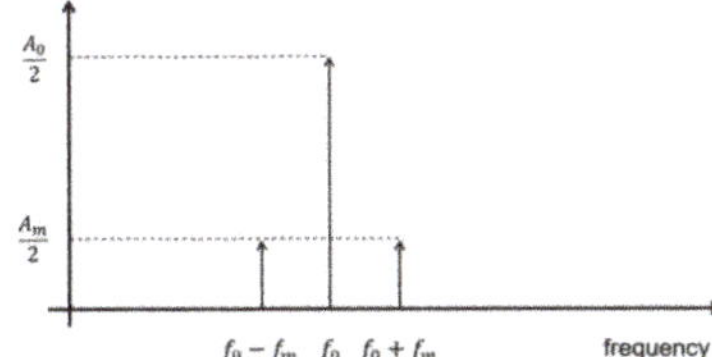

Figure 1. Positive-side amplitude spectrum of a sinusoidal signal amplitude-modulated by a sinusoidal signal.

According to [25], the frequency range of the modulation signal spans from 0.5 Hz to 33.33 Hz. This applies to 50 Hz systems. Considering 60 Hz systems, the phenomenon is modeled with a modulation frequency that ranges from 0.5 Hz to 40 Hz.

In a dual manner, signal modulation can be caused by the presence of interharmonics located around the fundamental frequency that behave as the two sideband components of the above-described phenomenon [27,28].

2.2. Modulations and Out-of-Band Signals in Standard PMU Testing

The synchrophasor standard [1] defines two classes of performance: P and M. The P class is suitable for applications that need speed and low latency, while the M class is suitable for applications that require a better rejection of disturbances. An entire test suite is defined for each class and performance specifications are given for each test in terms of TVE, frequency error (FE), and ROCOF error (RFE). The TVE represents the synchrophasor error and is defined as:

$$\text{TVE} \triangleq \frac{\left|\hat{p} - p_{ref}\right|}{\left|p_{ref}\right|} \tag{1}$$

that is the relative magnitude of the vector error (difference between the measured $\hat{p}$ and reference p_{ref} synchrophasors) with respect to the magnitude of the reference phasor. TVE is typically expressed as a percent value. As far as FE and RFE are concerned, their absolute values are used as indices. The standard defines limits for these three indices together, with additional limits for latency and step-response parameters.

Focusing on the PMU test conditions reported in the synchrophasor standard, two tests can be considered as related to voltage fluctuations, where the quantities of influence depend on the class of performance:

- The amplitude modulation test, in which the signal at nominal frequency has a time-varying cosinusoidal amplitude. This test signal is introduced to be representative of a dynamic condition that the PMU must be able to measure. Thus, the synchrophasor magnitudes estimated in this condition are expected to follow the signal amplitude evolution. According to [1], to guarantee the interoperability of PMUs from different manufacturers, the device performance must remain within the given TVE, FE, and RFE limits when the modulation frequency is up to 2 Hz, in the case of a p-class device, and 5 Hz for M-class PMUs.
- The out-of-band (OOB) interference test, in which the signal is affected by a single interharmonic component. Interharmonics are considered as disturbances and M class PMUs (the only PMUs for which OOB test specifications are given) must properly filter them out.

Besides these requirements, the guide [4] and the standard [1] define for the PMU an *in band* range (also indicated as passband in Table 3 of [1], which includes the modulation frequency range) and a stopband region (including the OOB interference frequencies). In a simplified way, the PMU passband range is determined based on both the nominal frequency f_0 and the reporting rate F_s as follows:

$$B_{PMU} \triangleq \left|f - f_0\right| < \frac{F_s}{2} \tag{2}$$

indicating the Nyquist bandwidth of the PMU with respect to its reporting rate (thus giving an idea of the maximum speed of dynamic phenomena the PMU can follow without aliasing). The complementary region of the spectrum is thus considered as a stopband region that includes disturbances to be rejected.

According to [1], synchronized phasor measurements should be evaluated with a reporting rate submultiple or multiple of f_0 (e.g., $F_s = 10, 25, 50$, and 100 frames/s, fps in the following, for $f_0 = 50$ Hz). PMUs can be also configured with a reporting rate lower than 10 fps, but, in this case, they are not subject to the dynamic requirements. For the special case, where $F_s = 100$ fps, all the frequencies in the range from 0 to $2 f_0$ are considered *in band*. In this case, according to [1], the OOB interferences are considered in the frequency range from second to third harmonic.

A typical reporting rate used in transmission grids is half the system frequency [29], which is a significant improvement compared to traditional monitoring technology based on SCADA. It is worth highlighting that the F_s value should be chosen according to the type of electrical phenomenon that is of interest, but it is also influenced by the constraints imposed by the monitoring and management architecture.

As an example, Figure 2 shows the different frequency bands for the case of an M-class PMU configured with $F_s = 50$ fps. In particular, the ranges represented with gray stripes in Figure 2, included in the *in band* range, highlight the large transition regions that are not involved in the tests considered in the current version of the synchrophasor standard [1], since no test signals present frequency components in these ranges.

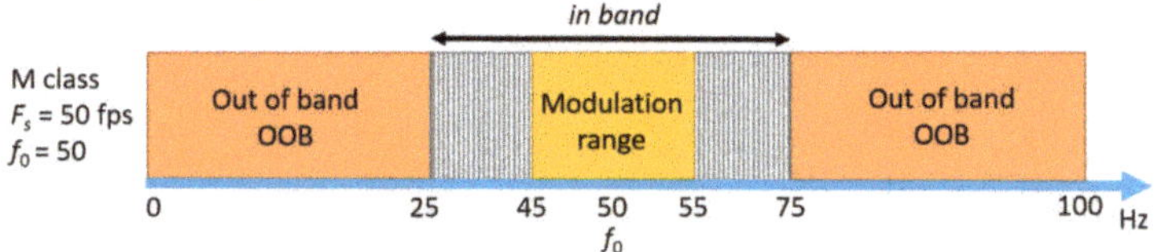

Figure 2. Out-of-band region, *in band* range, and modulation range for M Class with reporting rate 50 fps.

2.3. Voltage Fluctuations and PMU Specifications

Different situations may arise when voltage fluctuations are present in the PMU input signal, depending on the frequency of the phenomenon. If the fluctuation frequency exceeds the PMU Nyquist frequency, $F_s/2$ (see [1,30] for details), the two spectral components f_m shown in Figure 1 fall within the PMU stopband (OOB regions in Figure 2) and are thus considered as disturbances.

On the other hand, fluctuations up to the modulation frequency limit defined in [1] should be considered as amplitude modulations, i.e., dynamic phenomena that a PMU is requested to follow carefully.

Between these two ranges, there are frequency ranges (grey stripes in Figure 2) that are within the *in band* region defined in [1,3], but are not explicitly considered by any specific tests in either of the two ways. Therefore, PMU measurements may lead to different interpretations of flicker-generating voltage fluctuations, depending on the considered frequency ranges or on the PMU implementations and configurations, thus giving rise to possible debatable situations [24].

As already recalled, standard [1] suggests testing the PMU with amplitude and phase sinusoidal modulated input signals and sweeping the modulation frequency in a range whose upper limit varies from 2 Hz to 5 Hz, according to the PMU compliance class (which is usually configured in commercial PMUs).

However, the electrical phenomena observed in the real context of modern electric networks could be faster than 5 Hz. In [31], for instance, it is recalled that, during transient stability swings, oscillation frequencies can be in the range of 0.1–10 Hz, while in [32] it is specified that "The spectrum of frequency of interest with potential oscillatory activity goes from 0.1 Hz up to 50 or 60 Hz".

The frequency ranges involved in these phenomena, which can occur in the electric grids at every voltage level, cannot be adequately analyzed even with the maximum reporting rates currently defined for PMUs. For these reasons, [1] permits the use of other reporting rates. Even higher F_s are considered desirable [29] for applications trying to detect subsynchronous resonance. However, the increase of the reporting rate should be accompanied by a suitable increase of the modulation frequencies used for the bandwidth test. If this does not happen, the higher rates might not properly reflect the actual measurement capabilities of the PMU. In order to show this possible incongruence, Figure 3 points out the frequency band $f_m \pm F_s/2$ for different F_s values, along with the maximum modulation frequencies considered in the standard tests (compliance with the M class and P class), whose position does not change for different values of F_s.

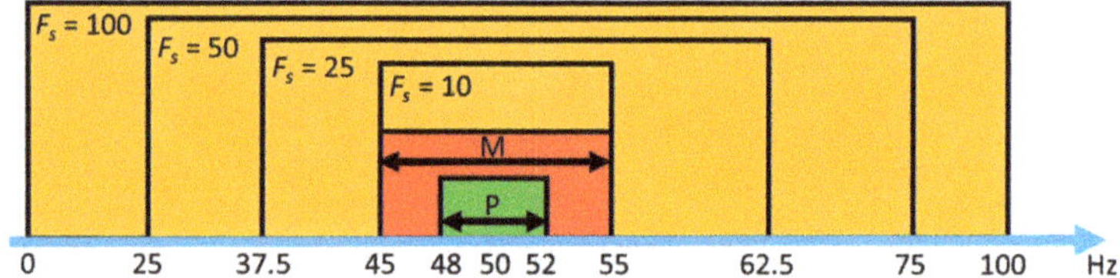

Figure 3. *In band* ranges for different reporting rates F_s and maximum modulation bandwidths required for the P class (green) and M class (red).

Finally, it should be taken into account that the behavior of the PMUs in the presence of these kinds of signals may strongly depend on the adopted measurement procedure, but PMUs' manufacturers usually do not provide enough information on the implemented algorithms. At the same time, PMUs are also specifically targeted to distribution systems which are usually claimed to be compliant with the only standards currently available and thus they are tested under the test conditions defined in those documents (i.e., [1–6]), which, as recalled above, were driven by the needs of the transmission systems. Triggered by all the above considerations, in the following section, an appropriate experimental characterization test setup is presented, designed with the aim of analyzing the PMU performance in the presence of possible voltage fluctuations and putting in evidence concerning the risk of misinterpreting the relevant measurement results.

3. Test Setup

To analyze the behavior of PMUs in the presence of voltage fluctuations with different modulation frequencies, three commercial devices with different possible configurations have been tested. The problem is not to evaluate the accuracy of the measurements, but to first understand the reference value with respect to which such accuracy must be evaluated.

The commercial PMUs have been tested with the highest selectable reporting rate for a 50 Hz system ($F_s = 50$ or 100 fps) to evaluate fast dynamic events. The schematic representation of the test architecture is described in Figure 4. The chosen PMUs are characterized by several features: in the following, the main characteristics of each device under test (DUT) are reported.

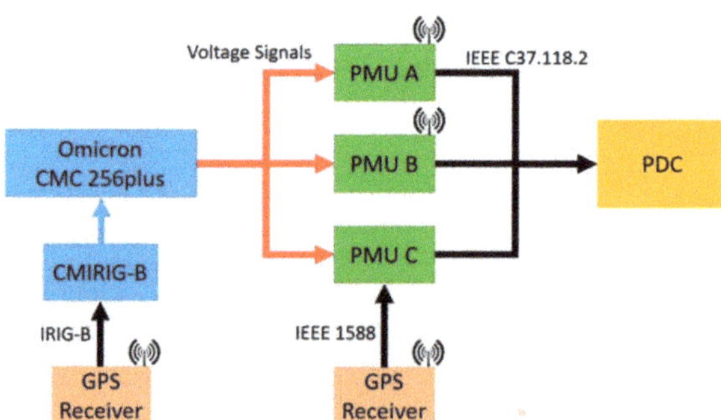

Figure 4. Test architecture.

PMU A is a device with a built-in GPS receiver able to measure the synchronised phasor up to 50 fps. The device is compliant with the 2005 version of the standard [4], and offers a wide range of configurations that might help in optimizing the measurement process. The two main selectable characteristics are the window type and window length, which define the characteristics of the weights employed to preprocess each sample record for synchrophasor extraction and thus directly impact on accuracy and latency of the measurement process. Any change or tuning of the parameters in the PMU algorithm can modify the performance of the device and make it suitable for a specific class. In the technical datasheet, a maximum TVE of 0.1% for all the contributions is reported, except for the estimation algorithm error. To underline different performances with the same reporting rate, the device has been tested with two different measurement configurations. For a more exhaustive comparison with the other devices, among the available configurations, the following ones have been used, inspired by those of the two algorithms suggested in the synchrophasor standard for p and M compliance classes:

- P-like configuration: Window type: triangular; window length: two cycles at nominal frequency.
- M-like configuration: Window type: Hanning; window length: seven cycles at nominal frequency.

PMU B is a programmable IED (Intelligent electronic device) characterized by a built-in synchronization provided by a GPS receiver. The device is fully compliant with [5] and presents two different measurement configurations, called P and M, respectively. The technical datasheet does not include any information about the implemented algorithm. The reported measurement accuracy is a maximum TVE less than 1%. The maximum reporting rate F_s available in the device is 50 fps.

PMU C is compliant with [5]. The synchronization is obtained with the IEEE 1588 protocol through an external GPS receiver with the functionality of grand master [33]. The device is characterized by an F_s up to 100 fps for the P configuration and up to 50 fps for the M configuration. The declared maximum TVE% is less than 1%.

The PMUs are tested using reference three-phase voltage signals generated by a power signal generator (OMICRON CMC 256 plus and CMIRIG-B) able to synchronize the phase angle of the generated signal, with respect to the UTC time reference. A GPS receiver Symmetricom with a time accuracy up to ±100 ns has been used to feed the CMIRIG-B module. The CMC 256 plus can also provide the PQ signals based on the tables specified in [25], which can be used for the calibration of a flickermeter. In particular, test signals that can be represented by the following voltage modulation equation are used:

$$v(t) = V\left(1 + \frac{1}{2}\frac{\Delta V}{V}\sin(2\pi f_m t)\right)\sin(2\pi f_0 t) \tag{3}$$

where f_0 is the nominal system frequency, while $\frac{\Delta V}{V}$ and f_m are the relative voltage change and the frequency modulation, respectively. Despite [25] defining different voltage changes for different test frequencies, a single level of relative voltage change is used in the following, for the sake of a clearer comparison. In particular, as suggested for the bandwidth test in [1], a 10% voltage amplitude modulation factor is used ($\frac{1}{2}\frac{\Delta V}{V} = 0.1$). The different modulation frequencies adopted in the tests are specified in [25] and reported in Table 1. A maximum modulation frequency of 24 Hz is reported in the results when F_s is equal to 50 fps, according to the Nyquist limit represented by Equation (2).

Table 1. Flickermeter mandatory (bold printed) and optional modulation frequencies for sinusoidal fluctuations [25].

Hz	Hz	Hz	Hz	Hz	Hz
0.5	4	7.5	8.8	13	**20**
1	4.5	8	9.5	14	21
1.5	5	**8.8**	10	15	22
2	5.5	9.5	10.5	16	23
2.5	6	10	11	17	24
3	6.5	7.5	11.5	18	**25**
3.5	7	8	12	19	**33 1/3**

The data from the PMUs under test are collected by a computer with PDC functionality. Afterwards, all the measurements and their time-tags are evaluated with a software developed in LabVIEW environment.

The synchronized phasors obtained from the DUT are evaluated using, as performance index, the maximum TVE% in one-minute measurement tests. The acquired data are evaluated with two different and complementary interpretations following the flowchart in Figure 5:

Case 1: The sinusoidal voltage changes are considered as signals of interest and thus the performance is evaluated as the ability of the PMU to follow the dynamic variations of the voltage amplitude for every modulation frequency.

Case 2: The sinusoidal voltage changes are considered as disturbances and the performance is evaluated as the rejection capability of the PMU.

FE and RFE values are also evaluated, and the outcome can be considered univocal, since all the measurements are compared with the reference values of 50 Hz and 0 Hz/s (for frequency and ROCOF, respectively), without any interpretation problems.

Besides the plots of the above defined errors, in the following, two concise indices will be used. Their definition is based on a threshold TVE% = 3%, which is the maximum error permitted under dynamic conditions in [1]. In particular, among the test modulation frequencies shown in Table 1, the highest value that allows a maximum TVE% < 3% in Case 1, which represents the actual PMU bandwidth, will be indicated as $f_{\max \text{ TVE3\%}}$. The minimum frequency value, again among the test modulation frequencies of [25], for which the maximum TVE% < 3% in Case 2, i.e., the minimum frequency of a disturbance the PMU is able to adequately reject, will be indicated as $f_{\min \text{ TVE3\%}}$. It is important to highlight that these values represent useful parameters for the comparison, but, since the PMU can even implement nonlinear processing routines, they do not fully represent the PMU filtering behavior and only the full test analysis can address the characterization in the presence of fluctuations.

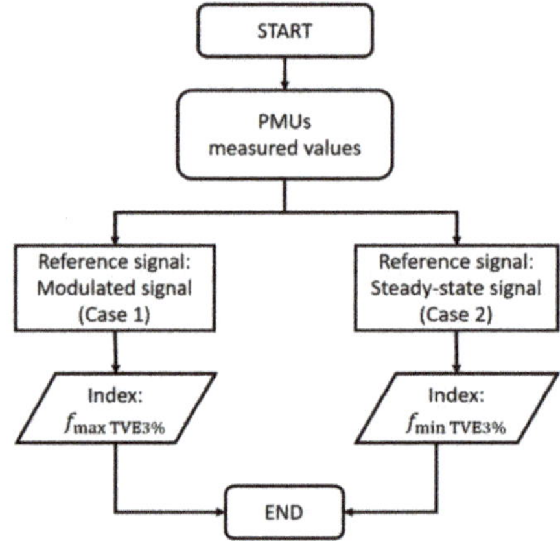

Figure 5. Flow chart of the evaluation process.

4. Test Results

In the following, the results of the experimental analysis for the aforementioned commercial PMUs are described in detail, focusing on both TVE performance and frequency measurement results.

4.1. TVE Results

Figure 6a reports the results of TVE% for Case 1 and Case 2 using the "P" configuration of PMU A. As recalled above, with F_S = 50 fps, the maximum frequency that can be tested to avoid aliasing is 24 Hz [34]. A horizontal red line at 3% is reported to allow defining $f_{\max \text{ TVE3\%}}$ and $f_{\min \text{ TVE3\%}}$. In this case, it is $f_{\max \text{ TVE3\%}}$ = 15 Hz, whereas $f_{\min \text{ TVE3\%}}$ cannot be defined for this test, since PMU A with

P-class settings is not designed to cancel interharmonics. This points out how important the definition of the measurement framework is, since the presence of fluctuations in the network is obviously independent of the PMU application context, and thus the behavior of a P-class PMU can become unpredictable if a dedicated characterization is not performed. Figure 6b shows the TVE for PMU A with M configuration. It results in $f_{\text{max TVE3\%}} = 5$ Hz, while $f_{\text{min TVE3\%}} = 9.5$ Hz.

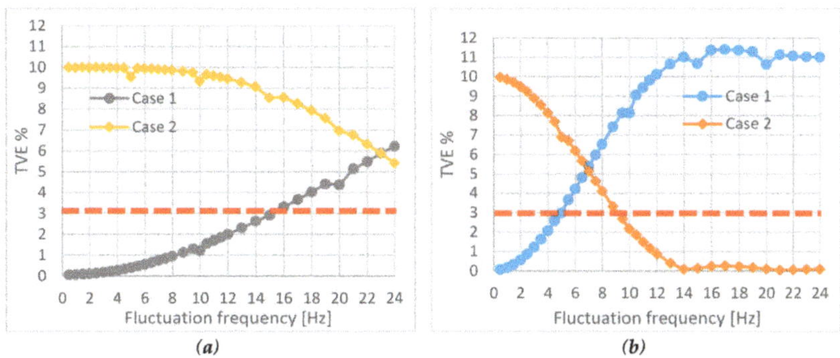

Figure 6. TVE results for PMU A with P (**a**) and M (**b**) configuration.

The results of PMU B with P configuration and $F_s = 50$ fps are shown in Figure 7a. It is clear that PMU B is able to also follow fast amplitude modulation signals up to 20 Hz ($f_{\text{max TVE3\%}} = 20$ Hz). Again, it is not possible to identify a value for $f_{\text{min TVE3\%}}$.

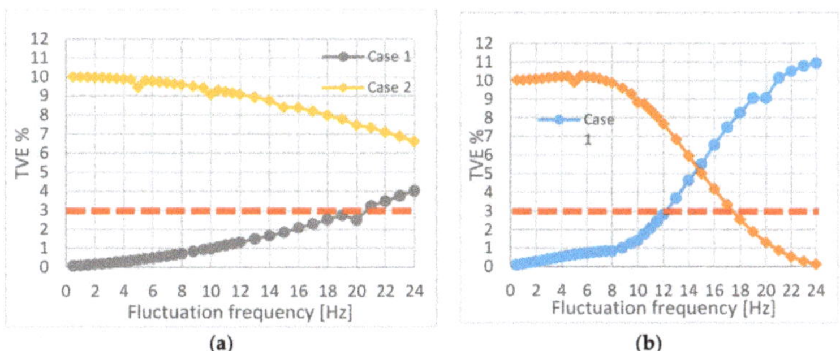

Figure 7. TVE results for PMU B with P (**a**) and M (**b**) configuration.

Unlike the P class, the M configuration (whose results for PMU B are shown in Figure 7b) is able to reject interharmonic interferences and, for this reason, it shows a much narrower passband. Fluctuations above 12 Hz ($f_{\text{max TVE3\%}} = 12$ Hz) cannot be correctly followed and the behavior in Case 2 entails, dually, a significant rejection of fluctuations above 18 Hz.

PMU C ($F_s = 50$ fps) with configuration P shows a behavior (see Figure 8a) like that of PMU A. In fact, it is $f_{\text{max TVE3\%}} = 15$ Hz, while a value for $f_{\text{min TVE3\%}}$ cannot be defined. With configuration M (Figure 8b), instead, PMU C has a behavior similar to PMU B: $f_{\text{max TVE3\%}}$ reaches a value of 13 Hz, while $f_{\text{min TVE3\%}}$ is equal to 18 Hz.

PMU C is the only DUT with an available F_s up to 100 fps for the P class, and Figure 9 thus shows the TVE for the highest reporting rate suggested in the standard [1]. The test frequency can thus reach 33.33 Hz, which becomes the $f_{\text{min TVE3\%}}$ (the corresponding TVE is 1.74%). For the other test frequencies, the behavior is the same as in Figure 8a, suggesting that the algorithm is the same as in the previous configuration.

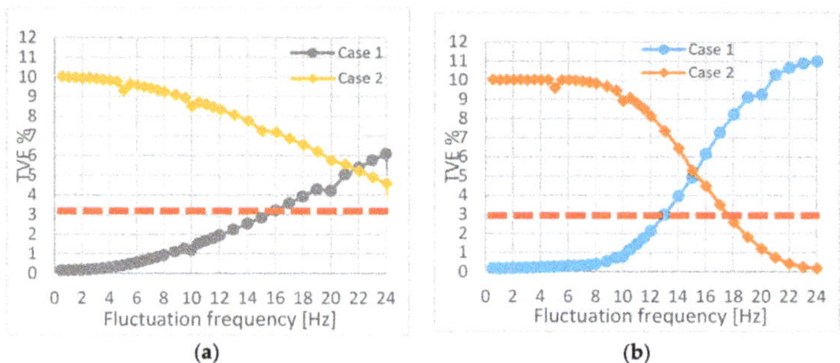

Figure 8. TVE results for PMU C with P (**a**) and M (**b**) configuration.

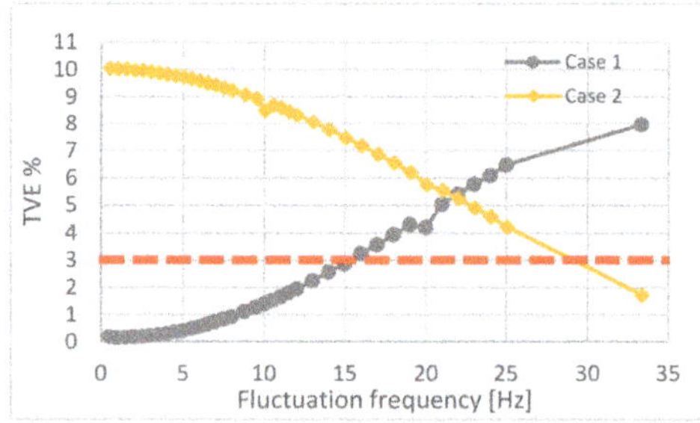

Figure 9. TVE results for PMU C with P configuration and F_s = 100 fps.

Table 2 summarizes the results in terms of $f_{\max\ \text{TVE3\%}}$ and $f_{\min\ \text{TVE3\%}}$. At F_s = 50 fps values for $f_{\min\ \text{TVE3\%}}$ are obtained only for the M class configurations, while at F_s = 100 fps, the PMU C also reaches this limit when configured as P class. The results of PMU C with configuration P and with the different reporting rates are similar. In this case, an increased value of F_s does not allow the ability to obtain an increased value of $f_{\max\ \text{TVE3\%}}$.

Table 2. Summarized results for the PMUs under test.

DUT	Configuration	$f_{\max\ \textbf{TVE3\%}}$ **(Hz)**	$f_{\min\ \textbf{TVE3\%}}$ **(Hz)**
PMU A	P	15	-
50 fps	M	5	9.5
PMU B	P	20	-
50 fps	M	12	18
PMU C	P	15	-
50 fps	M	13	18
PMU C 100 fps	P	15	33.33

4.2. Frequency and ROCOF Results

As for FE and RFE computation, the reference values are characterized by constant values of 50 Hz and 0 Hz/s for frequency and ROCOF, respectively. To evaluate the performance, it could be useful to analyze the results by means of the limits suggested by the synchrophasor standard for modulation tests. This can give an indication, but it is important to remember that the fluctuation frequencies can be outside the modulation range. The P-class configuration allows a value of $f_{\max\ \text{TVE3\%}}$ higher than the configuration M, thus meaning the possibility to follow faster fluctuations, but this result is accompanied by higher values of FE and RFE. For instance, during the tests, the maximum value obtained for $f_{\mathrm{max\ TVE3\%}}$ with a P-class configuration is 20 Hz for PMU B, but in this same case, the values of |FE| and |RFE| are the highest obtained.

Figure 10a,b report the values of FE for configurations P and M, respectively, of the PMUs under test as a function of the test modulation frequency. The values of errors for PMU A and PMU C are always below the limits imposed for the bandwidth test for F_s = 50 fps (FE limit is 0.06 Hz and 0.3 Hz for P class and M class, respectively). On the contrary, the FE of PMU B with configuration P increases with the increasing modulation frequency. Fluctuations thus have a strong impact on PMU B frequency measurements, which can easily become unacceptable even with much smaller voltage changes. With configuration M, instead, PMU B has lower accuracy for frequency measurements, which varies in a significant way with the fluctuation frequency. Finally, the FE results for PMU C with F_s = 100 fps are very similar to those achieved with F_s = 50 fps, confirming that the same algorithm is probably common to both reporting rates.

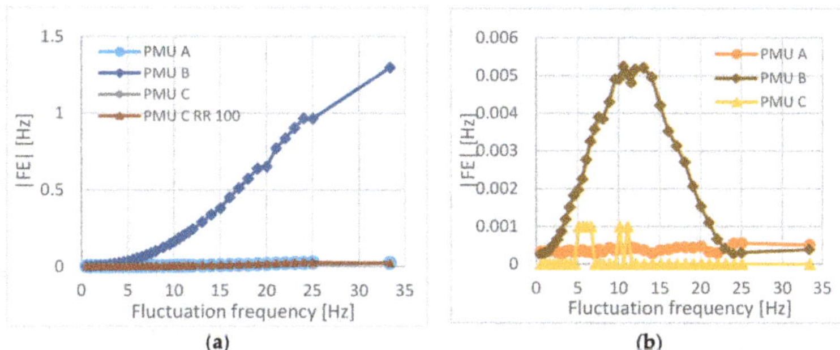

Figure 10. FE results for the PMUs with P (**a**) and M (**b**) configuration.

Analogously, Figure 11a,b show the results of the PMUs characterization in terms of RFE. The absolute RFE in Figure 11a increases with increasing modulation frequency for all the PMUs up to the Nyquist frequency, exceeding the P-class limit of 2.3 Hz/s defined in [1] for modulation tests. Moreover, also for RFE, the performance of PMU C (configuration P) remains almost the same when the reporting rate becomes higher than 50 fps. With the M configuration, the RFE is contained below the limit imposed by [1] (14 Hz/s, for F_s = 50 fps). Even if the PMUs show different measurement accuracies, the RFE trend in Figure 11b suggests that fluctuations are either followed or cancelled without dramatically affecting ROCOF measurement. This once more highlights how every measured quantity has a different behavior under voltage fluctuations and the device specifications are not sufficient to give a representation of measurement performance in the presence of common PQ events.

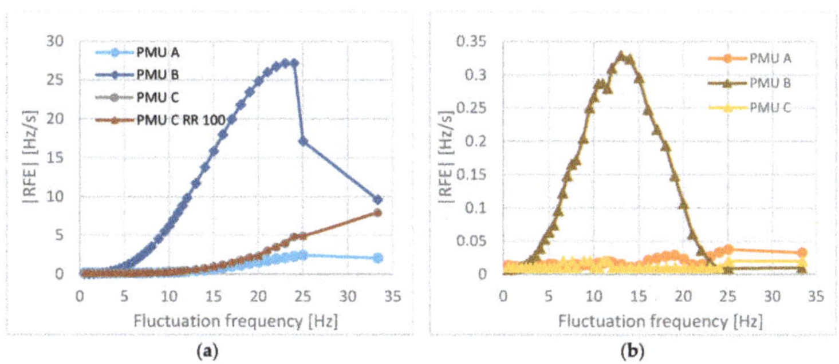

Figure 11. RFE results for the PMUs with P (**a**) and M (**b**) configuration.

5. Conclusions

The possibility of using high-performance measuring instruments, such as the PMUs, in a rapidly evolving context, such as that of electrical distribution networks, is attracting great interest from the operators of such systems. However, an effective use of these instruments requires that their performance be guaranteed in the conditions in which they will actually operate. This means, first of all, checking their behavior in the presence of voltage and current signals affected by significant power quality disturbances.

In this context, this work has presented and discussed the results of experimental tests carried out on some commercial PMUs in the presence of voltage fluctuations that give rise to the flicker. The performed characterization does not aim at highlighting the best or worst performance of the tested commercial devices, but at pointing out how possible different interpretations could be given to the PMU outputs in the same signal, depending on the quantity of interest. It emerged that the first step to solve the possible misinterpretations of the measurement results is to clearly define the objective of the measurement (e.g., to remove all non-fundamental frequency components or to follow the dynamic variation of a signal), which in turn depends on the requirements of the specific application. Only in this way, new and more suitable test conditions and performance evaluation criteria could be defined for PMUs targeted to distribution networks.

The results obtained, therefore, may represent a useful contribution, both for the preparation of a new synchrophasor standard specific for distribution networks, which pays the right attention to the performance of the devices in the presence of electrical phenomena typical of that type of network, and for PMU manufacturers, who will have clearer design criteria available, aimed at dealing with more realistic and often challenging conditions.

Author Contributions: Conceptualization, P.C., C.M., P.A.P. and S.S.; methodology, P.C., C.M., P.A.P. and S.S.; software, P.C.; validation, P.C. and S.S.; formal analysis, P.A.P.; investigation, S.S.; data curation, P.C. and P.A.P.; writing—original draft preparation, P.C.; writing—review and editing, P.C., C.M., P.A.P. and S.S.; supervision, C.M.

Funding: This research received no external funding.

Conflicts of Interest: The authors declare no conflict of interest.

References

1. IEEE Standard Association. *IEEE/IEC International Standard—Measuring Relays and Protection Equipment—Part 118-1: Synchrophasor for Power Systems—Measurements, in IEC/IEEE 60255-118-1:2018*; IEEE Power & Energy Society: Piscataway, NJ, USA, 2018; pp. 1–78. [CrossRef]
2. IEEE Standard Association. *IEEE Guide for Phasor Data Concentrator Requirements for Power System Protection, Control, and Monitoring, IEEE Std C37.244-2013*; IEEE Power & Energy Society: Piscataway, NJ, USA, 2013; pp. 1–65. [CrossRef]
3. IEEE Standard Association. *IEEE Guide for Synchronization, Calibration, Testing, and Installation of Phasor Measurement Units (PMUs) for Power System Protection and Control, IEEE Std C37.242-2013*; IEEE Power & Energy Society: Piscataway, NJ, USA, 2013; pp. 1–107. [CrossRef]
4. IEEE Standard Association. *IEEE Standard for Synchrophasors for Power Systems, IEEE Std C37.118-2005*; Revision of IEEE Std 1344-1995; IEEE Power & Energy Society: Piscataway, NJ, USA, 2006; pp. 1–57. [CrossRef]
5. IEEE Standard Association. *IEEE Standard for Synchrophasor Measurements for Power Systems, in IEEE Std C37.118.1-2011*; Revision of IEEE Std C37.118-2005; IEEE Power & Energy Society: Piscataway, NJ, USA, 2011; pp. 1–61. [CrossRef]
6. IEEE Standard Association. *IEEE Standard for Synchrophasor Measurements for Power Systems—Amendment 1: Modification of Selected Performance Requirements, in IEEE Std C37.118.1a-2014*; Amendment to IEEE Std C37.118.1-2011; IEEE Power & Energy Society: Piscataway, NJ, USA, 2014; pp. 1–25. [CrossRef]
7. Monti, A.; Muscas, C.; Ponci, F. *Phasor Measurement Units and Wide Area Monitoring Systems*, 1st ed.; Elsevier Academic Press: Cambridge, MA, USA, 2016. [CrossRef]

8. Barchi, G.; Macii, D.; Petri, D. Phasor Measurement Units for Smart Grids: Estimation Algorithms and Performance Issues. In Proceedings of the AEIT Annual Conference 2013, Mondello, Palermo, Italy, 3–5 October 2013; pp. 1–6. [CrossRef]

9. Castello, P.; Ferrari, P.; Flammini, A.; Muscas, C.; Pegoraro, P.A.; Rinaldi, S. A Distributed PMU for Electrical Substations with Wireless Redundant Process Bus. *IEEE Trans. Instrum. Meas.* **2015**, *64*, 1149–1157. [CrossRef]

10. Robson, S.; Tan, G.; Haddad, A. Low-Cost Monitoring of Synchrophasors Using Frequency Modulation. *Energies* **2019**, *12*, 611. [CrossRef]

11. Distribution Task Team, Synchrophasor Monitoring for Distribution Systems—Technical Foundations and Applications, 2018, NASPI-2018-TR-001. Available online: https://www.naspi.org/node/688 (accessed on 3 April 2019).

12. Shen, Y.; Abubakar, M.; Liu, H.; Hussain, F. Power Quality Disturbance Monitoring and Classification Based on Improved PCA and Convolution Neural Network for Wind-Grid Distribution Systems. *Energies* **2019**, *12*, 1280. [CrossRef]

13. Von Meier, A.; Stewart, E.; McEachern, A.; Andersen, M.; Mehrmanesh, L. Precision Micro-Synchrophasors for Distribution Systems: A Summary of Applications. *IEEE Trans. Smart Grid.* **2017**, *8*, 2926–2936. [CrossRef]

14. Cifredo-Chacón, M.-Á.; Perez-Peña, F.; Quirós-Olozábal, Á.; González-de-la-Rosa, J.-J. Implementation of Processing Functions for Autonomous Power Quality Measurement Equipment: A Performance Evaluation of CPU and FPGA-Based Embedded System. *Energies* **2019**, *12*, 914. [CrossRef]

15. European Committee for Electrotechnical Standardization. *Voltage Characteristics of Electricity Supplied by Public Distribution Systems*; European Standard CENELEC EN 50160; European Committee for Electrotechnical Standardization: Brussels, Belgium, 2010.

16. Pogliano, U.; Braun, J.P.; Voljč, B.; Lapuh, R. Software platform for PMU algorithm testing. *IEEE Trans. Instrum. Meas.* **2013**, *62*, 1400–1406. [CrossRef]

17. Liu, H.; Bi, T.; Yang, Q. The evaluation of phasor measurement units and their dynamic behavior analysis. *IEEE Trans. Instrum. Meas.* **2013**, *62*, 1479–1485. [CrossRef]

18. Castello, P.; Liu, J.; Muscas, C.; Pegoraro, P.A.; Ponci, F.; Monti, A. A Fast and Accurate PMU Algorithm for P+M Class Measurement of Synchrophasor and Frequency. *IEEE Trans. Instrum. Meas.* **2014**, *63*, 2837–2845. [CrossRef]

19. Toscani, S.; Muscas, C.; Pegoraro, P.A. Design and Performance Prediction of Space Vector-Based PMU Algorithms. *IEEE Trans. Instrum. Meas.* **2017**, *66*, 394–404. [CrossRef]

20. Romano, P.; Paolone, M. Enhanced Interpolated-DFT for Synchrophasor Estimation in FPGAs: Theory, Implementation, and Validation of a PMU Prototype. *IEEE Trans. Instrum. Meas.* **2014**, *63*, 2824–2836. [CrossRef]

21. Hu, W.; Zhang, Y.; Chen, Z.; Hu, Y. Flicker Mitigation by Speed Control of Permanent Magnet Synchronous Generator Variable-Speed Wind Turbines. *Energies* **2013**, *6*, 3807–3821. [CrossRef]

22. Lodetti, S.; Azcarate, I.; Gutiérrez, J.J.; Leturiondo, L.A.; Redondo, K.; Sáiz, P.; Melero, J.J.; Bruna, J. Flicker of Modern Lighting Technologies Due to Rapid Voltage Changes. *Energies* **2019**, *12*, 865. [CrossRef]

23. CIGRE. *Review of LV and MV Compatibility Levels for Voltage Fluctuations*; Working Group C4.111; CIGRE: Paris, France, 2016.

24. Castello, P.; Muscas, C.; Pegoraro, P.A.; Sulis, S. Analysis of PMU Response Under Voltage Fluctuations in Distribution Grids. In Proceedings of the IEEE International Workshop on Applied Measurements for Power Systems (AMPS), Aachen, Germany, 28–30 September 2016.

25. International Electrotechnical Commission. *Electromagnetic Compatibility (EMC)—Part 4-15: Testing and Measurement Techniques—Flickermeter—Functional and Design Specifications*; IEC Int. Std. 61000-4-15; International Electrotechnical Commission: Geneva, Switzerland, 2011.

26. Wiczyński, G. Estimation of Pst Indicator Values on the Basis of Voltage Fluctuation Indices. *IEEE Trans. Instrum. Meas.* **2017**, *66*, 2046–2055. [CrossRef]

27. De Rosa, F.; Langella, R.; Sollazzo, A.; Testa, A. On the interharmonic components generated by adjustable speed drives. *IEEE Trans. Power Del.* **2005**, *20*, 2535–2543. [CrossRef]

28. Langella, R.; Testa, A. Amplitude and Phase Modulation Effects of Waveform Distortion in Power Systems. *Electr. Power Qual. Util.* **2007**, *13*, 25–32.

29. Reliability Guideline, PMU Placement and Installation, December 2016. Available online: https://www.nerc.com/comm/PC_Reliability_Guidelines_DL/Reliability%20Guideline%20%20PMU%20Placement.pdf (accessed on 25 June 2019).

30. Stenbakken, G.; Nelson, T.; Zhou, M.; Centeno, V. Reference values for dynamic calibration of PMUs. In Proceedings of the 41st Annual Hawaii International Conference on System Sciences, Waikoloa, HI, USA, 7–10 January 2008.

31. Ma, J.; Zhang, P.; Fu, H.J.; Bo, B.; Dong, Z.Y. Application of Phasor Measurement Unit on Locating Disturbance Source for Low-Frequency Oscillation. *IEEE Trans. Smart Grid* **2010**, *1*, 340–346. [CrossRef]

32. Vanfretti, L.; Baudette, M.; Domínguez-García, J.-L.; Almas, M.S.; White, A.; Gjerde, J.O. A Phasor Measurement Unit Based Fast Real-time Oscillation Detection Application for Monitoring Wind-farm-to-grid Sub–synchronous Dynamics. *Electr. Power Compon. Syst.* **2016**, *44*, 123–134. [CrossRef]

33. IEEE Standard Association. *IEEE Standard for a Precision Clock Synchronization Protocol for Networked Measurement and Control Systems, in IEEE Std 1588-2008*; Revision of IEEE Std 1588-2002; IEEE Power & Energy Society: Piscataway, NJ, USA, 2008; pp. 1–300. [CrossRef]

34. Phadke, A.G.; Thorp, J.S. *Synchronized Phasor Measurements and Their Applications*; Springer: New York, NY, USA, 2017. [CrossRef]

Article

Power Quality in DC Distribution Networks †

Julio Barros *, Matilde de Apráiz and Ramón I. Diego

Department of Computer Science and Electronics, University of Cantabria, 39005 Santander, Spain;
matilde.deapraiz@unican.es (M.d.A.); ramon.diego@unican.es (R.I.D.)
* Correspondence: julio.barros@unican.es; Tel.: +34-942-201-355
† This paper is a technically extended version of a conference paper entitled "Definition and measurement of power quality indices in low voltage DC networks", by the same authors, presented at IEEE AMPS 2018, Bologna, Italy, September 2018.

Received: 31 January 2019; Accepted: 27 February 2019; Published: 5 March 2019

Abstract: This paper presents an overview of power quality in low-voltage DC distribution networks. We study which of the power quality disturbances in AC networks are also relevant in DC networks, as well as other disturbances specific to DC networks. The paper reviews the current status of international regulations in this topic and proposes different indices for the detection and characterization of the main types of power quality disturbances, presenting some results obtained in different laboratory tests in DC networks using different DC voltage shapes delivered by different DC power source types.

Keywords: low-voltage DC networks; power quality disturbances; power quality monitoring; DC power quality indices; voltage ripple

---

1. Introduction

The use of distributed generation with the increasing number of renewable energy systems that directly deliver DC power such as PV systems, wind generation, or battery storage, combined with the advance in DC technology enables a direct, more efficient and sustainable use of the energy, eliminating losses associated with energy conversion. The growing use of DC distribution systems in data centers, residential buildings, lighting, transportation, and other applications [1–8], has prompted the need for definition and standardization of the supplied voltage and the power quality requirements to enable the reliable operation of equipment in DC networks, as well as in AC public distribution systems.

Low-voltage DC distribution networks (LVDCs) are regulated in the European Union by the Directive 2006/95/EC [9]. This directive enables DC voltage to be used in electricity distribution systems up to 1500 V. Table 1 shows some of the most common DC nominal voltages used and examples of their main applications [10].

Table 1. Low-voltage DC distribution network (LVDC) voltages and applications [10].

LVDC Voltages	Application
3.6–5 V	Mobile phone charging
9 V	Battery
12 V	Car battery, lighting
24–36 V	Electric bikes, high-power cordless, truck battery
48 V	Telecom power
400 V	Data centers, office buildings, hospitals, electric vehicles fast charging
Up to 1500 V	Urban railways

During recent years, a number of IEC international standards have been developed for LVDC systems, including electricity generation, transmission, distribution, storage and applications such as lighting, household appliances, and electric equipment for transportation, among others. The most comprehensive overview of the IEC related work in LVDC can be seen in [10].

In spite of being power quality an important issue in the design of DC distribution networks, at present there is no standard norm for power quality in this type of distribution network.

Independent of the future development of LVDC systems, there is a need for definition and measurement of power quality indices in LVDC networks. In this sense, the international committees in charge of maintenance of European Standard EN 50160 "Voltage characteristics of electricity supplied by public electricity networks" [11] and of the IEC 61000-4-30 "Power quality measurement methods" [12] have started some preliminary studies in order to extend the scope of these two standards to low-voltage and medium-voltage DC public distribution networks in their future revisions.

At present, there are only a few proposals in the technical literature for definition and measurement of indices for relevant power quality disturbances in DC networks. Some of the first attempts in this subject have defined indices for harmonic/interharmonic distortion and for ripple evaluation [13,14]. Power quality indices in the time and frequency domains, such as average, median, and percentile variations of DC components, as well as peak-to-peak and r.m.s. variations, and low frequency distortion indices, have been defined for assessing power quality in DC microgrids [15]. Finally, other important power quality concerns in DC networks, such as fault currents, inrush currents and grounding, are considered in [3].

This paper presents an overview of power quality in LVDC distribution networks, studying the power quality phenomena in AC distribution networks that also apply to DC networks, as well as disturbances specific to DC networks, proposing some indices for their characterization. Section 2 reviews the different types of DC generation and DC voltage shapes and their characterization method. Voltage dips, voltage supply interruptions, rapid voltage changes, and voltage ripple in DC voltage are considered in Section 3 as some of the main types of power quality disturbances in LVDC networks, defining indices for their assessment. Section 4 reports some results obtained in different laboratory tests in the detection and assessment of these power quality disturbances, using different DC voltage shapes delivered by different DC power source types. Section 5 presents a summary of the power quality indices defined, and finally, Section 6 presents the conclusions.

2. DC Voltage Characteristics

There are two main types of DC generation: DC sources that produce ideal or quasi-ideal DC voltage, such as PV, DC generators, and battery storage; or AC/DC conversion (single-phase or multi-phase rectifying) that produces DC voltage with a ripple content caused by feed-through from the AC input section to the DC output. Figure 1 shows some examples of different voltage waveforms produced by different DC power sources [16].

As can be seen in Figure 1, DC voltage shapes can be very different, requiring precise definitions of DC voltage characteristics as well as methods for their measurements, both under steady state and non-steady state conditions. It is necessary to clearly define the expected characteristics of the electricity supplied in DC distribution systems under normal operating conditions.

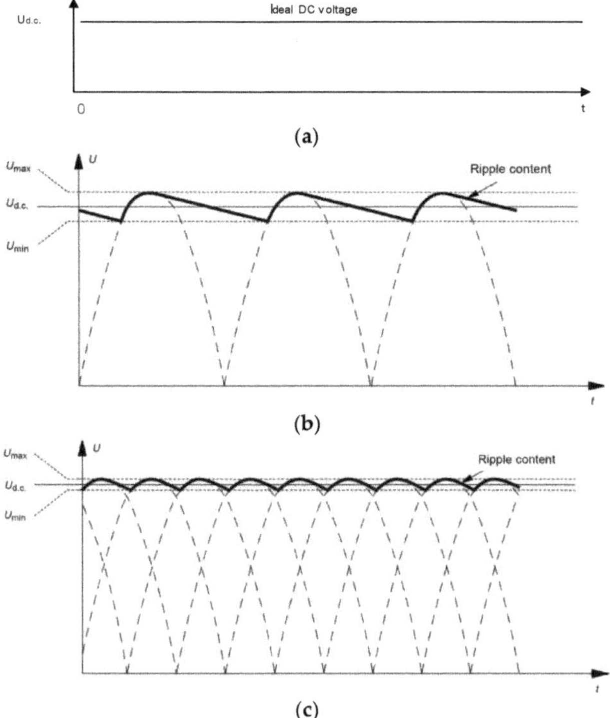

Figure 1. DC voltage shapes produced by different DC power sources: (**a**) Ideal DC voltage, (**b**) single-phase AC/DC conversion, and (**c**) three-phase AC/DC conversion.

3. Power Quality Disturbances in DC Networks

The key question concerning power quality in DC networks is whether power quality is relevant for DC networks. If so, which disturbances should be considered to define DC power quality and which indices should be defined for its accurate characterization?

As is the case with low-voltage and medium-voltage AC distribution networks, power quality is an important factor to ensure the correct performance of DC distribution networks, as well as a key factor for the correct and efficient integration of the new distributed energy resources into the network. A poor power quality in DC voltage supply can produce a malfunction or damage electrical and electronic equipment, affecting process operations with important economic and technical impacts.

Most of the power quality disturbances in AC power distribution networks apply to DC networks too. According to the preliminary work developed by the international standardization committees in charge of standards EN 50160 and IEC 61000-4-30, the power quality phenomena that should be assessed for LVDC distribution networks are:

- Supply voltage deviations
- Voltage unbalance
- Ripple/harmonics
- Voltage swells and voltage dips
- Voltage supply interruptions
- Rapid voltage changes and flicker

This section defines indices for the assessment of most of these power quality disturbances, that is, voltage swells and voltage dips, voltage supply interruptions, rapid voltage changes, and ripple/harmonics, taking as reference those previously defined in AC networks.

3.1. Voltage Dips, Voltage Swells and Voltage Supply Interruptions

As in the case of AC networks, voltage dips, voltage swells, and voltage supply interruptions are basically caused in LVDC networks by faults, sudden load changes, or by load switching. It is important to point out that these types of voltage event can be more detrimental in DC networks than in AC ones because of the lack of the reactive part of the impedance, which in the case of AC grids smooths these voltage events.

There are no standard requirements defined for tolerance of equipment to voltage dips, voltage supply interruptions, or voltage swells in LVDC networks, with the exception of the electrical and electronic equipment used in railway, aircraft and telecommunication applications [17–19].

According to European Standard EN 50155 electrical and electronic equipment is expected to operate reliably in the range 0.7–1.25 U_N (U_N: Nominal voltage) during a time t < 1 s, in the range 0.6–1.4 U_N during a time t < 0.1 s, and up to 10 milliseconds during a voltage interruption [17]. This time-dependent characteristic is shown in the voltage tolerance curve in Figure 2. No voltages within the compliant zone produce malfunctions of equipment.

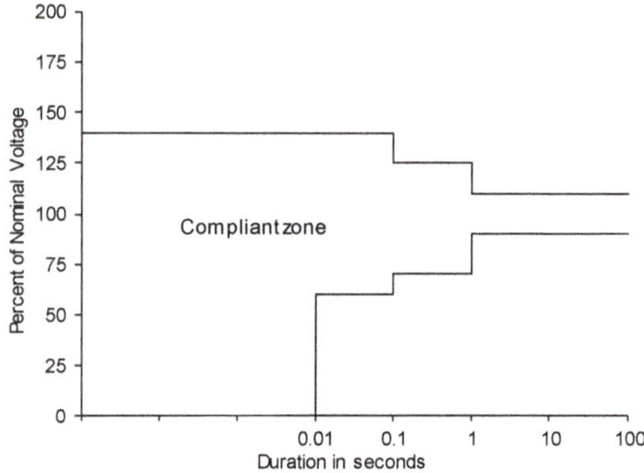

Figure 2. Voltage tolerance of equipment as defined in [17] for railway applications.

Other standards, such as IEC 61000-4-29 [20], define test methods for immunity to voltage dips, short interruptions, and voltage variations on the DC input power port of electrical or electronic equipment. This standard is applicable to low voltage DC power ports of equipment supplied by external DC networks.

Voltage dips/swells and voltage supply interruptions are characterized by two parameters in IEC 61000-4-30 for AC public distribution networks [12]: The residual voltage or the depth (or the maximum voltage in the case of a voltage swell) and the duration. The magnitude and duration of these voltage events can be computed in a similar way in LVDC networks, simply replacing the $U_{rms(1/2)}$ value defined in IEC 61000-4-30 (r.m.s. voltage computed over one cycle and refreshed each half-cycle) for their detection and evaluation by an equivalent DC magnitude. To this end, and solely for the purpose of detection and characterization of these voltage events, we propose the use of a DC voltage computed as the arithmetic mean value of the instantaneous DC voltages in a 20 millisecond sliding window ($U_{DC,20ms}$), computing a new DC mean value with each new voltage sample acquired. When this DC voltage magnitude is below/above a defined detection threshold, a voltage dip/swell is detected.

3.2. Rapid Voltage Changes

A rapid voltage change (RVC) is defined in AC networks as "an abrupt transition between two r.m.s. steady-state voltages and during which the r.m.s. voltage does not exceed the dip/swell thresholds" [12]. RVCs are a type of power quality disturbance that is also prevalent in DC networks.

RVCs can be produced in DC networks by switching operations in equipment, such as motor starting, generation units, fluctuating loads or load shedding, and they can affect the performance of electrical and electronic equipment, potentially even producing flicker. In the case of PV generation, and as previously reported [21], fast changes in solar irradiance could cause instantaneous variations in PV power output resulting in rapid voltage changes.

In AC public distribution networks and according to previous work [12], an RVC event begins when $U_{rms(1/2)}$ values corresponding to the previous 100/120 values (1 s) are over the RVC threshold of the arithmetic mean of these values, and ends when the $U_{rms(1/2)}$ values return to a new stationary-state condition. The RVC threshold should be set by the user, typically in the range from 1% to 6% of the nominal or declared voltage.

Two indices have been defined [12] for RVC characterization in AC networks: Magnitude and duration. The RVC magnitude ΔU_{max} is defined as the maximum voltage variation of the $U_{rms(1/2)}$ value during the event, and ΔU_{SS} is the absolute difference between the $U_{rms(1/2)}$ values between the two steady-states. In order to classify an event as a RVC, its magnitude ΔU_{max} should be less than $\pm 10\%$ of the nominal voltage or the declared input voltage; otherwise, the event should be classified as a voltage dip or as a voltage swell.

In the case of DC networks, two equivalent indices can be used for RVC event detection and characterization: Magnitude and duration. An RVC event is detected when the $U_{DC,20ms}$ value corresponding to the last second is over the RVC threshold and ends when the $U_{DC,20ms}$ value returns to a new stationary-state condition. The RVC magnitude in a DC network, ΔU_{DC}, can be defined as the maximum voltage variation of the $U_{DC,20ms}$ value during the event. Similar to the case of AC networks, if the magnitude of ΔU_{DC} is less than $\pm 10\%$ of the nominal voltage, the event should be classified as a voltage dip or as a voltage swell.

3.3. Voltage Ripple

Voltage ripple is a superimposed alternating voltage on a DC voltage that is defined in IEV 161-02-25 [22] as "the quantity derived by removing the direct component from a pulsating quantity."

The main sources of voltage ripple in DC networks are ripple produced by AC/DC conversion, ripple produced by battery charging, and ripple generated from equipment absorbing pulsating currents [16]. In the case of the ripple generated by AC/DC conversion, voltage ripple can be considered a continuous phenomenon with its magnitude decreasing and its frequency increasing with the number of rectifier units, as is shown in Figure 1. According to Standard IEC 61000-4-17 [16] "the frequency of the ripple is the power frequency or its multiple 2, 3 or 6," but other types of AC/DC converters produce different and more complex ripple spectra [14].

The ripple in a DC voltage should be estimated and limited in order to prevent malfunctions of electrical and electronic equipment connected to the networks. Voltage ripple can produce additional heating and increase losses and audible noise in audio circuits because its frequency and its harmonics are within the audio band; it can also interfere with TV displays.

Up to now, there are no clear limits for ripple magnitude, with up to $\pm 15\%$ of the nominal voltage in railway applications [17], or up to $\pm 10\%$ in DC power systems on ships [23], and no limits defined for the ripple frequency distortion. Standard IEC 61000-4-17 defines "test methods for immunity to voltage ripple at the DC input port of electrical and electronic equipment" [16].

Two main parameters are used in international standards for ripple characterization: Magnitude and frequency. In the case of ripple magnitude, two indices are used for its assessment: The "peak-ripple factor" and the "r.m.s.-ripple factor".

The "peak-ripple factor" is defined as "the ratio of the peak-to-valley value of the ripple content to the absolute value of the direct component of a pulsating quantity" [24]:

$$peak - ripple\,factor(\%) = \frac{V_{peak-to-valley}}{V_{dc}} 100\%$$

On the other hand, the "r.m.s.-ripple factor" is defined as "the ratio of the r.m.s. value of the ripple content to the absolute value of the direct component of a pulsating quantity" [25]:

$$r.m.s. - ripple\,factor(\%) = \frac{V_{ripple(rms)}}{V_{dc}} 100\%$$

An index similar to the total harmonic distortion (THD), used for harmonic distortion characterization in AC power system networks [26], can be defined in order to characterize the spectral components of the voltage ripple of a DC signal, the "ripple distortion factor" (RDF).

We define the RDF index as the square root of the sum of the ratio of the squares of all of the harmonic components of the ripple content G_n obtained using DFT analysis, up to a specific order in the range 2–3.6 kHz (to cover the frequency range defined by harmonic distortion related standards), with respect to the DC component G_0 as:

$$RDF = \sqrt{\sum_n \left(\frac{G_n}{G_0}\right)^2}$$

Other distortion factors, such as the "low-frequency sinusoidal distortion index" [11], can also be used, considering in this case all the frequency components provided by the DFT analysis with the selected frequency resolution, instead of using only the fundamental and the predominant harmonic components of the ripple content.

Another possible alternative is the use of different frequency ranges depending on the type and switching frequency of the power converters used for AC/DC generation, for a more accurate evaluation of the ripple, as has been proposed previously [13].

4. Experimental Results

This section presents some of the results obtained in the laboratory set-up developed for DC power quality analysis. The laboratory was made up by two different DC power sources: A regulated AC/DC three-phase rectifying power source with high ripple content and a quasi-ideal programmable DC power source with very low ripple content; and DC loads (resistive loads and DC motors) and a data acquisition system. Figure 3a shows the block diagram and Figure 3b the laboratory set-up when using the AC/DC three-phase rectifying programmable power source as the DC voltage source and a DC motor as the DC load.

The data acquisition system used for data acquisition and signal processing consisted of an LEM LV 25-P Hall effect voltage transducer with a frequency range from DC to 150 kHz; an NI CompactDAQ 9172 with a NI 9215, 4 input channels, 16-bit resolution, 100-kS/s/ch maximum sampling rate, and ±10 Volt maximum input range data acquisition board; and a laptop computer (see Figure 3a).

The sampling frequency used was 6.4 kS/s, high enough for the accurate characterization of the spectrum of the voltage ripple superimposed on the DC voltage. In each test a 10 s record of the DC voltage signal was taken, computing the DC voltage magnitude and different power quality indices depending on the test performed. The software was implemented using the LabVIEW 15 graphic programming environment.

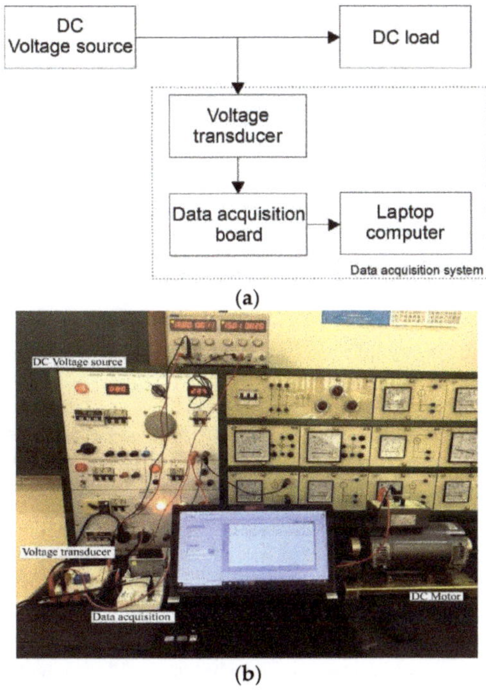

Figure 3. (**a**) Block diagram, and (**b**) laboratory set-up developed for DC power quality analysis.

Up to now, there is no standard measurement method for estimation of DC voltage in LVDC distribution grids. In the case of aircraft electric power systems, standard MIL-STD-704F [18] defines the steady-state voltage as "the time average of the instantaneous DC voltage over a period not to exceed one second," the same criteria as the standard used for shipboard electric installations [27]. In all the tests conducted in this paper, we have used a 200 millisecond period for estimation of the steady-state DC voltage (U_{DC}), the same time interval as the one used for the measurement of r.m.s. voltage in AC grids.

4.1. Voltage Dips and Short Interruptions

This section presents some results obtained in the detection and assessment of voltage dips and short interruptions in DC voltage supply using the sliding arithmetic mean value of the DC voltage previously defined in Section 3.1.

Figure 4 shows an example of a voltage supply interruption produced by the switching on/off of a 230 V, 0.69 kW, DC voltage source in our laboratory. The voltage event is detected when the DC voltage is below the voltage dip detection threshold. The detection threshold selected, as in AC networks, is 90% of the nominal voltage, considering the event as a voltage supply interruption when the DC voltage magnitude during the event is less than 5% of the nominal value, as previously proposed [11].

As can be seen in Figure 4, the voltage supply interruption starts at instant 1.24 s and ends at instant 2.60 s of the record, with a duration of 1.36 s. According to the voltage tolerance of equipment in Figure 2, the duration of this voltage supply interruption could cause the malfunction or the trip of equipment.

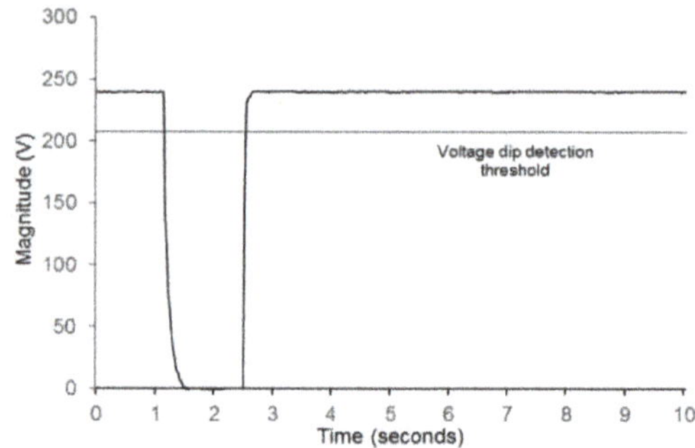

Figure 4. An example of a short interruption in a DC power source.

4.2. Rapid Voltage Changes

One type of RVC event that can appear in a DC distribution grid is associated with motor start-ups. In these voltage events and after a sudden transition, the voltage gradually recovers to a new value. A 230 V, 0.44 kW, 1500 rpm, DC motor, with different starting resistors connected to a 230 V, 0.69 kW, DC power source was used in our laboratory to produce RVCs of different magnitude and duration.

Figure 5 shows two different examples of RVC events. Different starting resistors were used to limit the starting current to safe values, producing RVCs of different magnitude and duration. The DC arithmetic mean voltage value corresponding to the last second, and recalculated with each new DC voltage sample acquired, as defined in Section 3.2, was used for RVC detection and characterization. A magnitude of 1% of the nominal voltage was selected as the RVC detection threshold (2.3 V in this case). The two RVC events in Figure 5 are 20.126 and 8.098 V magnitude (9.148% and 3.361% of the nominal voltage), with 0.18 s and 0.10 s duration, respectively.

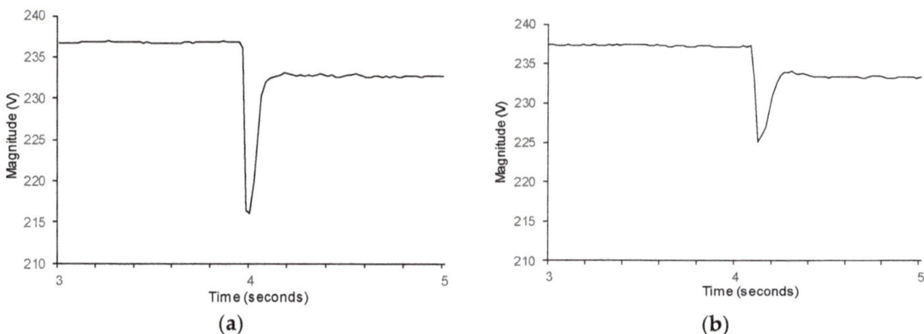

Figure 5. Examples of rapid voltage changes produced by DC motor starting.

4.3. Voltage Ripple

In this section, the results obtained in the assessment of the ripple produced by two different DC power sources in our laboratory are presented. In each case a 10 s record in the DC voltage was recorded, computing the peak-ripple factor, the r.m.s. ripple factor and the ripple distortion factor.

Figure 6 shows an example of the ripple produced by a 230 V, 0.69 kW, AC/DC three-phase rectifying programmable DC power source. Figure 6a presents 200 milliseconds of the DC voltage waveform, the ripple content is shown in Figure 6b, and its spectral analysis is presented in Figure 6c. The DC voltage (computed as the mean value during the 200 milliseconds period in the record), the peak-to-valley, and the r.m.s. values were 237 V, 50.478 V, and 14.735 V, respectively, whereas the peak-ripple factor and the r.m.s.-ripple factor (as defined in the previous section) were 21.299 % and 6.217%, respectively, of the DC voltage.

Table 2 and Figure 6c show the magnitude of the frequency components of the ripple content computed in the 200 millisecond window in Figure 6a, applying the DFT analysis with 5 Hz resolution. As can be seen from the results reported, the DC voltage component is 237 V, the fundamental component of the ripple content is 300 Hz, as expected in a three-phase AD/DC conversion voltage source when using 50 Hz AC input voltage with a magnitude of 14.46 V (6.1% of the DC voltage), and predominant harmonic components at 600 Hz, 900 Hz, 1200 Hz and 1500 Hz, with magnitudes of 1.99 (0.84% of the DC voltage), 1.06 (0.45%), 0.60 (0.25%), and 0.38 V (0.16%), respectively. The ripple distortion factor (RDF) of the signal in Figure 6a is 6.18% of the DC voltage.

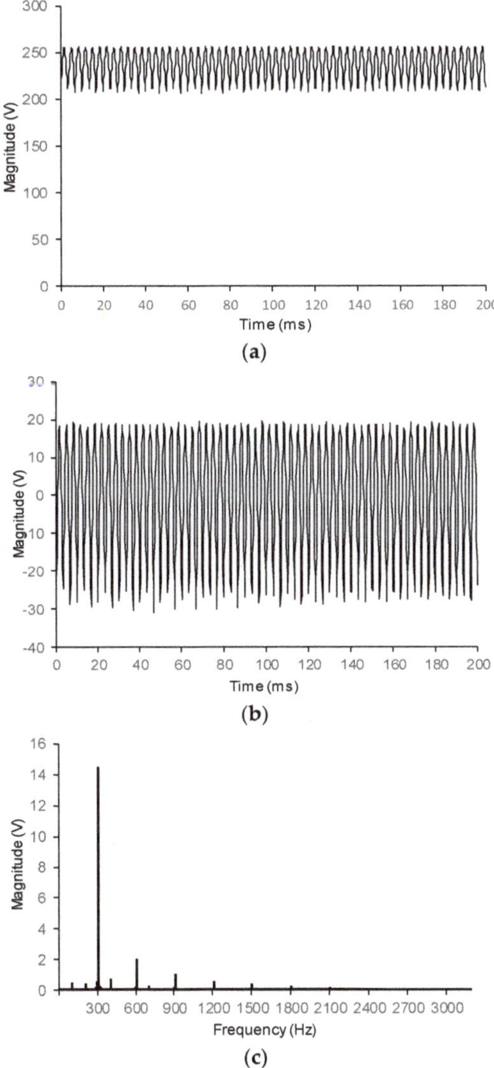

Figure 6. (**a**) DC voltage shape of an AC/DC three-phase rectifying DC power source, (**b**) ripple content, and (**c**) spectral components in voltage ripple excluding the DC component.

Finally, Figure 7 shows the DC voltage shape and the ripple content as produced, in this case, by a 600 V, 3 kW, programmable ideal DC power source. Figure 7a shows 200 milliseconds of a 400.06 DC voltage waveform produced by the power source, Figure 7b shows the ripple content and Figure 7c shows its spectral components. As can be seen, the DC voltage is quasi-ideal, with deviations from the output voltage of less than 1.5% of the selected voltage magnitude and very low ripple magnitude. The peak-ripple factor and the r.m.s.-ripple factor were 0.621% and 0.104%, respectively, of the DC voltage, and the ripple distortion factor (RDF) was only 0.022% of the DC voltage, with spectral components less than 0.1 V magnitude (0.25% of the DC voltage).

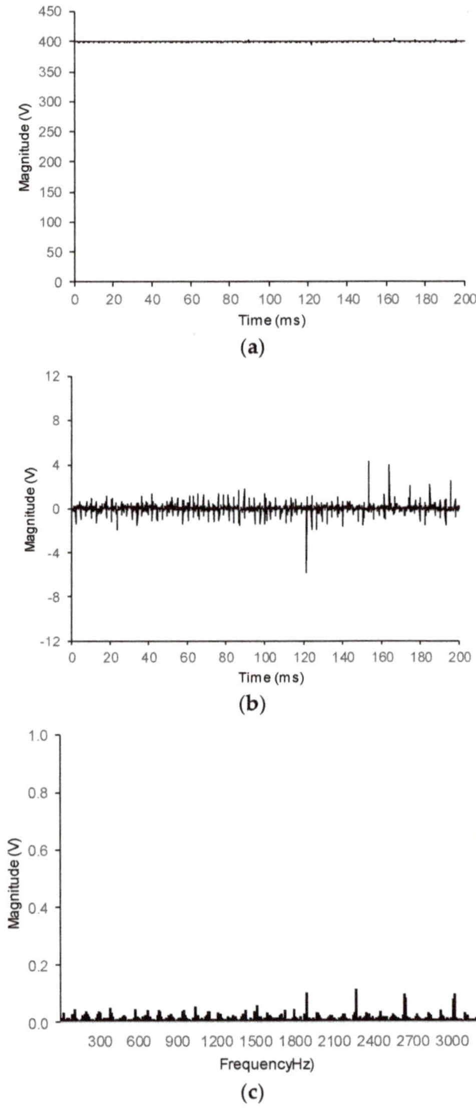

Figure 7. (**a**) DC voltage shape of quasi-ideal programmable DC power source, (**b**) ripple content, and (**c**) spectral components in voltage ripple excluding the DC component.

Table 2. Spectral analysis of the ripple content of the signal in Figure 6a.

Magnitude (V)	Magnitude (%)	Frequency (Hz)
237.00	-	DC
14.46	6.10%	300
1.99	0.84%	600
1.06	0.45%	900
0.60	0.25%	1200
0.38	0.16%	1500

5. Summary

Table 3 summarizes the indices used for characterization of power quality disturbances in LVDC distribution networks and their corresponding indices in AC networks.

Table 3. Power quality indices in LVDC distribution networks and their corresponding indices in AC networks.

Magnitude	AC	DC
Voltage	U_{rms}	U_{DC}
Voltage dips/swells/interruptions	$U_{rms(1/2)}$ Magnitude; duration	$U_{DC\text{-}20ms}$ Magnitude; duration
RVCs	$U_{rms(1/2)}$ Magnitude (ΔU_{max}); Duration	$U_{DC\text{-}20ms}$ Magnitude (ΔU_{DC}); duration
Harmonic distortion/ripple	Harmonic component (G_n) THD (Total Harmonic Distortion)	Harmonic component (G_n) RDF (Ripple Distortion Factor) Peak-ripple factor rms-ripple factor

6. Conclusions

This paper presents an overview of power quality in DC low-voltage distribution networks. The paper studies the definition and measurement of power quality indices in low-voltage DC networks. Voltage supply interruptions, rapid voltage changes, and voltage ripple content are considered in the paper. We reported the results obtained in different laboratory tests in the detection and measurement of these power quality disturbances using different DC voltage shapes delivered by different DC power source types, and proposed different indices for their characterization. From the results obtained, it can be concluded that the different power quality indices defined enable the correct detection and evaluation of the different DC power quality disturbances considered.

Author Contributions: Conceptualization, J.B. and R.I.D.; methodology, M.d.A., J.B. and R.I.D.; software, M.d.A.; validation, M.d.A., J.B. and R.I.D.; investigation, J.B.; writing—original draft preparation, J.B.; writing—review and editing, J.B., M.d.A, and R.I.D.; project administration, J.B. and R.I.D.

Funding: This research was funded by Comisión Interministerial de Ciencia y Tecnología (CICYT) of the Spanish Government, grant number ENE-2014-54039-R.

Conflicts of Interest: The authors declare no conflict of interest.

References

1. EPRI. *DC Power Production, Delivery and Utilization*; EPRI: Palo alto, CA, USA, 2006.
2. Baran, M.E.; Mahajan, N.R. DC distribution for industrial systems: opportunities and challenges. *IEEE Trans. Ind. Appl.* **2003**, *39*, 1596–1601. [CrossRef]
3. Whaite, S.; Grainger, B.; Kwasinski, A. Power quality in DC power distribution systems and microgrids. *Energies* **2015**, *8*, 4378–4399. [CrossRef]

4. Sanino, A.; Postiglione, G.; Bollen, M.H.J. Feasibility of a DC network for commercial facilities. *IEEE Trans. Ind. Appl.* **2003**, *39*, 1499–1507. [CrossRef]

5. Wunder, B.; Ott, L.; Szpek, M.; Boeke, U.; Weiβ, R. Energy efficient DC-grids for commercial buildings. In Proceedings of the 2014 IEEE INTELEC, Vancouver, BC, Canada, 28 September–2 October 2014. [CrossRef]

6. Waffenschmidt, E. Low voltage DC grids. In Proceedings of the 2013 IEEE INTELEC, Hamburg, Germany, 13–17 October 2013.

7. Parhizi, S.; Lofti, H.; Khodaei, A.; Bahramirad, S. State of the art in research on microgrids: a review. *IEEE Access* **2015**, *5*, 890–925. [CrossRef]

8. Kumar, D.; Zare, F.; Ghosh, A. DC microgrid technology: System architectures, AC grid interfaces, grounding schemes, power quality, communication networks, applications, and standardization aspects. *IEEE Access* **2017**, *5*, 12230–12256. [CrossRef]

9. European Commission. Low Voltage Directive, LVD 2006/95/EC European Union Directive, Brussels. Available online: https://eur-lex.europa.eu/LexUriServ/LexUriServ.do?uri=OJ:L:2006:374:0010:0019:EN:PDF (accessed on 31 January 2019).

10. International Electrotechnical Commision, LVDC: The Better Way. Available online: http://www.iech.ch/about/brocures/pdf/energy/iec_lvdc_the_better_way_en_lr.pdf (accessed on 31 January 2019).

11. *European Standard EN 50160: 2015. Voltage Characteristics of Electricity Supplied by Public Electricity Networks*; CENELEC: Brussels, Belgium, 2015.

12. International Electrotechnical Commission. *Electromagnetic Compatibility (EMC)—Part 4–30: Testing and Measurement Techniques. Power Quality Measurement Methods, IEC 61000-4-30, Ed. 3*; International Electrotechnical Commission: Geneva, Switzerland, 2015.

13. Caserza Magro, M.; Mariscotti, A.; Pinceti, P. Definition of Power Quality Indices for DC Low Voltage Distribution Networks. In Proceedings of the 2006 I2MTC, Sorrento, Italy, 24–27 April 2006. [CrossRef]

14. Mariscotti, A. Methods for ripple index evaluation in DC low voltage distribution networks. In Proceedings of the 2007 I2MTC, Warsaw, Poland, 1–3 May 2007. [CrossRef]

15. Ciornei, I.; Albu, M.; Sanduleac, M.; Hadjidemetriou, L. Analytical derivation of PQ indicators compatible with control strategies for DC microgrids. In Proceedings of the 2017 IEEE PowerTech, Manchester, UK, 18–22 June 2017. [CrossRef]

16. International Electrotechnical Commission. *Electromagnetic Compatibility (EMC)—Part 4–17: Ripple on d.c. Input Power Port Immunity Test, IEC 61000-4-17*; International Electrotechnical Commission: Geneva, Switzerland, 1999.

17. CENELEC. *European Standard EN 50155, Railway Applications. Electronic Equipment Used on Rolling Stock*; CENELEC: Brussels, Belgium, 2017.

18. Department of Defense. Interface Standard. MIL-STD-704F, Aircraft Electric Power Characteristics. Available online: https://prod.nais.nasa.gov/eps/eps_data/137899-SOL-001-015.pdf (accessed on 31 January 2019).

19. European Telecommunications Standard Institute. *ETSI EN 300 132-3-1, Power Supply Interface at the Input to Telecommunications and Datacom Equipment*; Part 3: Operated by Rectified Current Source, Alternating Current Source or Direct Current Source Up to 400 V; Sub-Part 1: Direct Current Source Up to 400 V; European Telecommunications Standard Institute: Valbonne, France, 2012.

20. International Electrotechnical Commission. *Electromagnetic Compatibility (EMC)—Part 4–29: Testing and Measurement Techniques. Voltage Dips, Short Interruptions and Voltage Variations on d.c. Input Power Port Immunity Tests, IEC 61000-4-29*; International Electrotechnical Commission: Geneva, Switzerland, 2000.

21. Trindade, F.C.L.; Ferreira, T.S.D.; Lopes, M.G.; Freitas, W. Mitigation of fast voltage variation during cloud transients in distribution systems with PV solar farms. *IEEE Trans. Power Deliv.* **2017**, *32*, 921–932. [CrossRef]

22. International Electrotechnical Commission. *IEC 60050—International Electrotechnical Vocabulary (IEV 161-02-25)*; nternational Electrotechnical Commission: Geneva, Switzerland, 1990.

23. International Association of Classification Societies. *Unified requirement UR E5—Voltage and frequency variations. Rev. 1*; International Association of Classification Societies: Hamburg, Germany, 2005.

24. International Electrotechnical Commission. *IEC 60050—International Electrotechnical Vocabulary (IEV 161-02-26)*; International Electrotechnical Commission: Geneva, Switzerland, 1990.

25. International Electrotechnical Commission. *IEC 60050—International Electrotechnical Vocabulary (IEV 161-02-27)*; International Electrotechnical Commission: Geneva, Switzerland, 1990.

26. International Electrotechnical Commission. *Electromagnetic Compatibility (EMC)—Part 4-7: Testing and Measurement Techniques—General Guide on Harmonics and Interharmonics Measurement, for Power Supply Systems and Equipment Connected Thereto, IEC 61000-4-7*; International Electrotechnical Commission: Geneva, Switzerland, 2002.
27. IEEE Std 45.1—2017. *IEEE Recommended Practice for Electrical Installations on Shipboard*; IEEE: New York, NY, USA, 2017.

Article

Application of Spectral Kurtosis to Characterize Amplitude Variability in Power Systems' Harmonics

Jose-María Sierra-Fernández [1,*], Sarah Rönnberg [2], Juan-José González de la Rosa [1], Math H. J. Bollen [2] and José-Carlos Palomares-Salas [1]

[1] PAIDI-TIC-168, Computational Instrumentation, University of Cádiz, Av. Ramón Puyol, 11202 Algeciras, Spain; juanjose.delarosa@uca.es (J.-J.G.d.l.R.); josecarlos.palomares@uca.es (J.-C.P.-S.)
[2] Electric Power Engineering, Luleå University of Technology, Forskargatan 1, 931 87 Skellefteå, Sweden; sarah.ronnberg@ltu.se (S.R.); math.bollen@ltu.se (M.H.J.B.)
*   Correspondence: josemaria.sierra@uca.es

Received: 30 November 2018; Accepted: 2 January 2019; Published: 8 January 2019

Abstract: The highly-changing concept of Power Quality (PQ) needs to be continuously reformulated due to the new schemas of the power grid or Smart Grid (SG). In general, the spectral content is characterized by their averaged or extreme values. However, new PQ events may consist of large variations in amplitude that occur in a short time or small variations in amplitude that take place continuously. Thus, the former second-order techniques are not suitable to monitor the dynamics of the power spectrum. In this work, a strategy based on Spectral Kurtosis (SK) is introduced to detect frequency components with a constant amplitude trend, which accounts for amplitude values' dispersion related to the mean value of that spectral component. SK has been proven to measure frequency components that follow a constant amplitude trend. Two practical real-life cases have been considered: electric current time-series from an arc furnace and the power grid voltage supply. Both cases confirm that the more concentrated the amplitude values are around the mean value, the lower the SK values are. All this confirms SK as an effective tool for evaluating frequency components with a constant amplitude trend, being able to provide information beyond maximum variation around the mean value and giving a progressive index of value dispersion around the mean amplitude value, for each frequency component.

Keywords: harmonics; constant amplitude trend; fourth-order statistics; detection; spectral kurtosis

---

1. Introduction

The current electric power system requires extensive control power electronics within all the stages, from the energy generation stage to consumption units (e.g., inverters, rectifiers, DC/DC converters, etc.) [1–4]. All those power electronics have their own switching frequency, introducing those frequencies in the resultant signal, joined with their harmonics. In addition, non-linear loads, magnetic cores' saturation, or power unbalance introduce frequencies that are multiples and non-multiples of the fundamental one (harmonics, inter-harmonics, and supra-harmonics), as well. Lower frequencies than the fundamental one (the so-called sub-harmonics) also appear.

Calculation techniques can be classified by the statistical order, based on the order of the moment used. Moments have many formulations due to their being few parameters, but the order is related to the maximum number of times input data are multiplied in the same term by itself (the same or other index). A differentiation is made among first- (no multiplication among them) and second-order methods (two elements are multiplied among them, the most common) and Higher-Order Statistics (HOS), where the order is three or more (three or more elements are multiplied among them).

In order to assess the power system distortion level, the traditional procedures have been designed on three main methodological lines: Discrete Fourier Transform (DFT), Wavelet Transforms (WT) [5,6], and recursive-based methods. The two former ones are purely second-order methods, and the latter incorporate many procedures, such as the Kalman filter [7], neural networks [8], the gravity search algorithm [9], the biogeography hybridized algorithm [10], the hybrid firefly algorithm [11], etc. The convergence of the recursive methods is related to signal conditions, so the accuracy changes. On the other hand, wavelet transform results are related to the mother wavelet used [12]. For all these reasons, DFT is the only technique selected by UNE-EN 61000-4-30 [13] and UNE-EN 61000-4-7 [14] for spectral measurement in Power Quality (PQ) evaluation. DFT still has the second-order characteristics, which makes it vulnerable to noise, and it cannot provide characterization beyond the second order.

In PQ, harmonics are measured as instantaneous and time-averaged values, according to the regulations. This work focuses on the detection of frequency components (sub-harmonics, harmonics, or inter-harmonics) with a constant amplitude trend, which implies a permanent distortion; otherwise, a transitory state may be present during the averaged time or the amplitude values may change. Therefore, DFT will be compared with Spectral Kurtosis (SK), a fourth-order technique (HOS), which shows a very good capacity for different stationary signals [15,16]. It has been previously used in PQ evaluation in [17–19] (in addition to other fields, such as insect detection based on vibration [20] or the main application up to now, which is fault detection in rotatory machines [21]), where SK was used for the detection of PQ events, as well as their characterization. In this work, the best configuration for DFT and SK will be studied, for the detection of frequency components with a constant amplitude trend.

The work is structured as follows: In Section 2, SK is studied, then a comparison is made between DFT and SK for detection in Section 3. Then, detection is tested with the current of the arc furnace in Section 4 and with the voltage of the power grid in Section 5. Finally, the conclusions are presented in Section 6.

2. Spectral Kurtosis

While the DFT and the classical power spectral density return amplitude values, without considering any kind of non-stationary information, the SK can target the presence of transients. Indeed, the SK returns the fourth-order statistical moment for each frequency component and, therefore, constitutes a measure of the statistical kurtosis. Thus, unlike the standard deviation, its value is not related to the averaged amplitude, but to the distribution peakedness.

For the SK calculation, the analysis vector must be split into a non-overlapping set of realizations (or segments), and the FFT is applied over them. According to the UNE-EN 61000-4-7, for spectral measurements in PQ, realizations must have a 0.2-s length (i.e., 10/12 cycles of 50/60 Hz). The results of DFT for all realizations are introduced in Equation (1), where the magnitude $X_i(m)$ estimates the mth frequency component under the ith realization index.

$$SK(m) = \frac{M}{M-1} \left[\frac{(M+1) \sum_{i=1}^{M} |X_i(m)|^4}{\left(\sum_{i=1}^{M} |X_i(m)|^2 \right)^2} - 2 \right] \tag{1}$$

The calculation is done for each frequency component, using the estimated amplitude at that frequency for each realization obtained with DFT, as indicated in [15].

To illustrate the procedure, a preliminary example signal (Figure 1) is first analyzed. It is based on a constant amplitude 50-Hz signal, with 230 V and a 2-s length, and different Signal-to-Noise-Ratios (SNR) of coupled random Gaussian noise are considered (10, 20, 30, 40, 50 dB; 50 in the figure and the others for the analysis). Two harmonics are introduced, without a constant amplitude. The fifth harmonic has 30% of the amplitude, in relation to the fundamental one, during the first half of the signal (1 s), and zero amplitude the rest of the time. The seventh has an amplitude of 30% in relation to the fundamental one during the last cycle of the realization. Figure 1 represents one cycle of each realization analyzed. Each realization has a length of 0.2 s, and 10 realizations are analyzed, obtaining a 2 s-length signal . Ten amplitude values per frequency are consequently obtained.

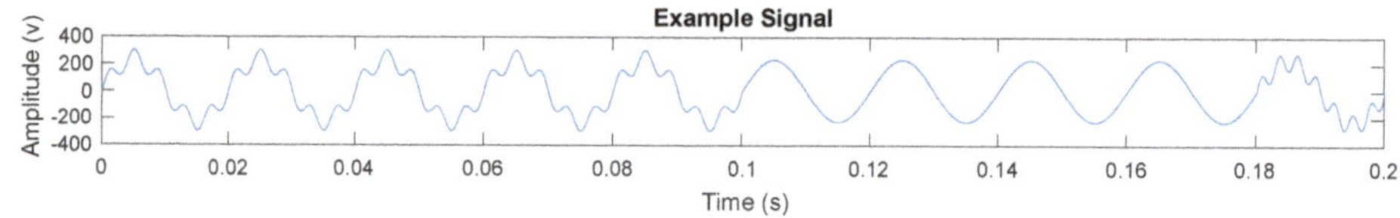

Figure 1. One cycle of each realization analyzed in the example signal for the explanation of Spectral Kurtosis (SK).

In Figure 2, a histogram of amplitudes is represented for each frequency: 50 Hz (fundamental), 250 Hz (fifth harmonic), and 350 Hz (seventh harmonic).

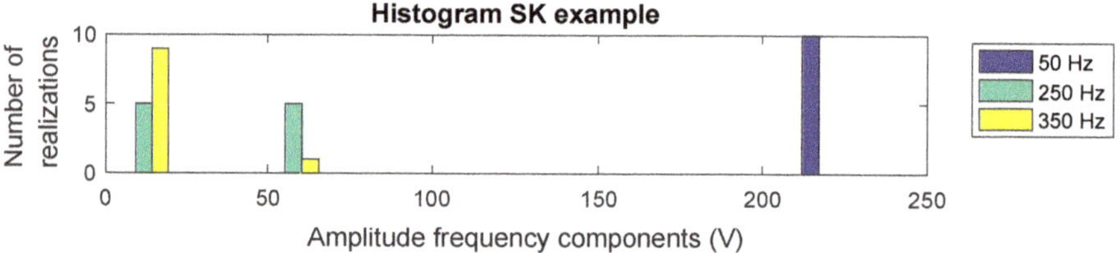

Figure 2. Histogram of amplitude values for each frequency introduced.

The histogram shows a constant amplitude for 50 Hz because the ten realizations have the same amplitude. For 250 Hz, half of the points (five) show an amplitude of 70 and the other half zero amplitude. For 350 Hz, only one point shows an amplitude of 70, and the others show zero amplitude. There are three different amplitude value dispersion situations. These amplitude values obtained from DFT are introduced into the SK formula. However, first, these three situations will be theoretically studied over the SK formula. First, if all amplitudes' values take the same value, the sum result of that value is multiplied by the number of elements summed. With this, the result of Equation (1) is as seen in Equation (2).

$$SK\,(constant) = \frac{M}{M-1}\left[\frac{(M+1)M|X_i(m)|^4}{\left(M|X_i(m)|^2\right)^2} - 2\right] = \frac{M}{M-1}\left[\frac{(M+1)M|X_i(m)|^4}{M^2|X_i(m)|^4} - 2\right] = $$
$$= \frac{M}{M-1}\left[\frac{(M+1)}{M} - 2\right] = \frac{M}{M-1}\left[\frac{(M+1-2M)}{M}\right] = \frac{M}{M-1}\frac{(1-M)}{M} = -1 \tag{2}$$

Therefore, the ideal returned value for a completely constant amplitude situation is -1; noise can affect this, but SK is very noise resistant. The second situation indicated is a signal with the half values with zero amplitude and half the values with an amplitude, which implies a great change (∞ from zero) in half the signal length. In this situation, the sum result is the summed element multiplied by half the number of the added elements. With that, the result of Equation (1) is as seen in Equation (3).

$$SK\,(equal) = \frac{M}{M-1}\left[\frac{(M+1)M/2|X_i(m)|^4}{\left(M/2|X_i(m)|^2\right)^2} - 2\right] = \frac{M}{M-1}\left[\frac{(M+1)M/2|X_i(m)|^4}{(M/2)^2|X_i(m)|^4} - 2\right] = $$
$$= \frac{M}{M-1}\left[\frac{M+1}{M/2} - 2\right] = \frac{M}{M-1}\left[\frac{M+1-2(M/2)}{M}\right] = \frac{M}{M-1}\frac{1}{M/2} = \frac{2}{M-1} \tag{3}$$

Now, the SK value is a near-to-zero value, nearer as more realizations are used. In the example considered, 10 realization are used, so $2/(10-1) = 0.2222$ is obtained. Finally, to simulate impulsive behavior, one point with a non-zero amplitude value and the rest with a zero amplitude value (which implies a punctual ∞ increment) are studied. In this situation, the sum result is a single value of the sum. With that, the result of Equation (1) is as seen in Equation (4).

$$SK\,(max) = \frac{M}{M-1}\left[\frac{(M+1)|X_i(m)|^4}{\left(|X_i(m)|^2\right)^2} - 2\right] = \frac{M}{M-1}\left[\frac{(M+1)|X_i(m)|^4}{|X_i(m)|^4} - 2\right] = $$
$$= \frac{M}{M-1}\left[M + 1 - 2\right] = \frac{M}{M-1}\left[M - 1\right] = M \tag{4}$$

Here it can be seen that the SK value for this situation is the number of realizations (actually, it is the maximum SK value that can be returned by SK), in the considered example with 10 realizations. All this was studied in [22]. SK values are shown in Table 1.

Table 1. SK values for the frequencies in the example signal.

Freq	50 Hz	250 Hz	350 Hz
SNR		SK	
10	−0.9998	0.2206	9.6754
20	−1	0.2219	9.9774
30	−1	0.2223	9.9945
40	−1	0.2222	9.9994
50	−1	0.2222	9.9998

Noise contamination affects the theoretical values calculated for non-noise-contaminated situations. However, for the five digits represented, for SNR of 20 dB or higher, the constant amplitude situation (50 Hz) is marked with −1; only an SNR of 10 dB affects the constant amplitude value.

In relation to the half values affected by the huge amplitude change (250 Hz), with less noise contamination (higher SNR), a value nearer to 0.2222 (the value calculated for a non-noise situation) is obtained. However, even for the 10-dB SNR situation, 0.2206 is obtained, which implies an error of less than 1%, and the values are the same as the ideal one for an SNR over 40 dB.

For impulsive behavior (only one point is affected by a huge amplitude increase) (350 Hz), as in the 250-Hz case, as less noise is applied, a value nearer to 10 is obtained. However, with an SNR of 20 dB, an error lower than 0.5% is obtained, and for an SNR of 50 dB, the error is lower than 20 ppm.

With all this, with an SNR over 20 dB, SK can detect with an acceptable error these three situations, and almost no error for an SNR over 50 dB. Specifically, the constant amplitude is the focus of this work. Additionally, SK has perfectly differentiated two spectral components with the same amplitude change (250 Hz and 350 Hz), but with different time behaviors, an impossible task for a second-order method.

This work is centered on the capacities of the SK for the detection of frequencies with a constant amplitude trend (−1 value). This feature is tested by generating variations around a constant level (unitary amplitude). Four signal lengths have been considered: 10, 20, 50, and 200 seconds (50, 100, 250, and 1000 realizations). Amplitude changes from 10–30% are applied over a length up to 20% of their total lengths. SK has been proven as a noise-resistant technique in the first case. However, in practice, the voltage signal obtained from a power system has an SNR ranging from 50–70 dB [23]. For that, hereinafter, an SNR of 50 dB will be used, due to it having highest contamination in this range. The results of the case are shown in Figure 3.

The horizontal axis shows the relative length of the changed amplitude segment in relation to the whole signal length. The vertical axis shows SK. In this graph, five groups of lines are observed, each one related to a different amplitude change, indicated in the graph on the right. All lines related to the same amplitude change looks like a single line, although they correspond to different signal lengths. This fact illustrates that SK is sensitive to the relative disturbance length in relation to the total signal length, and not the disturbance length. In addition, more amplitude change involves higher SK values. Fifteen percent is needed to get SK over −0.99 with a relative length of 12%. As higher amplitude change is observed, a smaller defective length would be needed to reach this level. For instance, only a relative length of 6% with an amplitude change of 20% is needed, or a relative length less than 4% with an amplitude change of 25%, or a relative length of 2% with an amplitude change of 30%.

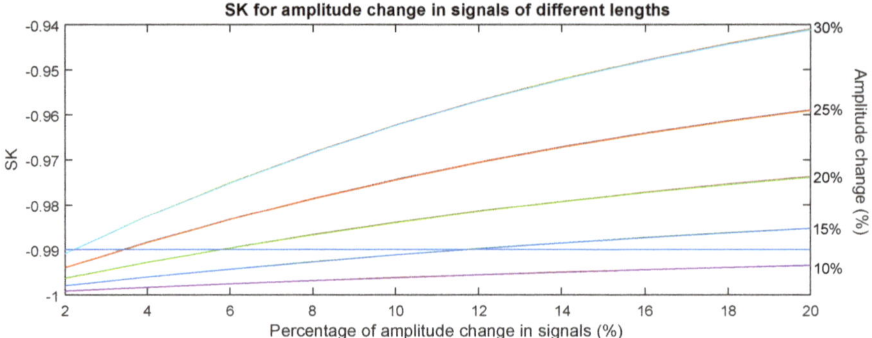

Figure 3. SK values for different amplitude changes and lengths.

3. SK vs. DFT in the Harmonic Evaluation

In this section, a synthetic signal, based on the one used in [24,25], is studied. This signal represents a typical industry waveform, which comprises the effects of power electronic devices, Variable Frequency Drives (VFDs), and arc furnaces. The signal under test has a 10-s length, with a constant amplitude, up to the ninth harmonic. In addition, temporal harmonic distortion has been introduced in harmonic of orders of 11, 13, and 15. The signal harmonic composition is detailed in Table 2. For signal generation, a sampling frequency of 2560 Hz is used, and Gaussian noise is coupled with a 50-dB SNR.

Table 2. Harmonic composition of the signal.

Harm Order	Amp (V)	Phase (°)	Evolution	Start (s)	End (s)
1st	1.5	80	Constant	0	10
3rd	0.5	60	Constant	0	10
5th	0.2	45	Constant	0	10
7th	0.15	36	Constant	0	10
9th	0.1	30	Constant	0	10
11th	0.3	0	Δ25%	0.9	1.1
13th	0.3	0	Δ25%	1	2
15th	0.3	0	Δ25%	0.9	1.2

A time domain signal is obtained, composed of multiple frequencies. SK is based on DFT, so first, DFT must be applied. As indicated before, the first step consists of the signal segmentation into 0.2 s-length realizations (50 realizations are obtained), according to the PQ spectral analysis in UNE-EN 61000-4-30 [13] and UNE-EN 61000-4-7 [14]. The amplitudes returned by DFT are shown in Figure 4. In the left graph, all frequencies involved in this study are shown, and in the right graph, a zoom over the changing amplitude components is applied.

Different levels are observed in the left parts of the graphic representation. All are almost constant, except three of them, which represent the components with frequencies of variable amplitudes. In the graph on the right, paying attention to the 11th harmonic, two points fall outside the normal value. However, the length of this disturbance is the same as the realization length (0.2 s). This is due to the fact that the change of the level starts in one realization and ends in another, so it is shown as two small changes of the level. The 13th and 15th harmonics have a longer changed amplitude length than a realization's length, so the maximum amplitude change can be read properly. In the 15th harmonic, first, half the realization has an amplitude change, and then, the complete realization has the proper amplitude change. In the 13th harmonic, five realizations with the proper amplitude change are given.

Even having the segment with the amplitude changed, the same length as a realization, the 11th harmonic has been observed as two realizations with the amplitude changed, with a lower amplitude change. This proves that DFT is sensitive to the disturbance starting point, due to the data segmentation performed. However, if we want to perform a detection of the amplitude evolution, this segmentation is needed. Moreover, as said before, this segmentation follows the regulations and standard requirements. In order to maximize the detection capacities, another DFT is calculated, but the starting point is delayed half a realization length. The 11th and 15th harmonics are analyzed again, and the results are shown for both DFTs (normal and delayed) in Figure 5.

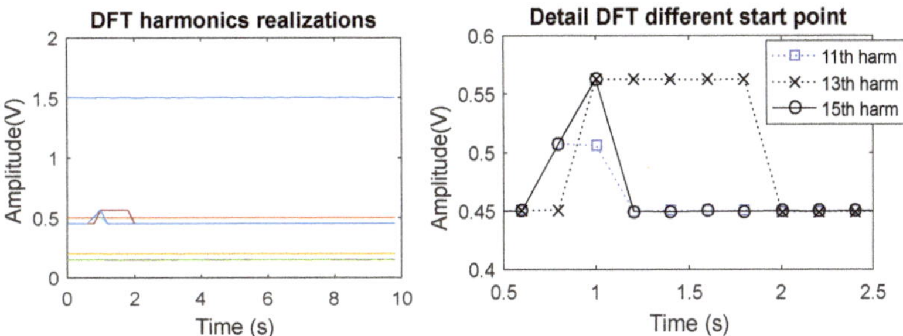

Figure 4. DFT for all harmonics introduced, including the fundamental one in the (**left**) graph, and the detail of the amplitude change harmonics in the (**right**) graph.

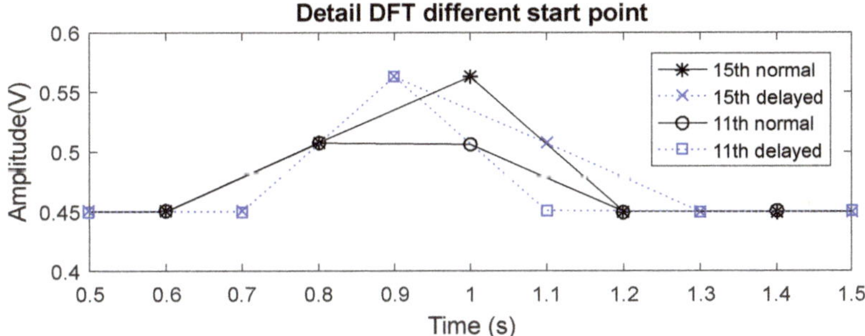

Figure 5. DFT for changed amplitude harmonics, for different starting points.

The 15th harmonic shows a similar response in the normal and delayed DFT. The only difference that can be observed is that in the normal DFT, the first realization is detected with half the amplitude change and the second one with the maximum amplitude, and the opposite occurs in the delayed DFT. Amplitude change starts in 0.9 s with a 0.3-s length. A 0.2-s length per realization is taken, starting at zero for the normal DFT and starting the delayed one in the second 0.1 (half the realization length). The starting points are in the middle of a realization for the normal DFT, but at the beginning of a realization for the delayed one.

However, the 11th harmonic shows a different response for the normal and delayed DFT. In normal DFT, it shows the same response previously seen, and in the delayed one the complete amplitude change in one realization can be seen, so delayed DFT reads the full amplitude change in only one realization. The amplitude change starting point links with the realization starting point of delayed DFT and with half the realization length of normal DFT.

With the double DFT, more information is obtained; even more in amplitude change segments with a small length, which can be split into two realizations. This creates a higher amplitude change read in one of the DFTs, due to the disturbance inside a realization and not being split. Therefore, for an amplitude stability analysis, it is recommendable to use the double DFT. However, up to this point, graphs against time have been obtained. In order to show a single value per frequency component, SK is applied. If we want to compare SK and DFT, single values for DFTs need to be obtained per frequency component. These values are the averaged amplitude value, maximum amplitude change, and maximum amplitude change relative to the averaged amplitude value, and they are shown in Table 3.

With the averaged amplitude value, the normal amplitude is seen, and it almost does not change, even in the longest amplitude change situations. The maximum amplitude change value is an absolute parameter, and the averaged amplitude value is needed in order to understand the change in relation to the normal situation (relative change). Relative amplitude change in the constant amplitude situation (third harmonic) shows a value of 0.01, due to noise. For amplitude changing components, a value of 0.25 is obtained for the 13th and 15th harmonics and delayed DFT of the 11th harmonic (the normal one shows a smaller relative change). It is 25% of the relative change introduced.

Table 3. Single value analysis for different harmonic orders, in DFT Normal (N) and Delayed (D).

Order	3rd		11th		13th		15th	
DFT	N	D	N	D	N	D	N	D
Avg	0.50	0.50	0.30	0.30	0.31	0.31	0.30	0.30
Max − Min	0.00	0.00	0.04	0.08	0.08	0.08	0.08	0.08
(Max − Min)/Avg	0.01	0.00	0.13	0.25	0.25	0.25	0.25	0.25
SK	−1.000	−1.000	−0.997	−0.994	−0.973	−0.976	−0.992	−0.992

All those indicators can detect amplitude change with a single DFT realization out of the indicated conditions, but cannot differentiate among one or more realizations. In this work, we want to study the constant amplitude trend, and with that aim, we need to know if the amplitude change occurs once or if it appears continuously over time. SK values show a value of almost −1 for the constant amplitude component, a value near −0.99 for a single realization being affected (0.2 s of 10 s implies 2%), and −0.97 for the amplitude change with a higher length (1 s of 10 s implies 10%). As seen in the previous section, SK can differentiate among short or long length amplitude changes, taking into consideration the amplitude change value as well.

In the 11th harmonic, the amplitude change starting point affects the SK value, so an SK analysis for different starting points must be done. As shown in Figure 3, different relative lengths for the amplitude change segment are considered. Now, two signal lengths are considered, 10 and 20 s. The amplitude change starting point is the starting point of the second realization of normal DFT. Therefore, normal SK sees the amplitude change in the complete realization, and delayed SK splits it. The results of this analysis are shown in Figure 6.

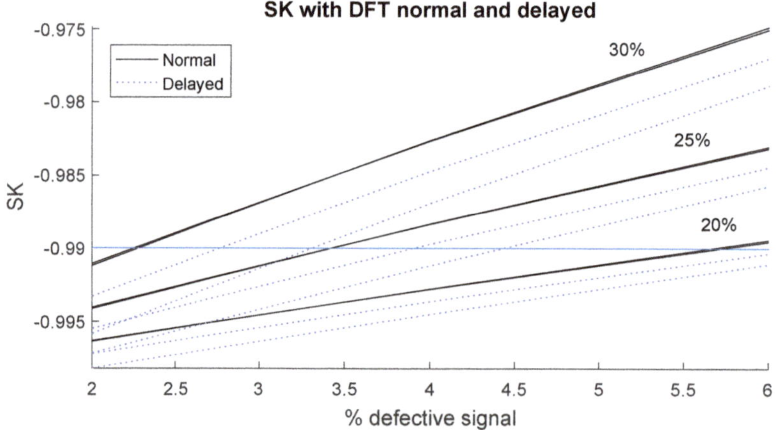

Figure 6. SK values for different amplitude changes and lengths, using double DFT.

In this figure, it can be observed that delayed SK, the one that splits the amplitude change segment into more realizations, returns lower SK values. Therefore, normal SK returns higher SK values in this situation. If the disturbance starting point changes, delayed SK can show higher values than normal SK. This shows that double SK ensures that the optimum SK value is obtained, associated with the worst situation. The solid line is associated with normal SK, and for both signal lengths, this line is overlapped. However, delayed SK, represented as a dotted line, takes a different position with each signal length. A great difference between the dotted line and solid line can even be seen, the absolute difference being lower than 0.05. Although the same relative length is used, as a higher total length is considered, less difference is observed. Realization length is constant, and the SK response difference is due to measuring one realization with the full amplitude change or two realizations with half the amplitude change. As more realizations are affected by the amplitude change, this difference has a lower impact.

Up to this point, only amplitude increases have been studied. However, if we want to study amplitude stability, increases or reductions can occurs in the amplitude value. Now, an amplitude increase and a reduction of up to 50% have been applied (now, increase or reduction is expressed as "amplitude change factor", which is the final amplitude/initial amplitude, reduction values below one being being increased values over one), over three different relative lengths (2%, 4%, and 6%) and using two different total lengths (10 and 20 s). The amplitude change starting point, as in the previous simulation, is the second realization of the normal SK. Normal and delayed SK values for this case are shown in Figure 7. An SNR of 50 dB has been considered.

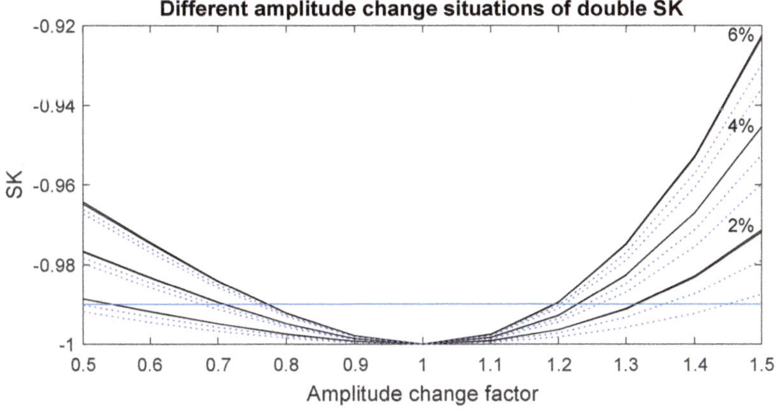

Figure 7. SK values for amplitude reduction or increase, using double DFT.

The SK value can be observed for amplitude reduction or increase. Smaller SK values are obtained when the amplitude reduces. As SK is the fourth-order moment, a value increase has more effect, due to it being magnified by the fourth power. As in the previous case, normal SK shows the same values for total length, and delayed SK shows slightly lower values. As more relative length is affected, higher SK values are given. In Table 4, the amplitude change needed for reach -0.99 SK values is shown, for the maximum value obtained from the double SK analysis.

Table 4. Amplitude change to reach a -0.99 SK value for different defective lengths.

Relative Length	Amplitude Up	Increment	Amplitude Down	Decrement
2%	1.31	0.31	0.54	0.46
4%	1.22	0.22	0.71	0.29
6%	1.19	0.19	0.77	0.23

As the relative length is affected by amplitude change increases, a smaller amplitude change is needed in order to obtain an SK value over -0.99. In this situation, SK under -0.99 is the condition to set a frequency component as a constant amplitude trend. As explained up to this point, SK will analyze amplitude changes, not only punctual, but along all of the signal, taking into consideration the duration of each amplitude change.

4. Arc Furnace Current Signal

As a real-life application, the electric current waveform from an arc furnace during the melting process is studied, for the three phases, in three time periods. During the melting process, an arc furnace requires a high current level, while it takes a similar level of power during this process. However, the current level is not properly constant, and that situation is a perfect example for the aim of this work. In this situation, low fluctuations are present, and the signal is detected as a constant amplitude trend. In Figure 8, the amplitude values for the current signal can be seen, for the fundamental frequency (50 Hz) measured with DFT. Each segment has a length of 40 s, so with a segmentation of 0.2 s, 200 realizations result per segment.

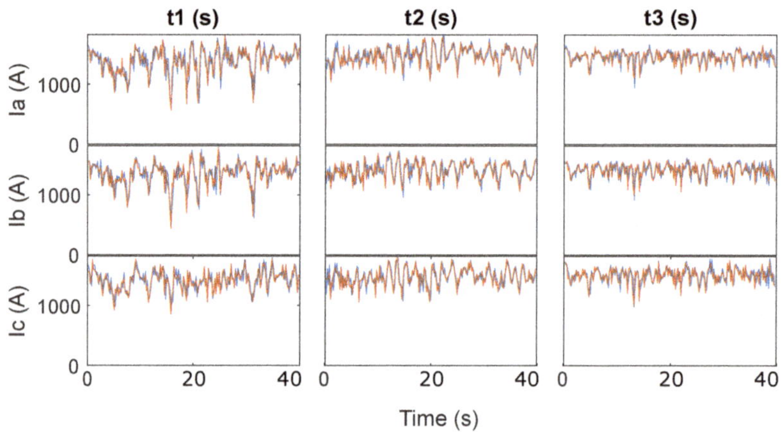

Figure 8. (Double) DFT amplitude measured for 50 Hz in the arc furnace current signal.

In this graph, oscillations in the fundamental component are observed. These oscillations are higher in the first time period, lower in the second one, and much lower in the third one. Moreover, Phases A and B show higher oscillations than Phase C in the first time period. As previously seen, these oscillations will be studied using normal and delayed SK. All SK values for fundamental frequency components are shown in Table 5.

Table 5. SK values for 50-Hz frequency components in the arc furnace current signal.

	t1		*t2*		*t3*	
SK	**N**	**D**	**N**	**D**	**N**	**D**
Ia	-0.916	-0.920	-0.963	-0.963	-0.982	-0.982
Ib	-0.923	-0.924	-0.960	-0.958	-0.976	-0.975
Ic	-0.948	-0.948	-0.960	-0.961	-0.976	-0.974

As explained before, an SK value of -1 is related to a constant amplitude trend, and higher SK values are obtained (more positive or less negative) as higher amplitude variations are observed. In Figure 8, higher amplitude variations are observed in Time Period 1, and due to those oscillations, values in the range of $[-0.948, -0.916]$ are observed, for all phases.

In the second period, a few level variations are observed, lower than in the first one. The SK values obtained are lower also in comparison to the ones obtained in the first period, all of them being in the range of $[-0.963, -0.958]$.

In the third time period, lower variations are observed than in the second one. The SK values obtained are again lower in comparison to those obtained in the second period, all of them being in the range of $[-0.982, -0.974]$.

In addition, if Time Period 1 is observed, as indicated before, lower amplitude variations are observed in Phase C than in Phases A and B. SK values for Phase C are -0.948, -0.948 (normal and delayed), and for Phases A and B, SK values are in the range $[-0.916, -0.924]$.

In this case, it is confirmed that as lower amplitude oscillations are observed in the frequency component studied, a lower SK value is obtained. In fact, all SK values are under -0.9, due to all amplitude values being near the averaged value. However, none of them obtain an SK value of -0.99 or lower, so no constant amplitude trend can be confirmed, even with the amplitude evolution observed in the third period, in Phase A, where amplitude values are really concentrated around the averaged value.

The SK value of zero is associated with noise (random oscillations due to noise). The third harmonic, which only reaches a maximum amplitude around 120 (when the fundamental frequency reaches an amplitude of 1600), implying 7%, is studied. It shows random oscillations. Amplitude variations for normal and delayed DFTs for the third harmonic are shown in Figure 9.

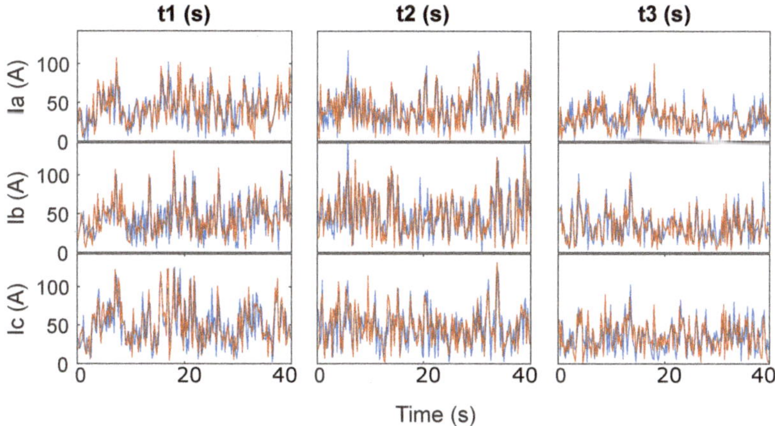

Figure 9. (Double) DFT amplitude measure for 150 Hz in the arc furnace current signal.

In Figure 9, these random amplitude oscillations can be observed, taking values from 0–120. Normal and delayed SK values are shown in Table 6.

Table 6. SK values for 150-Hz frequency components in the arc furnace current signal.

	t1		*t2*		*t3*	
SK	**N**	**D**	**N**	**D**	**N**	**D**
Ia	−0.194	−0.105	0.141	0.004	−0.004	0.234
Ib	0.025	0.025	0.074	0.060	0.263	0.082
Ic	0.014	0.015	−0.071	−0.107	0.044	−0.070

As explained before, random amplitude values are related to a zero SK value, as seen in this table. For all the time periods in all three phases, the SK values obtained are very near to zero. Then, in order to compare amplitude values' dispersion, histograms for the amplitude values will be represented.

A separate histogram is represented for each of the data used. As an example, Phase A is taken for all three time periods in both situations, fundamental frequency, and the third harmonic, resulting in six histograms. Over each histogram, the SK value obtained for that signal is indicated. For histograms, normal DFT is used and normal SK is indicated. Histograms are shown in Figure 10.

Now, in the fundamental frequency, it is easy to see how the values' concentration around the mean value is related to the SK values, returning lower values as concentration increases. In the third harmonic, it can be observed that amplitude values are not concentrated; they are distributed in a wide range, following a random distribution, associated with a kurtosis of zero (a near to zero value).

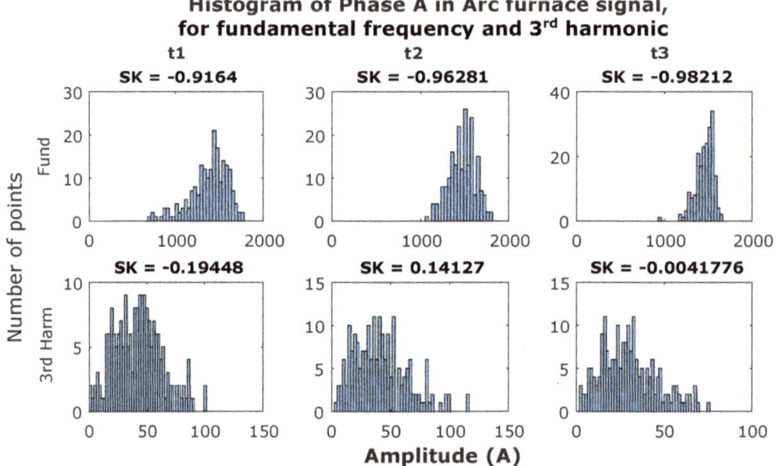

Figure 10. Histogram of Phase A of the arc furnace current signal.

5. Power Signal

An hour of the voltage waveform from the power grid in the research group laboratory facilities is analyzed. As explained before, a segmentation with 0.2 s, non-overlapped, has been done, and DFT has been calculated (which resulted in 18,000 realizations). The SK has been calculated over the results of that calculation. First, fundamental amplitude is shown for all segments in Figure 11, as the results of normal DFT.

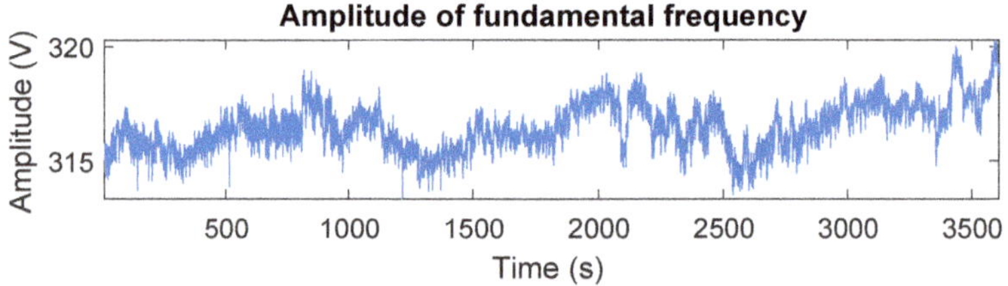

Figure 11. DFT amplitude measure for the fundamental frequency in the power signal.

Only small amplitude changes are observed in the fundamental frequency amplitude, taking amplitude values from 314 V–320 V, approximately 2% of variation. The nominal voltage level in Spain, by regulation, is 325 V in amplitude, with an allowed variation margin of ±10%. The values given are in the allowed range. The amplitude of a few harmonics are studied, due to their amplitude difference. They are plotted in Figure 12, a different figure than was used, due to the amplitude difference from the fundamental one.

The fifth harmonic is the one with a higher amplitude, and even it has an amplitude lower than 3% in relation to the fundamental amplitude. Even with these amplitude differences, SK evaluates amplitude changes in each frequency separately. The second and fourth harmonics have amplitude values centered on zero, with random oscillations. All other harmonics show a evolution around an average value, different for each frequency. They do not show a constant amplitude evolution; each one shows a different evolution, with low amplitude variation.

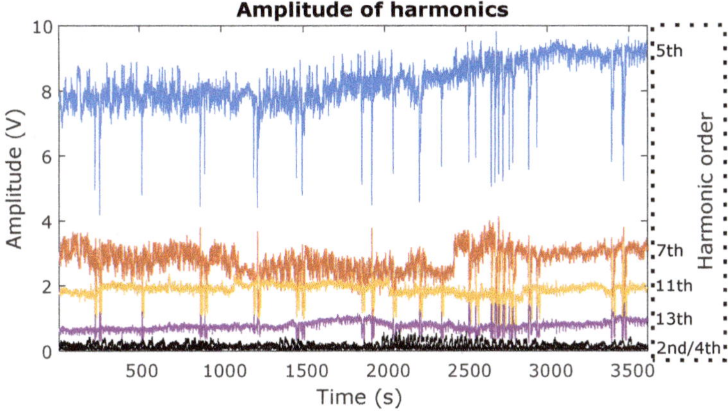

Figure 12. DFT amplitude measure for the harmonics of the fundamental frequency in the power signal.

In Table 7 the SK value for normal and delayed SK, the averaged DFT value (Avg), and the maximum (Max) and minimum (Min) value of the main data series (without the peaks) are shown for each studied frequency. Change is expressed in % in relation to the average amplitude for that frequency (Inc(%)).

Table 7. SK and DFT values for the studied frequencies in one hour of a power signal.

Harmonic	SK		DFT			
Order	N	D	Avg	Max	Min	Inc(%)
1.000	−1.000	−1.000	316.475	320.000	313.000	2
2.000	−0.080	−0.078	0.193	0.698	0.001	361
4.000	0.064	0.066	0.088	0.384	0.001	436
5.000	−0.975	−0.975	8.250	9.581	6.855	33
7.000	−0.919	−0.919	2.755	3.676	1.969	62
11.000	−0.938	−0.938	1.847	2.356	1.449	49
13.000	−0.848	−0.847	0.770	1.072	0.477	77

The SK values obtained for normal and delayed analysis are really similar, even in the second and fourth harmonic, which are only the noise levels. A long-term amplitude evolution was studied, and as there were no fast changes in the amplitude level, in this situation, there were no differences.

For the fundamental frequency, really small amplitude oscillations are observed (2% relative variation), so a constant amplitude trend can be considered. An SK value of −1 was obtained, so the analysis confirms this constant amplitude trend. As higher relative amplitude variations in relation to the averaged value were observed, higher SK values were obtained. This is confirmed comparing the relative variation and the SK value in Table 7 for all frequencies. The fifth harmonic has a relative amplitude change of 33% and an SK value of −0.975; the next one is the eleventh harmonic, with a relative change of 49% and an SK of −0.938; then, seventh harmonic, with a relative change of 62% and an SK of −0.919; and finally, the thirteenth harmonic, with a relative change of 77% and an SK of −0.848.

The second and fourth harmonic, the ones related to noise, showed high relative change values, due to their really small averaged amplitude values. These harmonics showed random amplitude variation. With random amplitude variations, a near to zero SK value was obtained.

A histogram was calculated per each analyzed frequency in this case, but the fundamental frequency. In order to understand dispersion in relation to the mean value, histograms starting with zero amplitude were calculated. When this was done for fundamental frequency, all points were concentrated in a single bar (not depicted). Histograms for all other frequencies are plotted in Figure 13, with the associated SK value over each one.

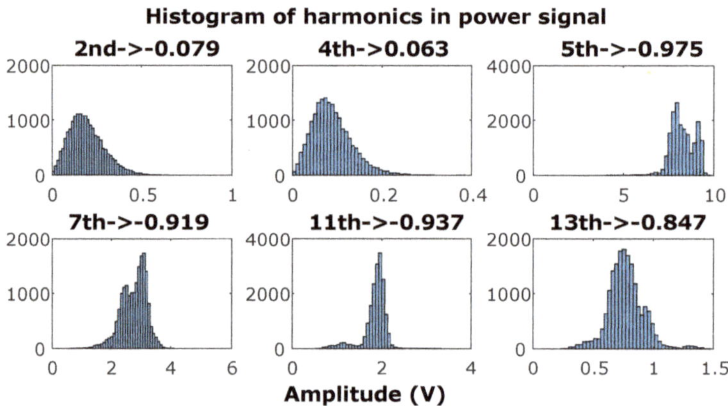

Figure 13. Histogram for the harmonics of the fundamental frequency in the power signal.

As in the arc furnace signal analysis, components with an SK value near zero are associated with non-concentrated values. In fact, these were the second and fourth harmonics, related to noise, which showed random amplitude oscillations. Lower SK values were obtained as the amplitude values were more concentrated around the mean value. This can be observed in the thirteenth (SK value -0.847) and seventh harmonic (SK value -0.919), where the point distribution presented two tails, so there were few values higher or lower than the mean value. The eleventh harmonic (SK value -0.937) only showed one tail to lower values in the point distribution, so few points took lower values. The fifth harmonic (SK value of -0.975) showed a point distribution without tails, so points were concentrated around the mean value, but not enough to be considered a constant amplitude trend.

6. Conclusions

In this research work, an estimator for the SK is used in order to characterize amplitude variability in power systems' spectra. More precisely, SK is studied to evaluate amplitude trends on each spectral component, focusing on the detection in constant amplitude trends.

The SK estimator shows flexible detection capacity, based on points' concentration around the mean value, and not if only one value is outside the threshold, which can cause false detections. This implies a detection of a big amplitude change with a short length or a small amplitude change with a long length. In this way, if SK indicates a value very near -1, a constant amplitude trend is detected, due to really small amplitude changes being detected.

The relation among the SK value and the amplitude values' dispersion, per each frequency component, is confirmed with two real-life signals. The first one is an electric current waveform form an arc furnace, taken at three different time periods of 40 s each. As the power level during the melting process is similar, the current level shows low variations. Amplitude value changes around the mean value are observed, and as postulated in this work, lower SK values are obtained when the amplitude values are more concentrated around the mean value. The second signal used is an hour of voltage signal taken from the power grid of the University of Cádiz. Even in a power system, with a nominal level, the amplitude value changes, and the 50-Hz frequency component shows low amplitude variations; it is set as a constant amplitude trend. SK analyzes each frequency component separately, so variations of each frequency component are only related to its mean value, allowing one to study in the same SK analysis frequency components with amplitude mean values greater than 100-times lower, in this case. In the study of these harmonics, it is confirmed again that points' dispersion around the mean value is related to the SK value, giving lower SK values as the amplitude values are more concentrated around the mean value.

With all this, it is shown and confirmed that SK is a good tool to detect a constant amplitude trend, being able to return information beyond maximum variation around the mean value and giving a progressive index of value dispersion around the mean value, for each frequency component. In addition, as frequencies are studied separately, a signal that shows different amplitudes at different frequencies can be examined without any problem (e.g., voltage or current harmonics).

Author Contributions: Writing original draft, methodology and visualization, J.-M.S.-F.; data curation, S.R.; funding acquisition, project administration, writing original draft, writing review and editing, J.-J.G.d.l.R.; conceptualization, supervision and validation, M.H.J.B.; investigation, J.-C.P.-S.

Funding: This research was funded by the Spanish Government with the research Project TEC2010-19242-C03-03 (Smart Inverter for Distributed Energy Resources - Higher-Order Statistics Applied to Power Quality (SIDER-HOSAPQ)), TEC2013-47316-C3-2-P (Smart Community Energy Management System-Advances Techniques for Power Quality Reliability (SCEMS-AD-TEC-PQR)), and TEC2016-77632-C3-3-R (Control and Management of Isolable NanoGrids: Smart Instruments for Solar forecasting and Energy Monitoring (COMING-SISEM)).

Acknowledgments: This research is supported by the Spanish Ministry of Economy, Industry and Competitiveness and the EU (Spanish Research Agency/European Regional Development Fund/European Union (AEI/FEDER/UE Spanish acronym, in English SRA/ERDF/EU)) in the workflow of the State Plan of Excellency and Challenges for Research, via the project TEC2016-77632-C3-3-R-Control and Management of Isolable NanoGrids: Smart Instruments for Solar forecasting and Energy Monitoring (COMING-SISEM), which involves the development of new measurement techniques applied to monitor the PQ in micro-grids. Furthermore, we express our gratitude to the Andalusian Government for funding the Research Group (Andalusian program of research, development and innovation, information and communication technologies (PAIDI-TIC-168)) in Computational Instrumentation and Industrial Electronics (ICEI).

Conflicts of Interest: The authors declare no conflict of interest.

References

1. Lei, X.; Lerch, E.; Povh, D.; Renz, K. Power Electronic Equipment for Increasing Capability of Transmission Links in Interconnected Power Systems. *IFAC Proc. Vol.* **1997**, *30*, 235–240. [CrossRef]
2. Chen, Z. 4-An overview of power electronic converter technology for renewable energy systems. In *Electrical Drives for Direct Drive Renewable Energy Systems*; Woodhead Publishing Series in Energy, Woodhead Publishing: Sawston, UK; 2013; pp. 80–105. [CrossRef]
3. Khan, O.; Xiao, W. Review and qualitative analysis of submodule-level distributed power electronic solutions in {PV} power systems. *Renew. Sustain. Energy Rev.* **2017**, *76*, 516–528. [CrossRef]
4. Shahbazi, M.; Khorsandi, A. Power Electronic Converters in Microgrid Applications. In *Microgrid*; Mahmoud, M.S., Ed.; Butterworth-Heinemann: Oxford, UK, 2017; Chapter 10, pp. 281–309.
5. Nath, S.; Sinha, P.; Goswami, S.K. A wavelet based novel method for the detection of harmonic sources in power systems. *Int. J. Electr. Power Energy Syst.* **2012**, *40*, 54–61. [CrossRef]
6. Mostavi, A.; Kamali, N.; Tehrani, N.; Chi, S.W.; Ozevin, D.; Indacochea, J.E. Wavelet based harmonics decomposition of ultrasonic signal in assessment of plastic strain in aluminum. *Measurement* **2017**, *106*, 66–78. [CrossRef]
7. Singh, S.K.; Sinha, N.; Goswami, A.K.; Sinha, N. Several variants of Kalman Filter algorithm for power system harmonic estimation. *Int. J. Electr. Power Energy Syst.* **2016**, *78*, 793–800. [CrossRef]
8. Ray, P.K.; Subudhi, B. Neuro-evolutionary approaches to power system harmonics estimation. *Int. J. Electr. Power Energy Syst.* **2015**, *64*, 212–220. [CrossRef]
9. Singh, S.K.; Kumari, D.; Sinha, N.; Goswami, A.K.; Sinha, N. Gravity Search Algorithm hybridized Recursive Least Square method for power system harmonic estimation. *Eng. Sci. Technol. Int. J.* **2017**, *20*, 874–884. [CrossRef]

10. Singh, S.K.; Sinha, N.; Goswami, A.K.; Sinha, N. Power system harmonic estimation using biogeography hybridized recursive least square algorithm. *Int. J. Electr. Power Energy Syst.* **2016**, *83*, 219–228. [CrossRef]

11. Singh, S.K.; Sinha, N.; Goswami, A.K.; Sinha, N. Robust estimation of power system harmonics using a hybrid firefly based recursive least square algorithm. *Int. J. Electr. Power Energy Syst.* **2016**, *80*, 287–296. [CrossRef]

12. Lachman, T.; Memon, A.P.; Mohamad, T.; Memon, Z. Detection of power quality disturbances using wavelet transform technique. *Int. J. Adv. Sci. Arts* **2010**, *1*, 1–13.

13. UNE-EN 61000-4-30. Testing and Measurement Techniques. Power Quality Measurement Methods. European Version in Spanish of IEC 61000-4-30. IEC Version. 2015. Available online: https://www.aenor.com/normas-y-libros/buscador-de-normas/iec?c=25344 (accessed on 7 January 2019).

14. UNE-EN 61000-4-7. Testing and Measuremet Techniques. General Guide on Harmonics and Interharmonics Measurements and Instrumentations, for Power Supply Systems and Aquipment Conneted Thereto. European Version in Spanish of IEC 61000-4-7. IEC Version. 2002. Available online: https://webstore.iec.ch/publication/4228 (accessed on 7 January 2019).

15. Antoni, J. The spectral kurtosis of nonstationary signals: Formalisation, some properties, and application. In Proceedings of the 2004 12th European Signal Processing Conference, Vienna, Austria, 6–10 September 2004; pp. 1167–1170.

16. Antoni, J. The spectral kurtosis: A useful tool for characterising non-stationary signals. *Mech. Syst. Signal Process.* **2006**, *20*, 282–307. [CrossRef]

17. Palomares-Salas, J.; De La Rosa, J.; Agüera-Pérez, A.; Sierra-Fernández, J. Smart grids power quality analysis based in classification techniques and higher-order statistics: Proposal for photovoltaic systems. In Proceedings of the 2015 IEEE International Conference on Industrial Technology (ICIT 2015), Seville, Spain, 17–19 March **2015**; pp. 2955–2959. [CrossRef]

18. De la Rosa, J.J.; Sierra-Fernández, J.M.; Palomares-Salas, J.C.; Agüera-Pérez, A.; Montero, A.J. An application of spectral kurtosis to separate hybrid power quality events. *Energies* **2015**, *8*, 9777–9793. [CrossRef]

19. Quirós-Olozábal, A.; González-De-La-Rosa, J.J.; Cifredo-Chacón, M.A.; Sierra-Fernández, J.M. A novel FPGA-based system for real-time calculation of the Spectral Kurtosis: A prospective application to harmonic detection. *Measurement* **2016**, *86*, 101–113. [CrossRef]

20. De la Rosa, J.J.G.; Moreno-Muñoz, A.; Gallego, A.; Piotrkowski, R.; Castro, E. On-site non-destructive measurement of termite activity using the spectral kurtosis and the discrete wavelet transform. *Measurement* **2010**, *43*, 1472–1488. [CrossRef]

21. Fan, Z.; Li, H. A hybrid approach for fault diagnosis of planetary bearings using an internal vibration sensor. *Measurement* **2015**, *64*, 71–80. [CrossRef]

22. Vrabie, V.; Granjon, P.; Serviere, C. Spectral kurtosis: From definition to application. In Proceedngs of the 6th IEEE International Workshop on Nonlinear Signal and Image Processing (NSIP 2003), Grado, Italy, 8–11 June 2003.

23. Tomic, J.J.; Kusljevic, M.D.; Vujicic, V.V. A New Power System Digital Harmonic Analyzer. *IEEE Trans. Power Deliv.* **2007**, *22*, 772–780. [CrossRef]

24. Ray, P.K.; Subudhi, B. BFO optimized RLS algorithm for power system harmonics estimation. *Appl. Soft Comput.* **2012**, *12*, 1965–1977. [CrossRef]

25. Singh, S.K.; Sinha, N.; Goswami, A.K.; Sinha, N. Optimal estimation of power system harmonics using a hybrid Firefly algorithm-based least square method. *Soft Comput.* **2017**, *21*, 1721–1734. [CrossRef]

 *energies*

Article

Reliability Monitoring Based on Higher-Order Statistics: A Scalable Proposal for the Smart Grid

Olivia Florencias-Oliveros [†], Juan-José González-de-la-Rosa [*,†], Agustín Agüera-Pérez [†] and José-Carlos Palomares-Salas [†]

Research Group PAIDI-TIC-168, Computational Instrumentation and Industrial Electronics (ICEI),
Area of Electronics, University of Cádiz, Higher Polytechnic School, Av. Ramón Puyol S/N,
E-11202 Algeciras, Spain; olivia.florencias@uca.es (O.F.-O.); agustin.aguera@uca.es (A.A.-P.);
josecarlos.palomares@uca.es (J.-C.P.-S.)
* Correspondence: juanjose.delarosa@uca.es; Tel.: +34-956-02-8069
† These authors contributed equally to this work.

Received: 18 November 2018; Accepted: 20 December 2018; Published: 25 December 2018

Abstract: The increasing development of the smart grid demands reliable monitoring of the power quality at different levels, introducing more and more measurement points. In this framework, the advanced metering infrastructure must deal with this large amount of data, storage capabilities, improving visualization, and introducing customer-oriented interfaces. This work proposes a method that optimizes the smart grid data, monitoring the real voltage supplied based on higher order statistics. The method proposes monitoring the network from a scalable point of view and offers a two-fold perspective based on the duality utility-prosumer as a function of the measurement time. A global PQ index and 2D graphs are introduced in order to compress the time domain information and quantify the deviations of the waveform shape by means of three parameters. Time-scalability allows two extra features: long-term supply reliability and power quality in the short term. As a case study, the work illustrates a real-life monitoring in a building connection point, offering 2D diagrams, which show time and space compression capabilities, as well.

Keywords: signal waveform compression; higher-order statistics (HOS); power quality (PQ); computational solutions for advanced metering infrastructure (AMI); smart grid (SG) applications

---

1. Introduction

Power quality monitoring (PQM) has become a major concern in recent years due to the increasing complexity of modern power systems within the integration of distributed energy resources in the electrical network [1]. The latest advances related to communication technologies, monitoring devices, and multiple data processing algorithms constitute significant changes that make it possible to introduce more monitoring points as part of the advanced metering infrastructure [2] at all voltage levels [3,4]. Consequently, the number of fixed PQ analyzers has increased, and the large amount of data is managed according to different measurement strategies, depending on the European country [5].

In this scenario, the methods to inform about the reliability of the system need to be updated and adapted to monitor both the reliability and the quality of the power supply in a large-scale context [6]. Nevertheless, the analysis is not easy if data lack organization and the appropriate format to be processed and the rules to be turned into knowledge [7]. Flexible and optimal representation is demanded in order to introduce methods at different scales such as the monitored area, site, the connection point under test, and the duration of the monitoring period in time (long and short term) [8]. The power quality triangle represents a unified result that outlines the different layers involved in different measurement campaigns [7] in order to extract conclusions within data information and make the correct decisions as a result of monitoring. As a consequence, companies could manage their assets more effectively by combining enterprise application integration (EAI) and PQ data [8,9].

In addition, the smart grid (SG) demands the empowerment of individual customers and prosumers taking part in the decisions in a new market model [10], where new energy profiles are appearing [11] and learning-based systems help empower customers to select companies according to the price and electricity fulfillment [12]. In this sense, it is crucial to fill in the gap between the energy delivered by utilities and how end users assess the quality of the energy. Indeed, the strategies to inform customers about reliability and PQ are crucial and therefore need to be considered by companies, even more when actual standards include contractual features related to power quality [2,13].

Regarding the time scale that best fits the requirements, the majority of the measurement systems cannot cope with all time intervals. Furthermore, not all instruments are suitable for all measurement location sites. For this reason, a variety of site and disturbance indices has been developed according to different thresholds. However, there is a lack of indices that account for waveform information, voltage, and current. Each index has a different purpose, which ranges from location to event type; consequently, it is difficult to established an objective comparison among all the indices.

Not very many tools enable navigation between reliability and power quality with relative ease. For this reason, scalable methods are breaking through in this new scenario; scalability allows representing different operating conditions ranging from short time to long time windows, providing a more comprehensive representation of the network status.

All in all, the present paper proposes a method to monitor the supplied voltage waveform (sinusoidal, periodic, and stationary) based on time-compression using higher-order statistics (HOS). The method translates the time domain information to the statistical plane, therefore proposing a new way to represent data in advanced metering infrastructure. HOS was previously used in power quality studies characterizing the waveform shape according to the deviations that arise from the non-stationary nature of data. Most of the literature shows that HOS has been applied efficiently to event detection [14,15]. Nevertheless, the aim of the present contribution is to explore the analytical capabilities of HOS in order to characterize the normal operating conditions of the network within a minimum period of a week. In this sense, the present work proposes three individual statistics (variance, skewness, and kurtosis) and a global PQ index to be combined so as to improve power quality assessment and graphical representation. The subsequent idea is to monitor the waveform shape by means of bi-dimensional scalable graphs, which, depending on the time interval, provide different insights into the point under test and the deviations with respect to the ideal supply. Additionally, the information characterizes reliability in the dual perspective of the utility-consumer. The proposed methodology has been applied in a monitoring campaign developed during a week and considering a single building connection point.

The paper is structured as follows: Section 2 reviews the trends in reliability and PQ monitoring, covering the challenges in future distribution networks. Section 3 describes the measurement method based on an example that illustrates the behavior of the voltage dynamics in the graphs, each of which are associated with the PQ index. The method is exposed in Section 3 through a real-life case-study, which performs a complete analysis of PQ during the whole week. Finally, conclusions and future work are drawn in Section 5.

2. Reliability and PQ Monitoring in Industry Practice

Technological advances both in architecture and communications in the SG, the EAI frame, and the regulatory aspects that trigger the requirements and recommendations regarding PQ monitoring are changing the methods and measurement equipment in the industry practice. This section summarizes the issues for PQ monitoring, including recommendations on how big data should be managed and organized.

2.1. Instruments and Communication

During the last three decades, instruments have incorporated new functionalities in order to perform advanced PQ measurements. The most relevant aspects associated with the instrumentation for reliability and PQ are summarized below.

Fixed instruments have been spread throughout the entire network thanks to the full- reliable last mile communication systems. In addition, the introduction of robust wireless communication protocols has led to the development of feasible AMR (automated meter readers) and DMS (demand side management). Current instruments incorporate communication interfaces, intelligent algorithms to process data, and storage capabilities.

Considering reliability, instruments measure system availability, outage frequency, and time responses under weather phenomena, overloading, wildlife effects, and other features. In addition, PQ measurements are being introduced in smart meters. Within this context, it has been reported in [16] that utilities and TSOs (transmission system operators) mainly use fixed PQ monitors, while DSOs (distributed system operators) handle portable analyzers. However, from the customer's point of view, the instrument should be more flexible, as the branch of customers' needs is wide. In this sense, it is expected that PQ functionalities can help the customer monitor power supply more dynamically [17]. Nevertheless, the flexible capabilities are a choice, without considering the customer requirements, and not a requirementwhen a measurement campaign is deployed [5].

Therefore, the criteria to select the instruments are based on the performance regarding the specific standards more than any other issues such as the support after purchase [16] and the customers' objectives. Indeed, new types of customers (prosumers) are appearing that require the introduction of friendly interfaces and practical representations of the network status that enable making prompt decisions and demands for supply breaches.

2.2. PQ Data Organization Tendencies: The Need for Compression

As a consequence of the increase in the number of measurement locations and the introduction of communication technologies in the SG, data management is a relevant issue in order to improve EAI [8,18].

The time-intensive activities of handling big data require more specific analysis techniques based on the introduction of automatic procedures capable of representing data in a compact way. Current instruments cannot provide the final user with this flexibility due to the lack of compression functionalities. In this sense, a mix of research goals between electrical power systems and signal processing methods that optimize electrical signal waveform compression is needed. Numerous efforts have been made during the last decade to shorten long time-series into a unique number as a result of post-processing and information sharing [19,20]. Extracting some relevant features capable of describing amplitude, frequency, and phase changes facilitates data storage and monitoring in the SG.

The main compression techniques used in electric signals have been accurately described in [20] and are divided into lossless and lossy techniques. The former includes image compression techniques, and the latter gathers discrete transformations and parametric coding and mixed solutions. Lossless compression algorithms (LCA) of archives in the smart grid and PQ data series are focused on solving the aggregation resulting from a continuous monitoring [21]. Among some works, we remark about the use of 2D representations for compact energy coefficients and avoiding data redundancy in the work [22]. Furthermore, based on wavelets, the work in [19,20] is remarkable, where an adaptive approach estimates the fundamental sinusoidal component and separates it from the transients. In these works, the limitations are based on the computational burden and the noise effects. These limitations can be solved using HOS [14].

Utilities can introduce this flexibility and navigate within the former reporting levels, which enables analyzing PQ data at the site level, adopting mitigation measures from the network perspective, or reporting PQ compliance if the focus is on the utility performance [8,16]. In this scenario, it has been claimed that both time and space compression are key aspects for all PQ reporting levels based on the PQ triangle [8]. Three types of PQ reports have been distinguished, namely, site, network, and utility, which involve time and space compression concepts. As will be seen hereinafter, compression is a concept to bear in mind in the goal of designing a PQ index.

2.3. PQ Global Indices

Dealing with global indices implies a series of issues to be borne in mind. First is scalability. Scalable indicators must deal with the time domain duality reliability and power quality assessments. Also important are the existing strategies to present monitoring results related to the network state in the normal condition and under the influence of electrical perturbations [2]. The main difference between these approaches is within the measurement interval: reliability deals with interruptions of more than five minutes, while PQ events concern short events of less than three minutes.

Secondly, global PQ indices were initially conceived to meet utilities needs (AMR, data processing, and monitoring) and were not customer oriented [23]. Thus, these concepts need to be reformulated, as the customer is acquiring an important role in power quality assessment within the framework of the SG. In this sense, researchers have proposed customer dissatisfaction indices (CDI), which are usually conceived of for network planning and defined as the probability of not having fulfilled the contractual conditions regarding the power supply, when a level of security is overcome [16,24,25]. A recent example of the customer-based reliability approach consists of a solution proposed for Finish distribution companies [26]. However, the perspective settling on PQ is lost within this conception, as the time instances are not really monitored, as a consequence of not having taken into account the time scalability concept and the compression strategies. This is the idea in the work [7], which separates disturbance indices used in standards, combining them in order to provide a unified PQ insight. However, the customer approach is not considered in monitoring issues.

Furthermore, it is worth mentioning a tendency to propose more intuitive indicators. The deviation index based on principal curves proposed in [27] is a generic index that can be used for any type of disturbance; it quantifies the deviation from the contractual conditions by making use of two time scales (reliability and power quality lengths) and new diagrams that avoid the time variable. However, this index has been tested only under the influence of short PQ events, and no disquisitions have been made regarding reliability.

Previous authors' contributions propose time-compression of the sinusoidal waveform of 50 Hz [14], in which the potential of HOS was shown. HOS are capable of mapping a stable triplet of values associated with each electrical disturbance type.

More work involving the trajectories of real HOS estimators and their deviations from the ideal 50-Hz sinusoidal waveform has been developed in [28], which proposes a method that computes deviations of the supplied energy in more intuitive graphs. That contribution was focused on specific PQ event detection such as sags and swells in short monitoring periods. Once again this is evidence of the fact that current systems neither include the possibility of navigation between different time scales, nor have they been tested in sessions where power quality and reliability are involved, i.e., it would be desirable to have an instrument capable of storing (compressed) data that will enable browsing from short to long time-scales.

In order to achieve the condition to provide a global index, the authors consider that it must enable navigation between different monitoring time-series, proving to be applicable to both long-term and short-term monitoring sessions, and this must lead to reliability and disturbance detection monitoring.

All in all, the PQ analysis tools lack temporal scalability, and there is no agreement regarding data processing and the whole measurement procedure. Consequently, there is a premier need for developing flexible methods to be deployed in instruments capable of adapting not only to the measurement points, but also presenting data using new concepts that are more understandable and valid through the entire chain of the electrical network and enhancing the EAI for the utility. In this sense, the proposed method is a time compression method conceived of to provide flexibility to the instruments that introduce PQ functionalities. Their main difference related to previous works [14,28] resides in the possibility to inform about the normal operating conditions by measuring the voltage waveform both in the long term and short term. A scalable and global PQ index is introduced and related to the bi-dimensional graphs that store the voltage information of the network status.

3. The Scalable Method Using HOS

The method based on HOS uses a global indicator of the quality of the waveform shape in the time domain. Based on a set of three 2D-HOS diagrams, which compare pairs of statistics (the measurands), it quantifies the non-ideal waveform characteristics (features) for a given time analysis window (measurement time), i.e., variance vs. skewness, variance vs. kurtosis, and skewness vs. kurtosis. Additionally, it is worth mentioning that the 2D-HOS traces provide complementary information that goes beyond the mere second-order values, like the RMS (root-mean square).

The main difference with respect to the existing indicators is that it gathers information related to the waveform shape, which has not been computed yet in former works. Indeed, while the current indices provide information regarding specific disturbances, this global index characterizes amplitude, frequency, and phase changes. The PQ index provides an average measurement of the voltage quality in the point under test, for a given time interval. Thus, for a given time frame, graphs complement the numerical information provided by the index, making it easy to relate the index value to the waveform shape. Furthermore, the scalability of the method can measure the voltage waveform ranging from 0.02 s (a cycle) to a week-long temporalwindow.

The current procedure of analysis is described as follows. As the time analyzing window sweeps the signal under test, the 2D diagrams are generated, and the following points are considered. Each point in a 2D-HOS diagram (a pair or coordinates) corresponds to one cycle in the time-domain; this is precisely the temporal resolution of the measurement method and the N:1 time compression.

The stable ideal (steady state) triplet is given by: variance = 0.5; skewness = 0.0; kurtosis = 1.5; and it describes the non-disturbed ideal sine wave ot the power supply. These ideal values are located in the origin of each coordinate in the three 2D-HOS diagrams.

In order to describe the PQ index, the following magnitudes have to be previously defined:

Δt: measurement interval.
s_{ij}: jth statistic associated with the ith period.
$\hat{s}_j$:   nominal jth statistic.
M: number of periods contained in Δt.
N: number of statistics for characterization.

Given a measurement points in the SG, the realization of the unified PQ index is initially conceived as a function of the specific deviations of each individual statistic with respect to its nominal value and is given by the generic expression in Equation (1):

$$PQ_{\Delta t} = f\left(|s_{11} - \hat{s}_1|, \ldots, |s_{ij} - \hat{s}_j|, \ldots, |s_{MN} - \hat{s}_N|\right). \tag{1}$$

In this paper, a version based on the sum of the deviations has been implemented; it is given by Equation (2):

$$PQ_{\Delta t} = \frac{\sum_{i=1}^{M} \sum_{j=1}^{N} |s_{ij} - \hat{s}_j|}{M}. \tag{2}$$

Certainly, the deviations of each statistic from its ideal values assess and quantify the waveform shape. The present research uses three deviation terms in Equation (2). Variance deviations are related to the amplitude; symmetry is associated with the skewness; and the kurtosis assesses the sinusoid aspect. Thus, more than paying attention to a certain type of electrical disturbance, or their trajectory in time, the method proposes an examination of the contractual waveform and lays down a monitoring procedure of the deviations from the ideal value and the measured statistic. The final expression for the index is described in Equation (3):

$$PQ_{\Delta t} = \frac{\sum_{i=1}^{M} |var_i - \hat{var}| + |sk_i - \hat{sk}| + |kur_i - \hat{kur}|}{M}.$$

(3)

Figure 1 illustrates the measurement procedure. The acquired signal, $u(t)$, is normalized to the maximum amplitude, providing $u_e(t)$, then the statistics are computed and located in the 2D trace. The 2D graph in Figure 1 shows how the diagram is traced as the time analysis window sweeps the whole time-domain register. In the end, N points are depicted (N signal periods), associated with the statistics pairs $(s_j, s - h)$ with j and $h \in [1, ..., M]$.

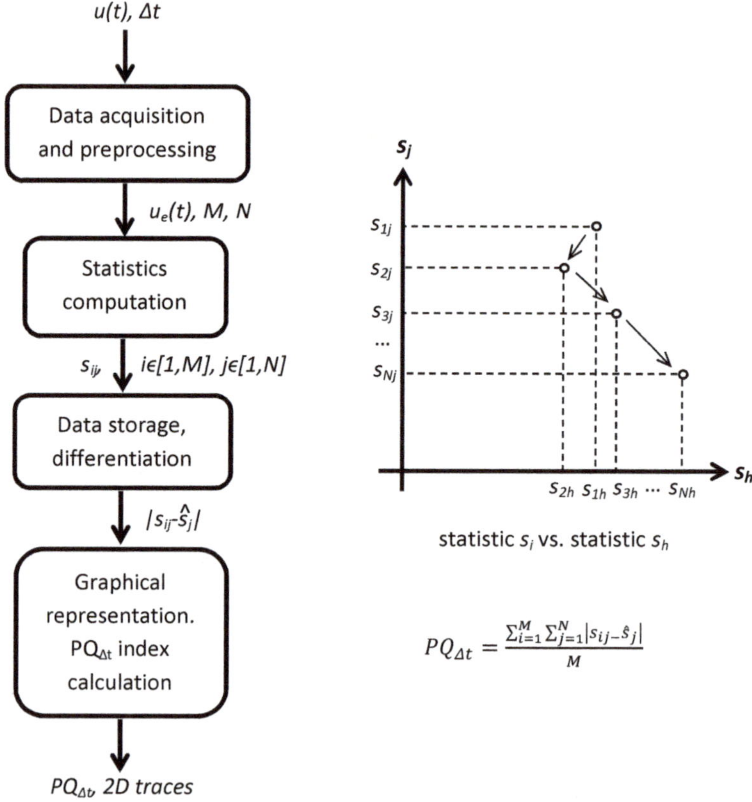

Figure 1. Measurement procedure along with the outputs: PQ index and 2D traces.

Hereinafter, the former concepts, associated with the method, are explained through Figure 2 with the aid of a real-life voltage-supply waveform (50 Hz, 230 V_{rms}) supposed to fulfill the compliance requirements, which was surveyed during one week (as part of a monitoring campaign) in the Polytechnic School of Algeciras (Spain). The temporal analysis has been segmented into six main intervals or observation periods (from left to right), each of which includes the previous one (overlapped) Figure 2. In the very first period of the series, it can be appreciated that the sinusoid waveform is not ideal at all. In the following ten cycles, the defects begin to be hidden, then it is followed by the sequences: 20-s, 1-h, 1-day, and 1-week segments. The complete one-week record is located at the top of Figure 2; just below the variance, the skewness and the kurtosis evolution during the surveyed week are depicted. The bottom three rows (fourth, fifth, and sixth) contain the variance versus skewness, variance versus kurtosis, and skewness versus kurtosis traces for each or the former time windows and aligned in these intervals. The *i*th-cycle is processed and generates a triplet (var_i, skw_i, kur_i); a pair of values corresponds to a point for each 2D graph; the origin of the coordinates represents the ideal situation; e.g., in the variance vs. skewness, the center is located in the coordinates (0.500, 0.000).

The span of the measurement is set to 0.07, according to the interval 0.45–0.52, inside which the variance fluctuations are well observed, so as to delve into the 2D graphs in Figure 2. The number of precision digits in the PQ has been set empirically to four, as it was observed that the fifth digit did not exhibit significant changes.

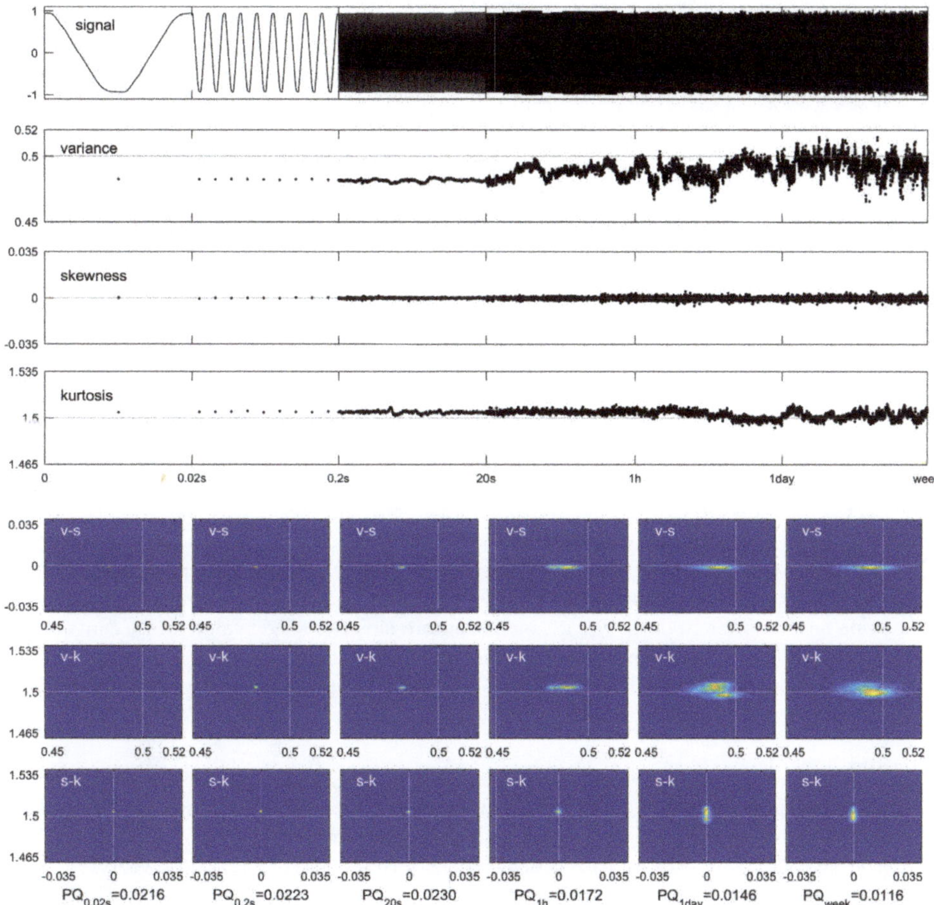

Figure 2. Monitoring of one week using the method based on HOS with a minimum resolution of one cycle. From top to bottom: complete power supply time-series; variance evolution, skewness evolution; kurtosis evolution; variance-to-skewness graph family; variance vs. kurtosis 2D graphs; skewness vs. kurtosis set of graphs. Each time interval is characterized by its associated PQ index.

The first period shows that the shape is not sinusoidal at all (Figure 2). In fact, the normalized power supply signal does not reach the top normalized values (flat-top) (± 1); a fact that is reflected in the variance, which keeps under the ideal reference of 0.500. The skewness is almost zero as the signal is equilibrated, and the kurtosis is slightly over 1.500, showing again a loss in the sinusoidal shape. The coordinates (x,y) of each pair of values quantify each of the PQ index terms.

During the next nine periods, the situation is very similar, with the exception of little variance and few kurtosis fluctuations, which show a dispersion in the probability 2D graphs. The next 20 s and the successive measurements are scattered round the first-cycle triplet values. In the next hour, the differences in variance are much more noticeable, the probability cloud being displaced towards the ideal values. A zone of this probability cloud corresponds to the 20 s analyzed in the previous graphs, since they are included in the represented hour. This fact can be observed in the probability cloud (cluster), which has moved to the right in the var. vs. kur. plane. This displacement is general for the rest of the elapsed time intervals.

During the following day, two different, and clearly marked, nuclei were observed, eliciting different behaviors Figure 2. The first one was characterized by a low variance, and its associated kurtosis was located above the ideal value, which is indicative of a shape degradation. The lower cloud centroid was nearer the ideal voltage supply state; the previously-analyzed hour was included in the first nucleus.

Looking at the time-series for the variance and the kurtosis, it can be seen that when the variance was above its ideal value, the same happened to kurtosis in general (Figure 2). During the last stretches of the day, the power supply voltage was recovered, a fact which is reflected in the second cluster of measurements. This part is associated with the night period, when the building is closed and the energy demand is minimum. It can be said that during the last intervals in this day, some measurement cycles would have reached the ideal values of variance and kurtosis (0.5; 1.5).

Finally, and concerning the whole analysis of the week, again, the profile of the behavior is confirmed; making it clear that the analyzed segments usually were much worse than the general behavior in the week. In summary, the proposed tool was able to provide characterizations of the waveform state in a similar framework, independently of the time interval.

A PQ index was defined on this basis, which gathered the deviations observed in these graphs, so that information could be compressed into a single value. However, some interesting remarks are inferred from the study of the time intervals. As the time interval length increased, the PQ value decreased. This was due to the averaging effect and the selection of a low quality starting point for explanation purposes; e.g., $PQ_{week} = 0.0116$ was much better than $PQ_{0.02s} = 0.0216$ and $PQ_{0.2s} = 0.0223$. The introduction of the advanced index in PQ analyzers should consider the time evolution line in order to compare current measurements to historical ones that has been registered at the same time in different days. The PQ index compression to a single value can be convenient in simple instruments or applications that do not demand high storage.

The next section analyzes the temporal structure of the supply trace by exploiting the proposed visualization tool and its associated index.

4. Results

This section goes into the PQ analysis regarding the previously-monitored week, from which we show that depending on the windows' length, the focus changes from reliability (wide windows) to PQ (shorter windows). Figure 3 shows the 2D voltage traces corresponding to the whole week, from which two relevant days, the best and the worst, have been selected so as to compare the trace evolution from the one-day frame to the 0.02-s frame (equivalent to ten cycles). The first column corresponds to the one-week results; then, the trace evolution from one day to 0.2 s is shown.

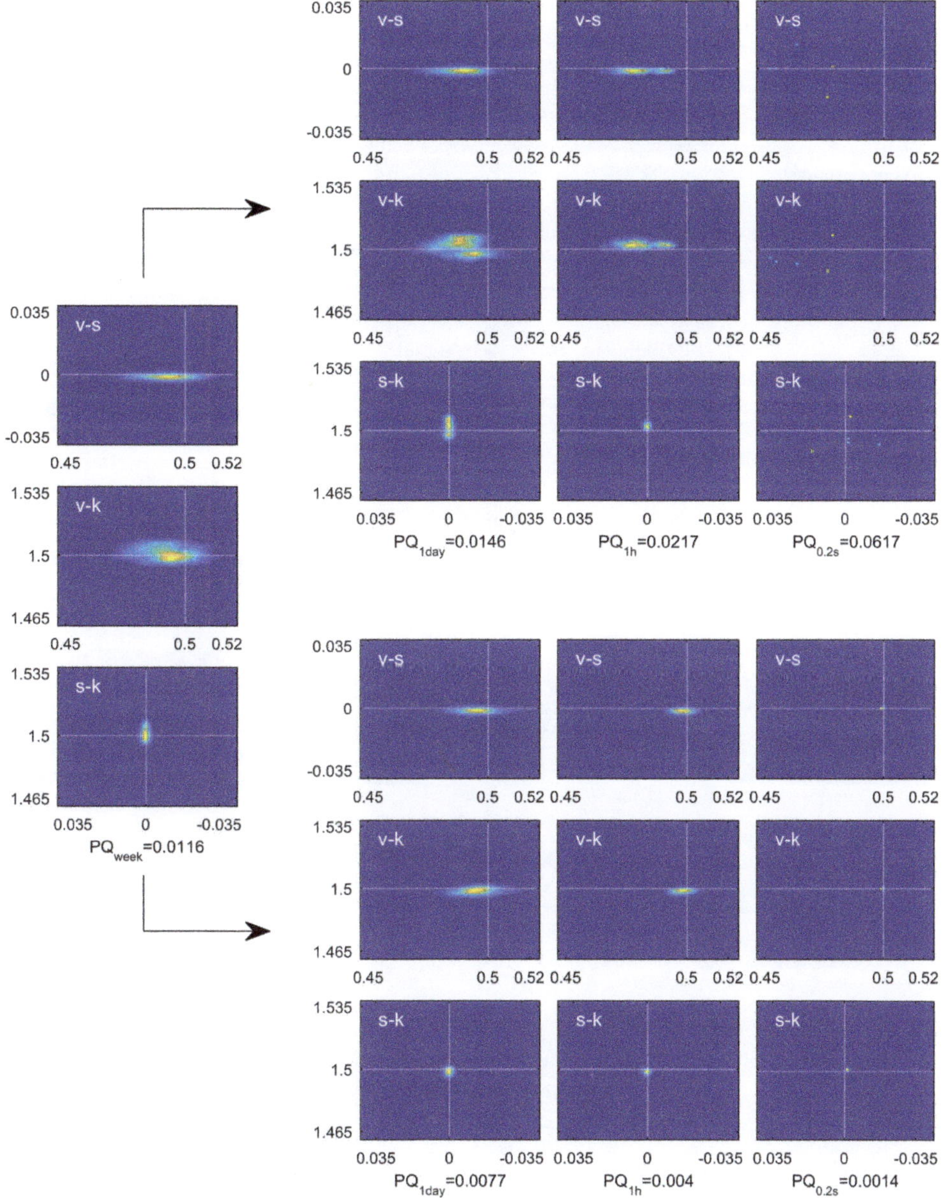

Figure 3. Scalability of the proposed method: two singular days within the analyzed week. The 3×3 2D-HOS graph matrix corresponding to the best (worst) day is located below (above). Within the worst (best) day, the worst (best) hour and the worst (best) 0.02 s are selected.

Indeed, this Figure 3 shows the detail of the PQ temporal organization. This tree structure can be easily followed from left to right as the temporal width decreases, from reliability (one week surveyed) to PQ indications (0.2 s monitored). Thus, it can be said that reliability patterns (long width time windows) can be confirmed as the aggregation of different PQ states (shorter windows). Therefore, as the monitoring scale decreases, the focus goes from the concept of reliability to the PQ event detection strategy. This fact illustrates one of the contributions of this measurement strategy: several analyses can be performed in a unified framework, allowing post comparisons among different time-scales. For instance, from the pattern in the one-week monitoring period, the worst and the best day can be distinguished, top and bottom 3×3 matrix block of 2D-HOS graphs, which exhibit extremely different behaviors. The general week pattern, resulting from the aggregation of shorter-time states, allows exploring the whole PQ history (from week to 0.2 s) following multiple sequences of frames, also enabling direct comparisons between frames of the same or different time scales, which provide information regarding deviations experimented in raw power supply data in the surveyed intervals.

Making quantitative remarks, for the worst day (the upper graph), the PQ increased as the time window was shortened. Paying attention to the heart of the figures, first of all, it is shown that the HOS plane allowed extracting details in the voltage signal supplied by the contract.

Figure 4 shows the non-stationary evolution of the PQ index during a day, with an average value of $PQ_{day} = 0.0138$. As a general remark, it is shown that the PQ depends on the period of the day.

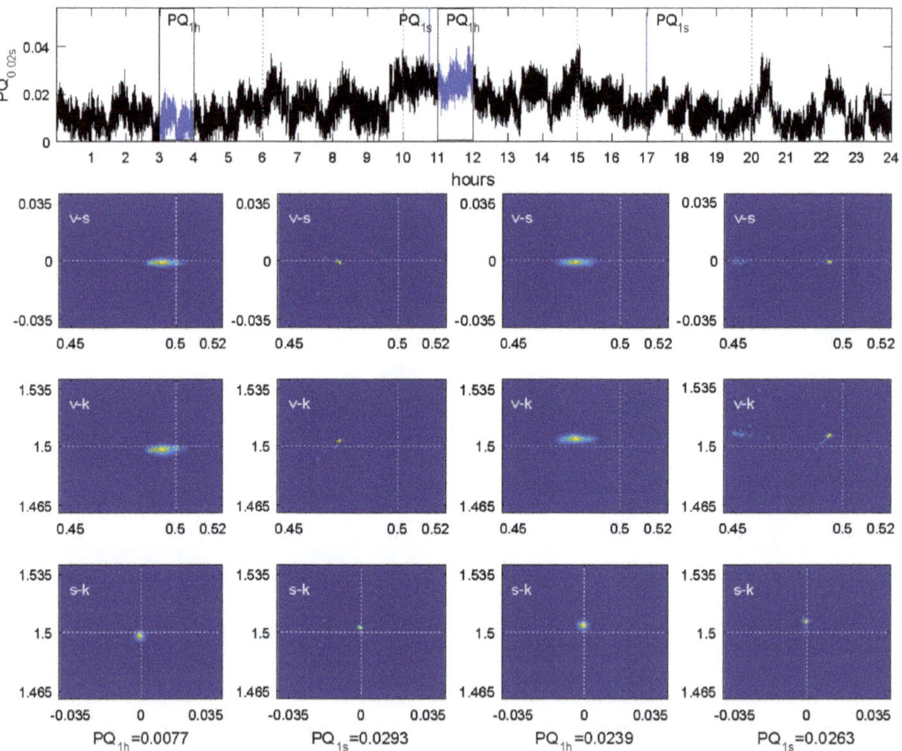

Figure 4. $PQ_{0.02s}$ values during 24 h, from which different intervals can be analyzed.

Hereinafter, the analysis of the top PQ series in Figure 4 begins. First of all, a general analysis regarding the tendency of the curve was performed. Between 00H00 and 06H00, oscillations occurred just above the ideal value of zero, with an average of $PQ_{6-h} = 0.0094$. Then, in the interval 06H00 and 10H00, the average value increased with the electricity demand, $PQ_{4-h} = 0.0148$. Within the time stretch 10H00–15H00, the quality became worse with a $PQ_{5-h} = 0.0205$, which were the central hours of the observed day. Then, in the interval 15H00–20H00, a soft decrease took place as a direct consequence of activity ceasing. The same was observed in the period 20H00–24H00.

Secondly, four time stretches have been selected to illustrate more specific behaviors: two relevant hours and two relevant seconds. For each one, its associated triplet of 2D-HOS graphs is drawn just below. The two-hour periods (03H04 and 11H12) enable explaining the dynamic of the trace through the day, making use of the PQ time series and the 2D HOS traces. The PQ_{1-h} was far different between the selected hours: $PQ_{1-h} = 0.0073$ vs. $PQ_{1-h} = 0.0237$, respectively; more stable (near the ideal zero value) in the first and significantly worse (above zero) in the second hour under test.

The maximum $PQ_{0.02s}$ value was reached at $PQ_{0.02s} = 0.05$, within the analyzed second, between 10H00 and 11H00. This maximum may indicate the presence of at least one event. Thus, the statistical parameters deviated from the ideal values, even out of the tendency within this time interval. A singular event was detected in the HOS plane and was part of a series of events that marked a tendency of a high PQ.

The same maximum was reached within the second interval 15H00–20H00. This time, it seemed as if the PQ evolved closer to the ideal value. This can be observed in the v-k graph, where two clouds can be differentiated; the one in the left associated with the event group and the second one nearer the origin, where the ideal values are located. The demand intervals 06H00–14H00 and 15H00–20H00 can be extracted from the HOS graphs. They are also reflected in a 24-h monitoring solution based on the PQ index Figure 5.

As a general remark, this paper has proven that the PQ index gathered in a single value the formal behavior of a signal and that the HOS space offered much more information about the events' occurrence. The combined use of the index and the 2D graphs improved the cycle-by-cycle detection. In addition, different monitoring objectives could be planned (e.g., testing PQ with different events, detecting only events that overpass a threshold or triggering the analysis before instructions or critical moments).

Finally, in order to show a strategy extracted from the whole of the raw data, Figure 5 represents the hourly trend of the index, during several days in the whole week. It is seen that while for a non-working day, the PQ value was always near zero, the five working days exhibited a quite different behavior, reaching maximum values.

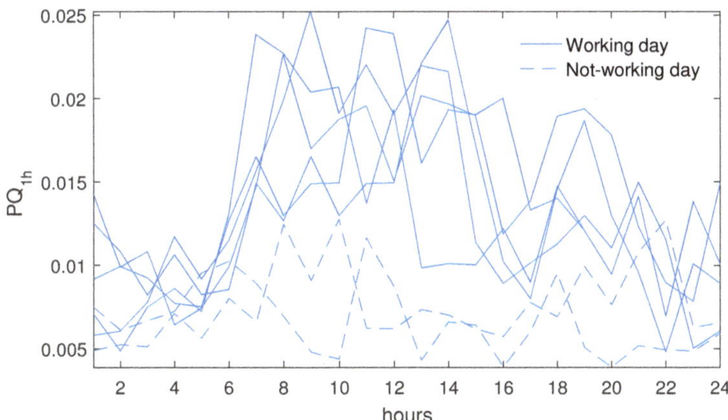

Figure 5. PQ tendency during 24 h, considering the whole week. Each PQ value is the result of an averaging of individual PQ-cycle measurements during one hour.

5. Conclusions and Future Work

The HOS-based method characterizes the voltage waveform shape under normal operating conditions during long-term monitoring sessions. The statistics help compress the waveform information into three parameters that inform about the quality of the waveform in the point under test.

The main novelty of this method lies in its capability of time compression. Scalability allows browsing between reliability and power quality monitoring objectives, which is a desired property in the SG context. Furthermore, the method improves the visualization of the network status based on new graphs that help report the waveform information.

The validation of the method during one week (within a wider monitoring campaign), is aligned to the minimum measurement period according to the norm IEC 61000-4-30:2015. Indeed, differences can be found between working and non-working days and are clearly outlined in the graphs, along with the probability of PQ events. In this line, the authors propose the use of the PQ index and 2D graphs as a complementary tool to the existing ones as part of the advanced metering infrastructure.

The method is valuable for both the utility and the customer sides. Once the network operates outside the contractual operating conditions, the PQ index can establish a threshold that sets the limits for normal operations and triggers any further analysis or data storage.

The proposed method not only allows the temporal characterization of PQ, but also would enable PQ prediction, which is certainly one of the objectives within the current SG research scenario.

Future work consists of three main lines: first, to develop a power index forecasting unit that could be integrated with energy management systems, adding valuable information to that provided by the existing indices; secondly, to broaden our method, including more terms in the index and more 2D graphs, derived eventually from the new advances in the analysis of power quality monitoring; finally, machine learning techniques could be applied in order to estimate the grid state, the space compression of several supply points, and the three-phase behavior in the point under test when some events propagate.

Author Contributions: Funding acquisition, J.-J.G.-d.-l.-R.; methodology, A.A.-P. and O.F.-O.; project administration, J.-J.G.-d.-l.-R.; supervision, A.A.-P., J.-J.G.-d.-l.-R., and J.-C.P.-S.; visualization, A.A.-P. and O.F.-O.; writing, original draft, O.F.-O. and J.-J.G.-d.-l.-R.; writing, review and editing, O.F.-O. and J.-J.G.-d.-l.-R.

Funding: This research was funded by the Spanish Ministry of Economy, Industry and Competitiveness and the EU (AEI/FEDER/UE) via the research project TEC2016-77632-C3-3-R.

Acknowledgments: This research is supported by the Spanish Ministry of Economy, Industry and Competitiveness and the EU (AEI/FEDER/UE) in the workflow of the State Plan of Excellency and Challenges for Research, via the project TEC2016-77632-C3-3-R-COntrol and Management of Isolable NanoGrids: Smart Instruments for Solar forecasting and Energy Monitoring (COMING-SISEM), which involves the development of new measurement techniques applied to monitoring the PQ in micro-grids. Furthermore, we express our gratitude to the Andalusian Government for funding the Research Group PAIDI-TIC-168 in Computational Instrumentation and Industrial Electronics (ICEI.)

Conflicts of Interest: The authors declare no conflict of interest.

References

1. Bollen, M.; Bahramirad, S. Is there a place for Power Quality in the Smart Grid? In Proceedings of the 16th IEEE International Conference on Harmonics and Quality of Power(ICHQP), Bucharest, Romania, 25–28 May 2014; pp. 713–717.
2. Kilter, J.; Meyer, J.; Elphick, S.; Milanović, J.V. Guidelines for Power Quality Monitoring—Results from CIGRE/CIRED JWG C4.112. In Proceedings of the 16th IEEE International Conference on Harmonics and Quality of Power (ICHQP), Bucharest, Romania, 25–28 May 2014; Volume 1, pp. 703–707.
3. Bollen, M.; Baumann, P.; Beyer, Y.; Castel, R.; Esteves, J.; Fajas, S.; Friedl, W.; Larzeni, S.; Trhulj, J.; Villa, F.; et al. Guidelines for good practice on Voltage Quality monitoring. In Proceedings of the 22nd IEEE International Conference on Electricity Distribution (CIRED), Stockholm, Sweden, 10–13 June 2013; Volume 1, pp. 1–4.

4. Kilter, J.; Meyer, J.; Howe, B.; Zavoda, F.; Tenti, L.; Milanovic, J.; Bollen, M.; Ribeiro, P.; Doyle, P.; Gordon, J. Current practice and future challenges for power quality monitoring-CIGRE WG C4.112 perspective. In Proceedings of the IEEE 15th International Conference on Harmonics and Quality of Power, Hong Kong, China, 17–20 June 2012.

5. McDaniel, J.; Friedl, W. Benchmarking of Reliability: North American and European Experience. In Proceedings of the 23th International Conference on Electricity Distribution, Lyon, France, 15–18 June 2015; pp. 15–18.

6. Meyer, J.; Klatt, M.; Schegner, P. Power quality challenges in future distribution networks. In Proceedings of the 2nd IEEE PES International Conference and Exhibition on Innovative Smart Grid Technologies, Manchester, UK, 5–7 December 2011; Volume 1, pp. 1–6.

7. Gosbell, V.J.; Perera, B.; Herath, H. Unified Power Quality Index (UPQI) for Continuous Disturbances. In Proceedings of the 10th IEEE International Conference on Harmonics and Quality of Power (ICHQP). Proceedings (Cat. No.02EX630), Rio de Janeiro, Brazil, 6–9 October 2002; Volume 1, pp. 316–321.

8. Braun, J.; Gosbell, V.J.; Robinson, D. XML Schema for Power Quality Data. In Proceedings of the 17th IEEE International Conference on Electricity Distribution (CIRED), Barcelona, Spain, 12–15 May 2003; Volume 1, pp. 1–5.

9. Kenner, S.; Thaler, R.; Kucera, M.; Volbert, K.; Waas, T. Comparison of smart grid architectures for monitoring and analyzing power grid data via Modbus and REST. *EURASIP J. Embed. Syst.* **2017**, *2016*, 1–13. [CrossRef]

10. Li, N.; Chen, L.; Dahlen, M.A. Demand Response Using Linear Supply Funtion Bidding. *IEEE Trans. Smart Grid* **2015**, *6*, 1827–1838. [CrossRef]

11. Rathnayaka, A.J.D.; Potdar, V.M.; Dillon, T.; Hussain, O.; Kuruppu, S. Analysis of energy behaviour profiles of prosumers. In Proceedings of the IEEE International Conference on Industrial Informatics, Beijing, China, 25–27 July 2012.

12. Apostolopoulos, P.; Tsiropoulou, E.; Papavassiliou, S. Demand Response Management in Smart Grid Networks: A Two-Stage Game-Theoretic Learning-Based Approach. *Mob. Netw. Appl.* **2018**, *23*, 1–14. [CrossRef]

13. IEC. IEC 61000-4-30. Electromagnetic Compatibility (EMC), Part 4: Testing and Measurement Techniques. Section 30: Power Quality Measurement Methods. Available online: https://webstore.iec.ch/publication/21844 (accessed on 20 December 2018).

14. Agüera-Pérez, A.; Palomares-Salas, J.C.; De-la Rosa, J.J.G.; Sierra-Fernández, J.M.; Ayora-Sedeño, D.; Moreno-Muñoz, A. Characterization of electrical sags and swells using higher-order statistical estimators. *Measurement* **2011**, *44*, 1453–1460. [CrossRef]

15. De-la Rosa, J.J.G.; Agüera-Pérez, A.; Salas, J.C.P.; Sierra-Fernández, J.M.; Moreno-Muñoz, A. A novel virtual instrument for power quality surveillance based in higher-order statistics and case-based reasoning. *Measurement* **2012**, *45*, 1824–1835.

16. Milanović, J.; Meyer, J.; Ball, R.; Howe, W.; Preece, R.; Bollen, M.H.; Elphick, S.; Čukalevski, N. International Industry Practice on Power-Quality Monitoring. *IEEE Trans. Power Deliv.* **2014**, *29*, 934–941. [CrossRef]

17. Alam, S.M.S.; Natarajan, B.; Pahwa, A. Distribution Grid State Estimation from Compressed Measurements. *IEEE Trans. Smart Grid* **2014**, *5*, 1631–1642. [CrossRef]

18. Zhang, J.; Zheng, X.; Wang, Z.; Guan, L.; Chung, C.Y. Power System Sensitivity Identification-Inherent System Properties and Data Quality. *IEEE Trans. Power Syst.* **2017**, *32*, 2756–2766. [CrossRef]

19. Ribeiro, M.V.; Romano, J.M.T.; Duque, C.A. An Improved Method for Signal Processing and Compression in Power Quality Evaluation. *IEEE Trans. Power Deliv.* **2004**, *19*, 464–471. [CrossRef]

20. Tcheou, M.P.; Lovisolo, L.; Ribeiro, M.V.; da Silva, E.A.B.; Rodrigues, M.A.M.; Romano, J.M.T.; Diniz, P.S.R. The Compression of Electric Signal Waveforms for Smart Grids: State of the Art and Future Trends. *IEEE Trans. Smart Grid* **2014**, *5*, 291–302. [CrossRef]

21. Kraus, J.; Stepan, P.; Kukacka, L. Optimal data compression techniques for Smart Grid and power quality trend data. In Proceedings of the IEEE 15th International Conference in Harmonics and Quality of Power(ICHQP), Hong Kong, China, 17–20 June 2012.

22. Gerek, O.N.; Ece, D.G. Compression of power quality event data using 2D representation. *Electr. Power Syst. Res.* **2008**, *78*, 1047–1052. [CrossRef]

23. Florencias-Oliveros, O.; De-la Rosa, J.J.G.; Agüera-Pérez, A.; Palomares-Salas, J.C. Discussion on Reliability and Power Quality in the Smart Grid: A prosumer approach of a time-scalable index. In Proceedings of the International Conference on Renewable Energies and Power Quality (ICREPQ18), Salamanca, Spain, 21–23 March 2018.

24. Bollen, M.; Holm, A.; He, Y.; Owe, P. A customer-oriented approach towards reliability indices. In Proceedings of the 19th IEEE International Conference on Electricity Distribution (CIRED), Vienna, Austria, 21–24 May 2007; Volume 1, pp. 1–4.

25. Holm, A.; Pyvänäinen, J.; Owe, P.; Paananen, H.; Bollen, M.H. Customer dissatisfaction index (CDI): Pilot use in network planning. In Proceedings of the 20th IEEE International Conference on Electricity Distribution (CIRED), Prague, Czech Republic, 8–11 June 2009; Volume 1, pp. 1–4.

26. Lassila, J.; Kaipia, T.; Haakana, J.; Partanen, J.; Bollen, M. Customer-based reliability monitoring criteria for Finnish distribution companies. In Proceedings of the 7th International Conference Electric Power Quality and Supply Reliability (PQ2010), Lyngby, Denmark, 26–27 September 2010.

27. Ferreira, D.D.; de Seixas, J.M.; Cerqueira, A.S.; Duque, C.A.; Bollen, M.H.; Ribeiro, P.F. A new power quality deviation index based on principal curves. *Electr. Power Syst. Res.* **2015**, *125*, 8–14. [CrossRef]

28. Florencias-Oliveros, O.; De-la Rosa, J.J.G.; Agüera-Pérez, A.; Palomares-Salas, J.C. Power quality event dynamics characterization via 2D trajectories using deviations of higher-order statistics. *Measurement* **2018**, *125*, 350–359. [CrossRef]

 *energies*

Article

A Memory-Efficient True-RMS Estimator in a Limited-Resources Hardware

Jose-Maria Flores-Arias *, Manuel Ortiz-Lopez, Francisco J. Quiles Latorre, Francisco Jose Bellido-Outeiriño and Antonio Moreno-Muñoz *

R&D Group 'Industrial Electronic Instrumentation', Universidad de Córdoba, 14071 Córdoba, Spain; el1orlom@uco.es (M.O.-L.); el1qulaf@uco.es (F.J.Q.L.); el1beouf@uco.es (F.J.B.-O.)
* Correspondence: jmflores@uco.es or el1flarj@uco.es (J.-M.F.-A.); amoreno@uco.es (A.M.-M.)

Received: 5 April 2019; Accepted: 30 April 2019; Published: 5 May 2019

Abstract: More and more human activities are increasingly dependent on the power quality energy they are supplied. In particular, those loads that have an electronic switching mode power supply (SMPS) and microcontrollers or microprocessors are very susceptible to de-energization of the AC line as voltage monitoring takes place on the DC end and may not have enough time to cope with a voltage event involving energy loss. There have been many proposals that analyze power quality or current consumption, even those using limited resources hardware or the classic formulas of discrete root mean square (RMS). In this proposal, an approximation to the problem by means of an RMS value estimator that uses as a base a microcontroller of basic range and low economic cost and algorithms of calculation of very low computational cost that elude complex arithmetic operation to controllers, such as powers or roots, is presented. The results of the experiments to which the proposal was subjected show its ability to provide an estimation of the RMS value of voltage with sufficient precision or an event alarm in less time than the options analyzed. The validation tests and functional comparison of the prototype which were carried out support its viability.

Keywords: RMS voltage estimation; low computational cost; limited resources hardware; power event detection; energizing warning

---

1. Introduction

Reliable and clean electric power conditions many human activities. Electric power is deemed 'reliable' if the electric supply is available and dependable. It is also called 'clean' if the AC current and voltage waveforms that are supplied essentially fit a sinusoidal wave shape at a frequency of 50 Hz or 60 Hz depending on countries where users live.

Regarding its quality and reliability, electric power is deemed 'computer grade' if it is specially conditioned and controlled ultra-clean electric power. By contrast, 'dirty' electric power is characterized by a significant distorted waveform which contains undesirable components like pulses, off-frequency sinusoidal waves, and other undesirable deviations from the sinusoidal desired waveform. These components can be transient events, they can repeat periodically at AC mains frequency, or they can cycle at higher or lower frequencies related to that, i.e., harmonic or subharmonic components [1].

Outages (i.e., power interruptions, including blackouts) can also cause serious consequences. Computer data can be lost; a shaft on a large, idle, rotary machine can bow or sag; molten metal can solidify; heating systems can shut down; protection systems can fail; a hospital's life-support equipment can become inoperative; and a chemical processing plant can experience runaway reactions.

Dirty power, steady-state undervoltages (brownouts) and overvoltages can lead to electrical and mechanical equipment damage.

Power quality and reliability (PQR) problems affecting the voltage in quantity or magnitude that occur in home or business facilities involve transients, sags, swells, surges, outages, harmonics, and impulses [2]. Short-term power quality disturbances include dropouts, sags, and swells. Dropouts are momentary power losses typically occurring within one electrical cycle. Sags are depletions of power caused when heavy loads start. Swells are power surges caused by sudden load disconnects. Sags and swells are typically seen as voltage changes lasting 2.5 s (150 cycles for a 60 Hz system) or less upon sudden changes in power demand.

They can disrupt sensitive loads or cause harm to installed equipment. Electronic equipment especially is highly susceptible to all three problems.

Before electronic loads were common, simple surge arresters, power supply transfer switchgear, and voltage regulators assisted most equipment and systems to tolerate common power problems and perform adequately. However, electronic loads are increasing and consume more than half the power generation capacity in the developed world. These loads are sensitive to power quality problems. Furthermore, in much of the electronic circuitry in use today, electronic power supplies with switching frequencies as high as 200 kHz have replaced older, linear (sinusoidal) power supplies. Switch-mode power supplies lower power quality. Today, while the miniaturization of modern electronic devices makes reliable, clean power even more crucial, these same electronic devices have become a major contributor to the dirty power problem.

Typically, power-conditioning equipment combines voltage regulation with power system isolation and shielding. Uninterruptible power supply (UPS) systems are generally an excellent way to control power distortions, but their primary purpose is to improve power reliability, not to clean poor power quality. Among those which exist there are currently available many solutions to voltage regulation [3,4]. Among them, UPS and AC/DC power factor correction (PFC) converters with universal voltage supply range and equipped with surge suppressors, which very often include a power digital manager module at the DC side and frequently may be found in very sensitive domestic electronic equipment. However, electrical appliances are one of the most complex sets in the Smart Grid definition of sensitive equipment; there also are adjustable and sheddable loads [5] managed with different strategies.

Electric power supply problems are inappropriately considered service quality problems and the sole responsibility of utility companies. However, a user's distribution circuit arrangement, system, and equipment can contaminate the power supply and cause power supply interruptions. Thus, utilities, equipment manufacturers and system designers all share responsibility for power supply conditions.

Those effects are less critical in appliances powered by switching mode power supply (SMPS) where the front side power converter filters and smooths most of the voltage level and frequency electrical disturbances within the range of the universal input supply range [6]. However, power abnormalities and waveform distortions cause control, data processing, and diagnostic systems to operate erratically and can lead to faults in microcontrollers and, consequently, operation downtime and process rejections with variable operating costs.

Furthermore, electronic control equipment needs fast and reliable voltage brownout detectors [5,7] in order to equip the adequate protection or compensation strategies or to detect voltage sags as fast as possible and trigger adequate alarm or control signals [8].

Finally, the Smart Energy Profile 2 (SEP 2) standard promoted by the Consortium for Smart Energy Profile 2 Interoperability (CSEP) enables management and control of smart grid devices in the home [9]. In this scenario, a fast and accurate RMS value measurement or estimator algorithm for any electrical variable which could be easily embedded in either home area network metering devices or smart appliances is also needed. Wider ARMs and sensor networks cannot be avoided either.

If the de-energizing process were only detected when the SMPS outlet was out of range, the affected intelligent load would not be able to adapt its run mode to deal with it.

Every PQR event may be defined according to the relationship between its residual voltage level and duration, as reflected in the standards EN 50160 [10] and IEEE 1159-2009 [11].

Moreover, the International Working Group (JWG) C4.110, sponsored by CIGRE (the International Council on Large Electric Systems), CIRED (the International Conference on Electricity Distribution), and UIE (the International Union for Electricity applications), has established a description of the different properties and characteristics of voltage dips based on dividing a dip into transition segments and event segments [12].

This works firstly aims to the automatic in advance detection of de-energizing dip, brownout, or outage events related to the AC voltage at the supply inlet by means of its phasor module measurement instead of monitoring the converter DC outlet.

One of the mean advantages of using micro-controlled SMPS is the monitoring of the power converter that leads to a supervisory function that produces an idle state or a soft switch-off if necessary. Many of those systems are based on the DC voltage level in the DC bus at the converter outlet.

This means that if a DC undervoltage occurs at that point, the output filter capacitor had been supplied all the energy it had stored for correct functionality and no operational return point could be achieved because of the recharging time gap needed to re-energize it. Consequently, the microcontroller faults and the appliance program resets.

However, the good news is that the AC mains can degrade their power quality several cycles before the power converter becomes de-energized.

A brief overview of the state of the art (SoA) on signal quality metering using microcontrollers shows the use of a widespread variety of hardware provided with different resource levels.

There are diverse approaches to the measurement of the quality of the electrical signal. Some of these use the RMS value of the signal and others use other parameters on which they carry out more complex transformations like the wavelet transform. Some of them are also implemented using limited resources systems, as described in [13], which employs a limited resources system based on an ATmega 328P microcontroller (Microchip, Chandler, AZ, USA). This system measures voltage, current, active power, and power factor over various loads and wirelessly transmits them to the main brain for processing. Voltage and RMS current are calculated to evaluate the energy consumption in each load, but it does not detect voltage drops. In [14] the ESP8266 module is used for *THDi* (Current Total Harmonic Distortion) via FFT (the Fast Fourier Transform), and the current RMS value for each load is given via the RMS method. The values of these parameters are transmitted to a server and implemented through a Raspberry Pi, using the MQTT (Message Queuing Telemetry Transport) protocol so that the user can monitor the energy quality signal and know the electrical faults of a load or the network. This monitoring system has the disadvantage of not being able to detect faults in real time. In [15] the power quality parameters of a single phase are measured and the arc-fault series are sensed, but the voltage drops are not able to be detected in real time. It uses a limited resources Microchip (Chandler, AZ, USA) dsPic33 processor to calculate the wavelet transform and the pass fundamental frequency using the zero-crossing technique, as well as RMS values and power. In [16] the V_{rms} for a specific number of signal periods is calculated, meaning it can detect neither voltage drops nor surges in real time. Its objective is to achieve an accuracy of 0.1% in the measurement of the RMS value with an analog-to-digital converter (ADC) of 10 bits so it is constrained to use the oversampling technique and therefore does not optimize the time of calculation to provide an RMS value. The system uses a single limited resources, low power CMOS 8-bit microcontroller based on an AVR-enhanced RISC (reduced instruction set computer) architecture.

In [17] some signal quality parameters such as V_{rms}, I_{rms}, active power, power factor, and *THDi* are measured and sent through the power line to a concentrator that makes decisions based on the consumption strategy adopted by the user. Therefore, it does not detect voltage drops. An STM32F407VG microcontroller is used to acquire and process the signals, and send them via a PLC (Power Line Communications) modem to a Raspberry Pi, model B.

In [18] the system uses the wavelet transform to detect voltage drops. It uses a sampling time of 156 μs, and although the calculation time is less than the acquisition time (32 μs), the final time needed to detect the voltage drop is much greater, on the order of 5 ms, since it needs to acquire 32 samples. This system uses a DSPIC processor.

The second aim of this work is to check the feasibility of a proposal of an RMS estimator built using limited resources hardware and to validate its reliability and accuracy to prove that even limited resources microcontrollers which are not intended for arithmetic functions are able to perform and to achieve an estimation capable of providing as fast as possible an RMS phasor module value to be used in a voltage control loop or in a de-energizing alarm system.

Section 2 describes the fundamentals of the RMS calculation. Section 3 approaches the methodology used in this proposal from the system functional blocks to the algorithm implementation including low computational complex arithmetic to estimate the phasor module. The paper finishes with the description of the experiments performed and the consequent results in Section 4 from which the conclusions are drawn in Section 5, where future work is additionally presented.

2. Backgrounding on Root Mean Square Value Calculation

2.1. Discrete RMS Value Determination

The most common way to describe or analyze power quality events is by the measurement of the RMS value of voltage or current. According to [19], the RMS value of an electrical magnitude (e.g., voltage) is determined by the formula:

$$V_{rms} = \sqrt{\frac{1}{T} \int_0^T v^2(t)\mathrm{d}t} \tag{1}$$

The discretization of this expression results in [13]

$$V_{rms} = \sqrt{\frac{1}{2N} \sum_{n=1}^{N} v_n^2} \tag{2}$$

RMS values can be computed continuously whether they are updated each time a new sample is obtained but generally these values are updated with a certain time interval (each cycle or half cycle) in a discrete mode [20]. The IEC standard [19] proposes the $U_{rms(1/2)}$ discrete measurement method for determining the voltage level during the event.

From the point of view of a customer, the IEC 61000-4-30:2015 standard [19] characterizes a sag/swell at the supply line by means of its duration and remaining voltage. The most common procedure for measuring this minimum/maximum voltage value during the event is the voltage RMS calculation described above. Once this value is measured, the event duration may be determined by comparing this value with the established threshold and hysteresis levels.

On the one hand, the RMS method is simple, well known, and easy to implement. On the other hand, the limitations in the use of RMS values in power quality analysis from the point of view of the IEC standard have been clearly overviewed [21,22]. In a single-phase, the major drawbacks that affect both the performance feasibility and the accuracy of results are:

- the dependency on the window length and the time interval for updating the values; the longer the measuring window length the less correct the RMS value estimation for short duration and less severe voltage events
- the dependency on the starting event point-on-wave for calculating the RMS value faster or slower
- the commonly unpredictable large number of processing steps that complex operations like squares or squared roots need, which determines a faster processor clock frequency to avoid sampling lagging or processing gap overtaking.

As a consequence, limited performance in the detection of voltage events and in the estimation of their magnitude and duration is produced.

2.2. Using the Moving Window to Evaluate the Voltage Magnitude

In contrast to the $U_{rms(1/2)}$ discrete measurement, in the RMS voltage magnitude evaluation by means of a moving window [23,24] the values are continuously calculated from the input voltage samples. This method expresses the energy content of the signal and provides a convenient measure of the magnitude evolution. Assuming that the window contains N samples per evaluated signal period, the RMS value at sampling instant k can be calculated by:

$$V_{rms}[k] = \sqrt{\frac{1}{N} \sum_{n=1}^{N} v_n[k]^2}$$

(3)

Thus, the RMS value can be computed each time a new sample is obtained, which requires a large size/amount of data memory if these data have to be stored. In the $U_{rms(1/2)}$ method, the values are updated each half cycle of the supply signal, meaning the required memory is significantly reduced. The discrete function that illustrates this evaluation method according to Equation (3) is shown in Figure 1. The N-values length data vector used in this calculation is completely refreshed after N samples to prevent buffer overflow.

The discretizing of the supply signal relies on several methods for sampling:

- Zero-cross synchronized sampling [25], where the electrical signal's zero-across synchronous detection is equal to synchronous sampling with an invariable interval. This method has no error of measurement and only presents errors related to the sampling process or the analog to digital conversion. This synchronous sampling error must be taken into account when sampling non-sine waves. An extension of this method using oversampling is used in [25] in order to measure true RMS values under harmonic conditions.
- Quasi-synchronous sampling [25]. This means that the alternative signal is discretized as a zero-cross synchronized sampling sequence after low-pass filtering the supply signal. The associated errors involve power calculation due to the displacement angle introduced during the filtering process.
- Non-synchronous sampling used in [26] for true RMS measurement based on the calculation of cycle double-end sampling values.

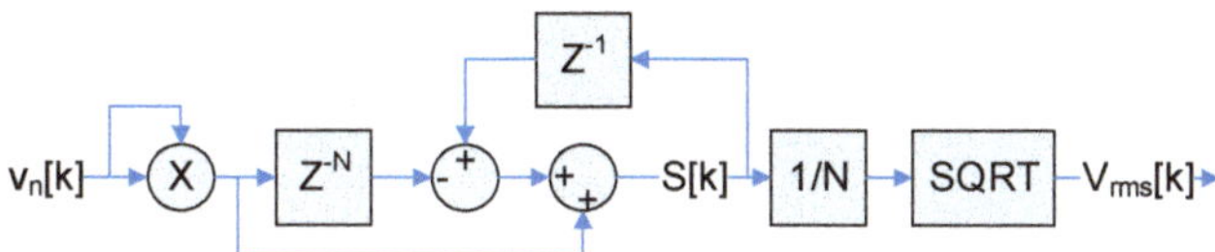

Figure 1. Z-domain moving window root mean square (RMS) evaluation function.

3. Methodological Approach

As stated above, sensitive electronic equipment needs voltage dip detection algorithms which are as fast and reliable as possible in order to become immune to this type of perturbation [27]. The IEC 61000-4-30 $U_{rms(1/2)}$ determination is not fast enough to achieve this critical requirement; this is, in fact, not its purpose. A simple comparative simulation study was performed in PSIM 11, updating [28], to compare how these aforementioned methods result in an RMS voltage magnitude value, with the results shown in Figure 2. Here, the response time of several sampling rate continuous RMS calculations, a supply period discrete RMS calculation, and $U_{rms(1/2)}$ for a voltage signal (Vs) are compared. The significant differences between the two compared methods may be easily checked:

the moving window method depicted in the traces is labeled as VCrms_x for four, eight, and 16 samples per half a period, respectively, and the IEC method presented in the traces is named VDrms_1T and Urms1_2_6100_4_30.

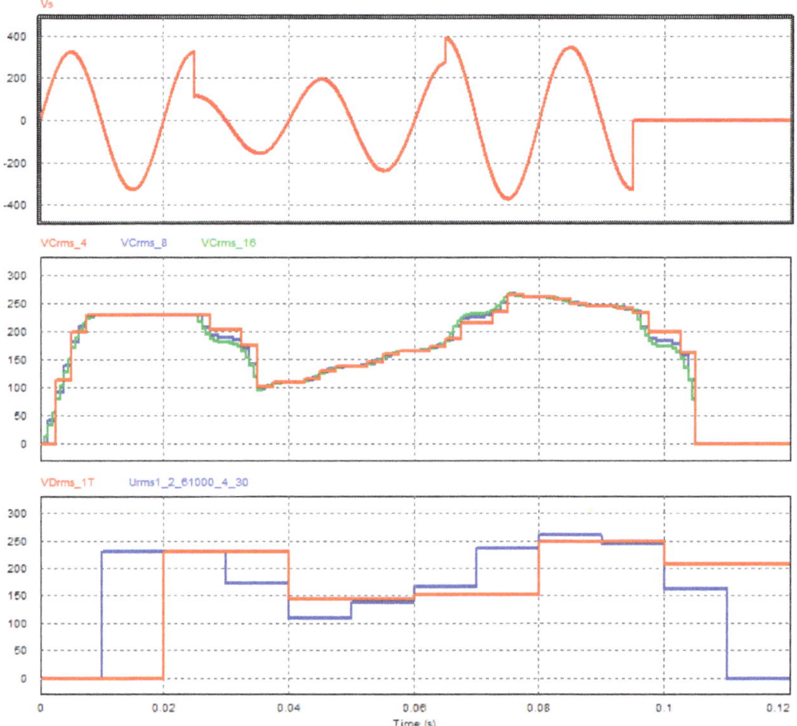

Figure 2. Response time of the discrete and moving window selected metering procedures versus IEC 61000-4-30 $U_{rms(1/2)}$.

As stated before, the main objective of our proposal is to follow the changes in the voltage magnitude as closely as possible during the sag/swell event. The more instant voltage values considered or employed in a window, the closer the actual RMS value is [17,18].

3.1. HW Proposal for an RMS Evaluator Prototype

The aim of this work is to propose a memory-efficient algorithm for metering the RMS voltage that runs on limited resources hardware. The proposed and built meter has the block diagram described below (Figure 3).

The voltage transducer block was based on an LEM® LV-25xx (LEM®, Milwaukee, WI, USA) [29] circuit that carries out galvanic isolation and scaling functions. Passive element values in its connection circuit were chosen for the nominal supply line voltage. A voltage transducer block is used to adapt the signal amplitude to the microcontroller input range. Once the level has been scaled, the bipolar signal is converted to a unipolar one with a general-purpose precision rectifier.

The precision rectifier is the standard full wave version of this function built with limited resources general-purpose operational amplifiers and standard diodes and resistors. This block provides the absolute value of the scaled supply signal and adapts the signal to be measured to the microcontroller ADC range. So far, nothing new has been set in practice.

The estimator proposed in this paper was built in a general-purpose limited resources 8–bit arithmetic logic unit (ALU) PIC that included five multiplexed 10-bit ADCs.

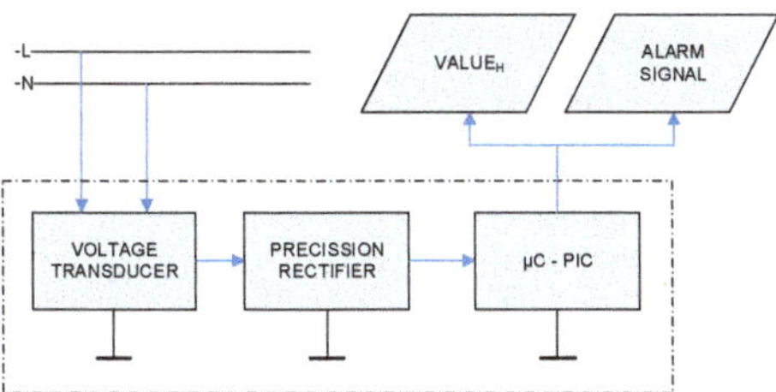

Figure 3. Structure of the proposed volt-estimator. PIC is a family of microcontrollers (µC) made by Microchip Technology, Chandler, AZ, USA.

3.2. Fitting the Main Program in the Microcontroller: HW/FW Limitations

The main program flow diagram implemented in the PIC is shown in Figure 4. Note that the first operation to be performed by the main function was to create the look-up-table (LUT) with the set of voltage values and their respective squared values according to the estimation range, as explained in Section 3.3 below.

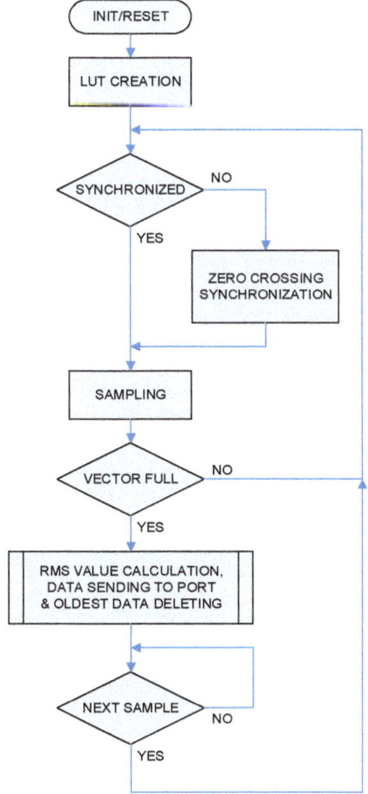

Figure 4. Flow chart of the PIC main program.

The selected microcontroller block was a Microchip® 18Fxx2 PIC [30] that performs the following functions:

- Signal sampling. In accordance with the Nyquist theorem, the sampling frequency must be more than twice the signal frequency in order to avoid the aliasing effect, which has been established with a synchronous sampling frequency of 800 Hz that provides an 8-sample vector per rectified scaled signal period.
- Analog to digital conversion is performed in a 10-bit ADC port.
- Zero-crossing synchronization mechanism.
- RMS voltage measure core function, which is described in a specific chapter.
- The RMS voltage value or the undervoltage alarm sending to an output port.

3.3. Selecting the RMS algorithm for Voltage Estimation

As stated above, limiting the time for a square or a squared root calculation may be the Achilles heel in this step, due to both operation spend for an a priori unknown number of processing steps, i.e., squaring a number might be highly dependent on the number size, and rooting another one could require many steps before the result converges. For instance, the time gap used to perform the Babylonian algorithm for square rooting is highly dependent on convergence speed and this dependence becomes uncertainty.

Given that our purpose is to obtain the most accurate possible value in the least time possible with the hardware employed, the focus of this procedure was on modifying and simplifying the RMS value calculation algorithm.

All the data were declared as unsigned integers.

The selected method for voltage value or event detection is an evolution of voltage RMS magnitude continuous evaluation using a moving window. As in the original method, the supply signal is discretized and processed using Equation (3) being rewritten in the following way:

$$V_{rms}[k] = \sqrt{\sum_{n=1}^{N} \frac{v_n[k]^2}{N}} \tag{4}$$

Using this method, the squared values of the sampled vector are divided by the vector length prior to the summation. Hence, by choosing an 8-data vector length, the instruction of dividing an integer by eight becomes a faster operation than the division by a non-power of two. In such a way, the 16-bit size of the data will not be overflown by the results and, therefore, will optimize their size. Thus, the discrete time function described above in Figure 1 was able to be restructured as shown in Figure 5.

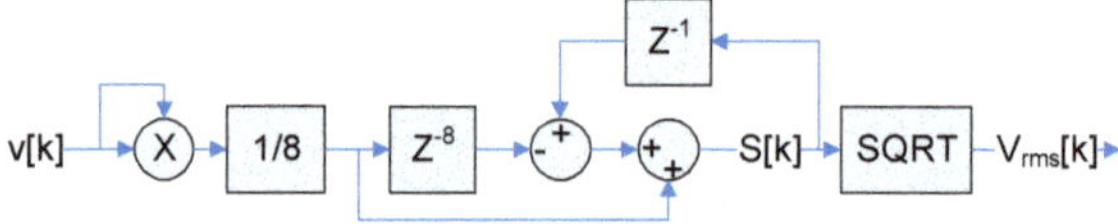

Figure 5. New Z-domain moving window function for RMS evaluation.

Working with powers of two values allows simplifying all the calculus to be done as well. Thus, only the eight most significant bits (MSBs) of the 10-bit data from the ADC were considered, as displayed in Figure 6. Otherwise, the squared value of a 10-bit datum would be that of a 20-bit one, with the result that the microcontroller's basic data size would be exceeded and there would be a need for a more than double register size register to define the result.

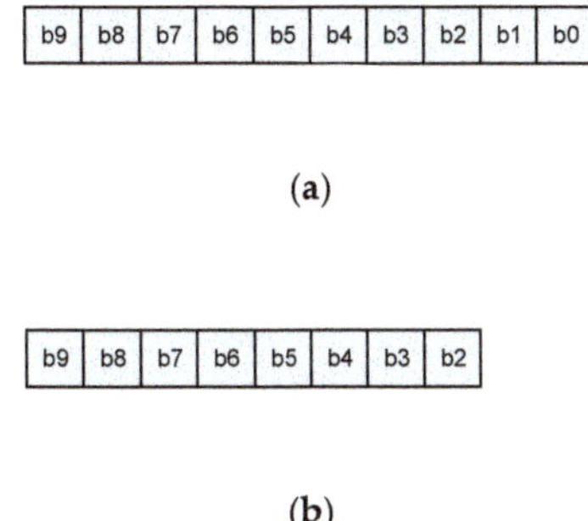

(a)

(b)

Figure 6. Only the eight most significant bits (MSBs) of the binary datum (**b**) from the ADC output (**a**) are used.

The RMS-calculation algorithm starts using these eight MSBs from the ADC output (Figure 7a) as the datum's index pointer of its squared value (Figure 7b) stored in a kind of lookup table. This squared datum is 16-bits long.

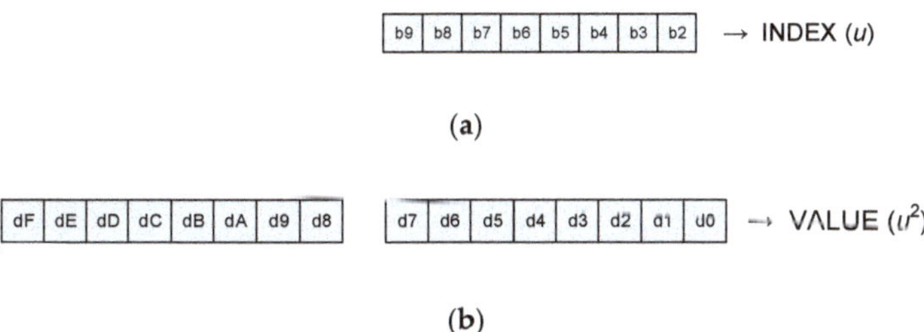

(a)

(b)

Figure 7. The eight MSBs from the ADC output (**a**) are the index pointer of its squared value (**b**).

Any additional operation would surpass the established limit of 16 bits, so if it was divided by eight, the data vector length, prior to the summation the result should fit within a 13-bit value. This division is displayed in Figure 8. Consequently, the sum of eight 13-bit terms will never exceed a 16-bit size.

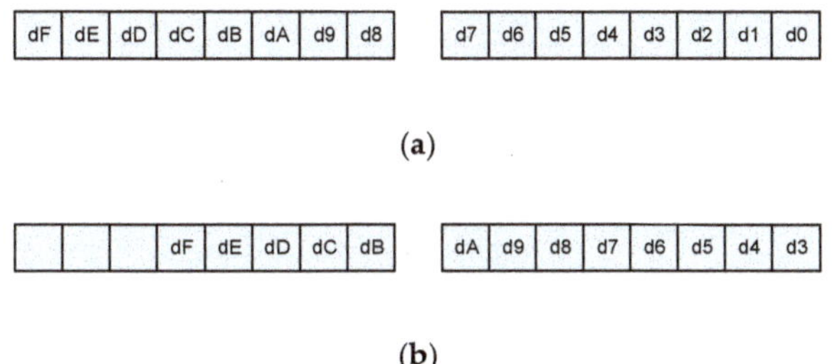

(a)

(b)

Figure 8. The squared item (**a**) divided by eight produces data shifted three positions to the right (**b**).

Once the data vector has been summed, the square root is as easy as finding the address of the data vector that matches or is close to the sum result. Both the data and the summation sizes are depicted in Figure 9.

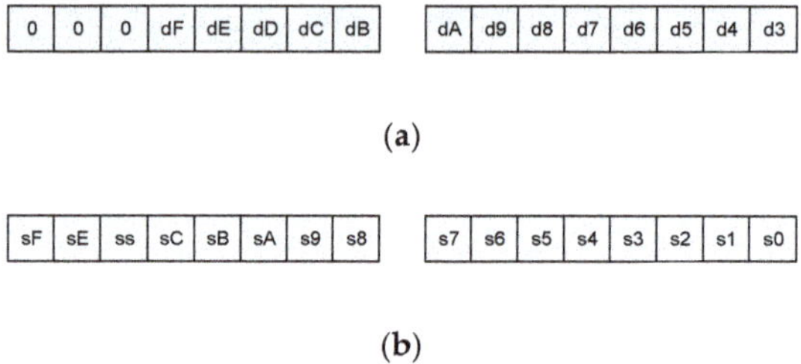

Figure 9. Adding eight 13-bit data (a) gives a result of up to 16 bits (b).

The selected method used to find the square root value of the summation is the bisection method, which is also called the binary search method. This method allows for the obtaining of a valid solution in a maximum of eight, which is the solution length in bits, comparison steps and, as a consequence, the solution can be obtained in a very short time with a very low computational cost. Figure 10 shows the relationship between the summation value and its position index to get the solution.

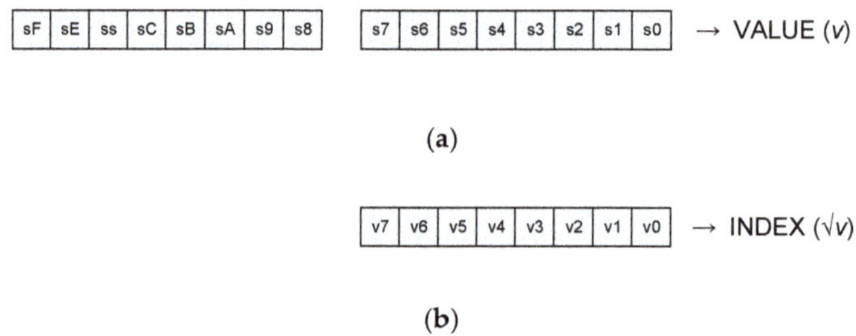

Figure 10. The address/index-pointer (b) where the result (a) is stored is identified by means of the bisection method.

This solution can be inputted into a communication port, compared to an alarm threshold level, or used as an input argument in other functions, such as a discretized voltage control loop.

The LUT for the RMS calculation algorithm is illustrated in Figure 11.

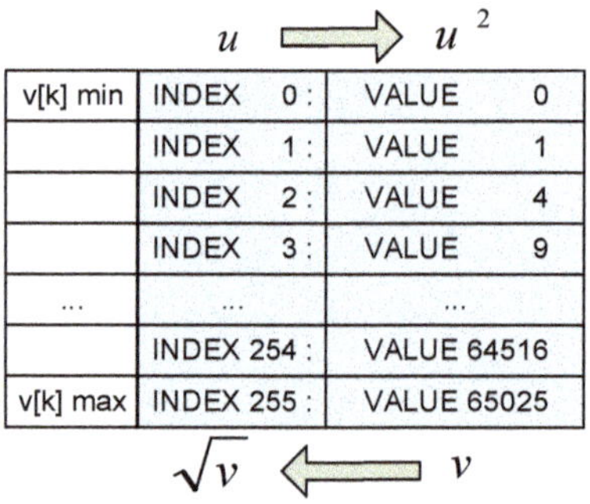

Figure 11. Both square and square root operations were performed using a look-up-table (LUT).

Finally, the relationship between the proposed moving window function for RMS evaluation and the register size can be clearly seen in Figure 12.

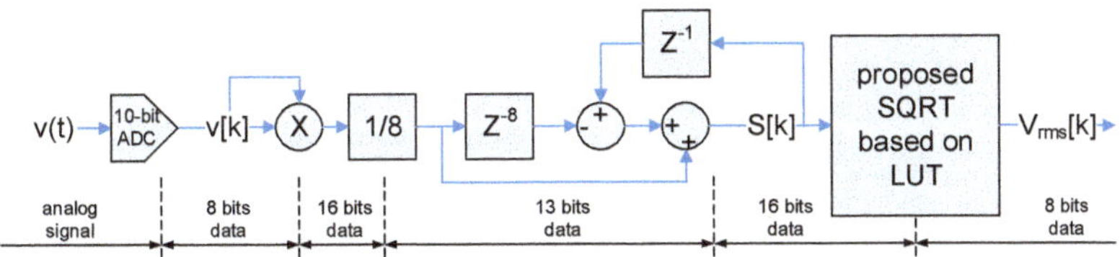

Figure 12. Proposed final moving window function for RMS evaluation with corresponding data bit lengths.

All the processes in the algorithm can be easily written in assembly PIC code or C/C++ language, minimizing and optimizing the program size and its execution time.

4. Performed Experiments

A set of different experiments were carried out with the prototype under testing in order to achieve its overall performance. For that purpose, static and dynamic tests were designed and the obtained results may indicate the proposal's abilities.

Firstly, on the one hand, for these experiments, the voltage transducer block scaling factor was established in 1:100 to fit the desired measuring range upper limit to 500 V and adapt it to the microcontroller ADC input range, which was set as up to 5 V. This upper limit was set because it is just over 150% of the supply voltage peak value, i.e.:

$$1.5 \cdot \sqrt{2} \cdot 230 + 2\%$$

On the other hand, there were only 8 bits considered from the 10 ADC output. Thus, the ADC resolution was set to 19.6078 mV per step, i.e., each bit in the RMS value corresponded to 1.96 volts with the referred scaling factor.

In the second stage, the speed of event detection capability and a performing algorithm evaluation regarding some measured data series were carried out.

Finally, a comparison experience was undertaken. A similar system was built with another limited resources microcontroller running a standard RMS basic algorithm with a 1 kHz sampling frequency in order to highlight its measuring uncertainty with respect to our proposal.

All the steps described above are covered in the following subsections.

4.1. Checking the Accuracy

To trial the accuracy of the proposed algorithm, the RMS estimator was correlated with the TRMS voltmeter of a Fluke ScopeMeter® (Fluke®, Everett, WA, USA) series 120 in order to check its accuracy. A set of steady-state metering tests were implemented for a series of supply line voltages from, approximately, 0.1 pu to 1.25 pu provided by an autotransformer in the experiment topology shown in Figure 13.

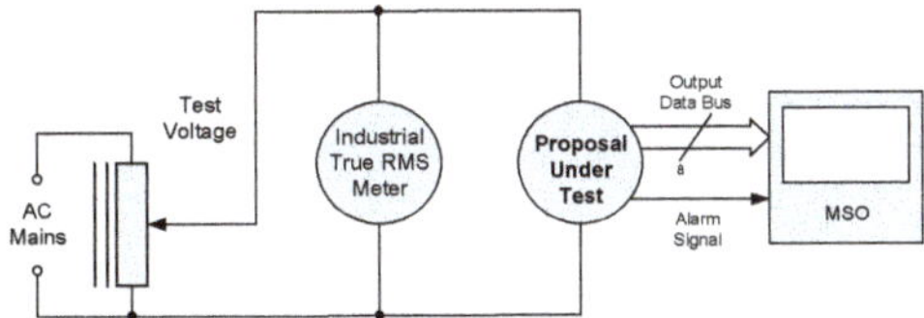

Figure 13. Circuit for characterizing the meter static accuracy. MSO means mixed-signal oscilloscope.

The correlation results are displayed in Figure 14.

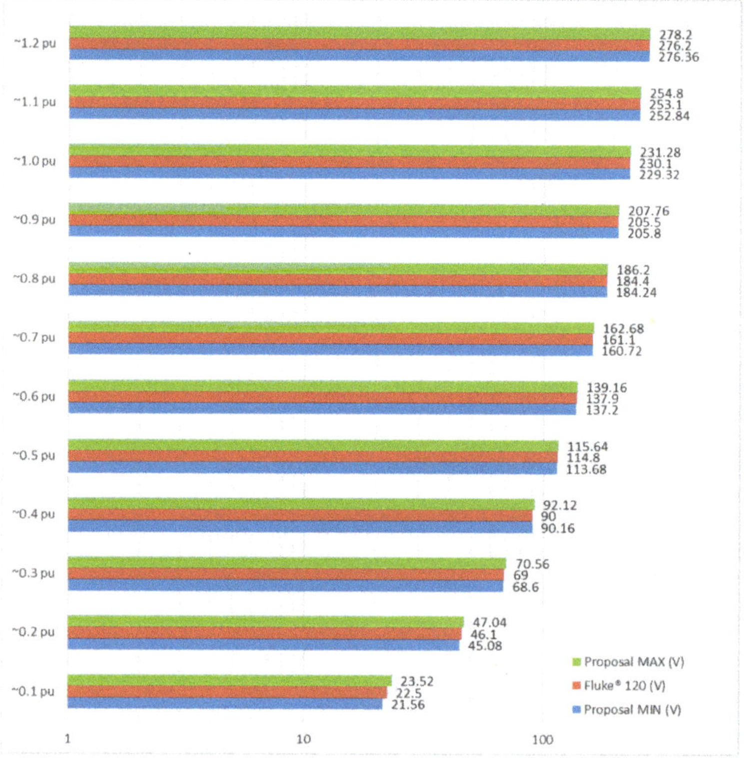

Figure 14. Correlated voltage measured in the proposal versus Fluke® 120.

Note in Figure 14 that the absolute difference between the maximum and minimum values provided by this proposal is the discretization step (1.96 V) in all cases and the relative error quickly reduces as the gap between the actual value and the estimation limits remains limited while the value of the phasor module increases. Figure 15 depicts the relative error for both the maximum and the minimum phasor module value estimations regarding the real objective value measured with the reference true-RMS digital multimeter.

4.2. Dynamic Response: Event detection and Present Value Estimation

Once the meter was characterized and its accuracy tested, more experiments were carried out to check the meter dynamic response. The circuit used for this set of tests is shown in Figure 16. The worst case analyzed, a 0.7 pu sag beginning at the zero-crossing supply signal, resulted in a time of 6 ms for the sag detection (a level below the 90% of the nominal value) and 8 ms for a valid remaining voltage metering, as shown in Figure 17.

Note that due to the event which took place at the null instant voltage value (signal zero crossing), the first value capable of modifying the summation is taken in the following sample, 1.25 ms later).

4.3. Comparing the Proposed Estimator Accuracy to Standard Measuring Methods

Once the initial accuracy test was successfully checked under lab conditions, several validation processes used to prove its capability for detecting de-energizing events with sets of captured field data were performed.

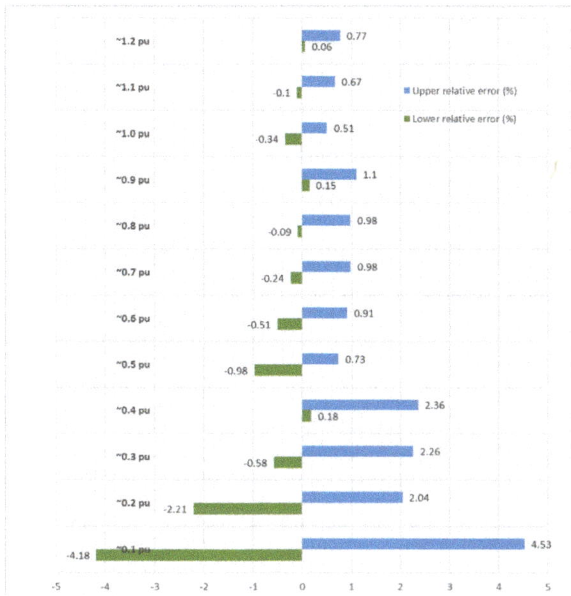

Figure 15. Experimental relative error margins of the proposed RMS estimator.

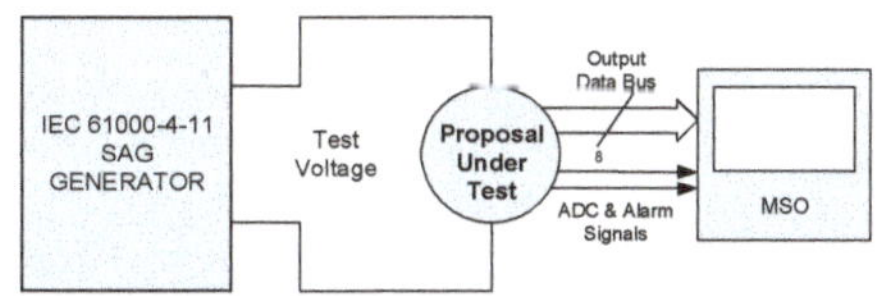

Figure 16. Circuit to determine the estimator's dynamic response.

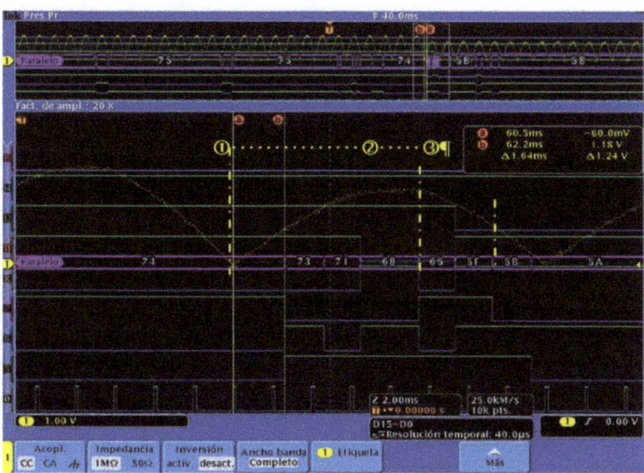

Figure 17. Oscillogram traces from one experiment to determine the event detection dynamic response of the proposed RMS. The dashed-dotted lines mark the start of the event (1), the <0.9 pu level detection (2), and the valid sag value (3).

First, registered data sets from AC mains in home appliances were analyzed using the algorithm used in [31]. Herein, in contrast to what was stated in the previous subsection, the experiment omitted zero crossing synchronization but achieved sufficient accuracy.

In the second stage, processing data from a 3-phase substation outlet were contrasted with data obtained from the already installed instrumentation at the facility.

In Figure 18a the time series of the 3-ph voltages with a clear brownout are shown. The measured data from a 96 sample per period double-precision data vector moving window from a DSP (digital signal processor) metering device (grey color and the legend 'VxMW48'), lines RMS values obtained using the IEC 61000-4-30:2015 method for half a period (orange color and the legend 'Vx61k4-30'), and the data provided by this proposal (blue color and the legend 'Proposal') are depicted in Figure 18b–d, respectively, for each AC phase.

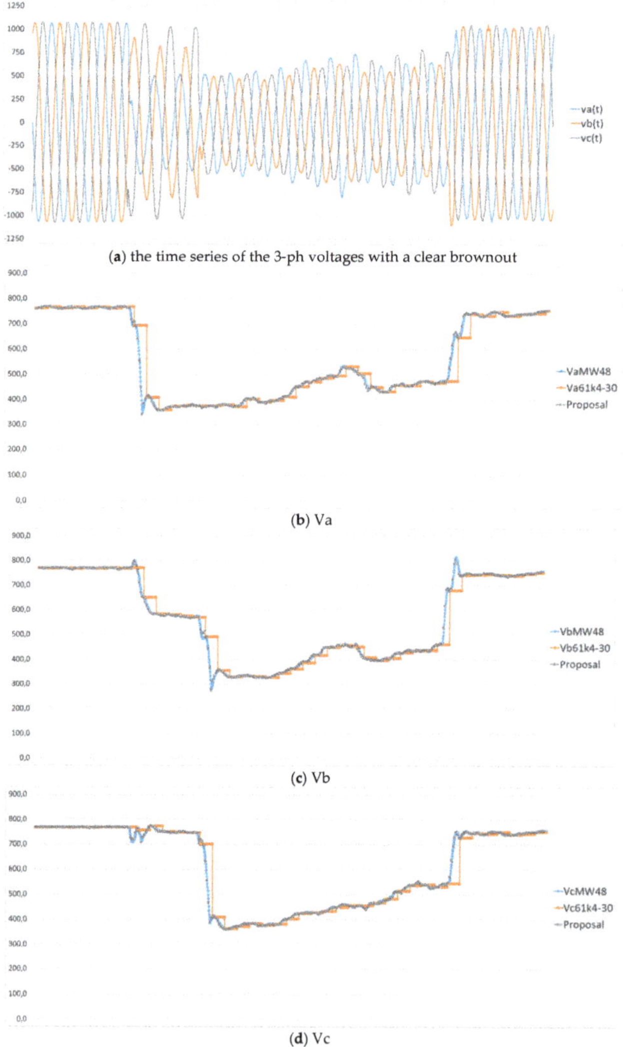

Figure 18. AC substation mains time waveforms and RMS values results. All y-axis values are in volts.

4.4. Contrast Experiment: Expanding the Samples Vector and Employing the Whole Math Function Library

To perform this experiment, a general purpose ATmega328P microcontroller-based prototype, as given in [22], was built and configured with a clock frequency of 16 MHz. The selected sampling frequency was 1 kHz, i.e., a 20 samples per AC voltage signal period, and the programmed procedure fit what is described in Figure 1 and in Equation (3).

The contrasting demonstration was limited to the dynamic response since the static accuracy had been verified as being as good as that from the proposed system. As has been mentioned in previous sections, no zero-crossing synchronization was performed in this experiment.

Of the set of experiments carried out, the one corresponding to a sag of 25% is shown. The depicted screens in Figure 19 represent the two extreme cases recorded within this. In this figure, the blue trace corresponds to an active low alarm related to surpassing the 10% undervoltage limit.

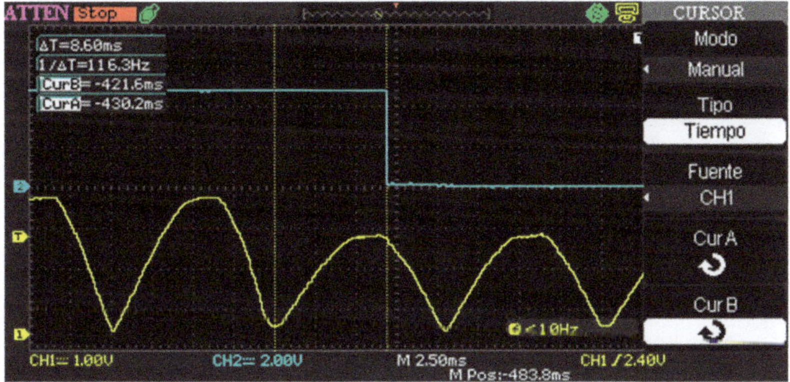

(**a**) Good result.

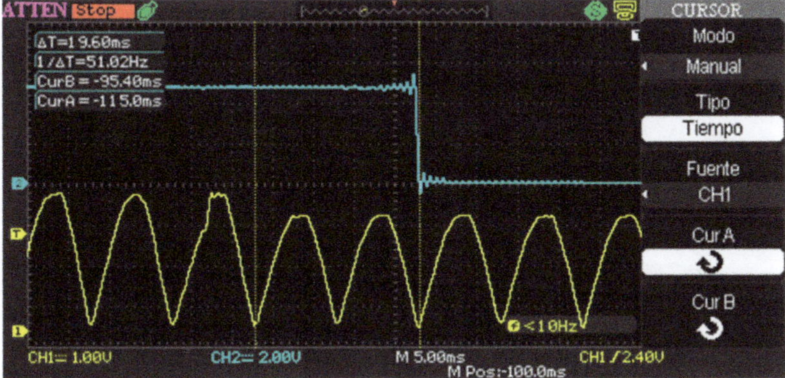

(**b**) Poor result.

Figure 19. Two different oscillograms of instant waveform values registered of various same-level de-energizing event.

Note that the shortest recorded time to reveal a 0.75 pu sag (Figure 19a) is pretty close to our proposal, though the longest one (Figure 19b) triples ours. This is due to the uncertainty in terms of duration both in the second power and in the square root operations.

5. Conclusions and Future Work

In this paper, an RMS voltage estimator that eludes the inherent uncertainty of complex arithmetic operations related to the discretized RMS algorithm has been presented.

Even though there are more accurate and faster methodologies present in the literature, the one proposed in this paper is built using limited resources hardware with a high-efficiency program which minimizes the system requirements and prevents data memory overflow. Therefore, the phasor module estimator presented here limits its response time for providing a valid estimation and it can be implemented in simple microcontrollers that do not have high-performance arithmetic units or specific resources for digital signal processing. Its ability to operate as either a sag/swell detector, a true RMS meter in a protective alarm or idling system, or at a not too demanding voltage loop of a regulator has been assessed and verified by comparing its performance to a low-cost embedded system based on an ATmega328p microcontroller, one of the most basic of those provided by the company Microchip.

The proposal aims in the near future to carry out its implementation in a very low-cost FPGA (field-programmable gate array) that only has basic logical resources such as LUTs, adders, flip-flop, and embedded block RAM, etc., such as, for example, the iCE40 LP/HX/LM family and the integration of voltage and current phasors estimation on a single device. This would allow for the embedding of this feature into a smart home appliance.

Author Contributions: J.-M.F.-A. conceived the proposal idea; J.-M.F.-A. and A.M.-M. performed the algorithm design and its modeling and simulation; M.O.-L. and F.J.Q.L. defined the HW requirements and selected the microcontrollers; F.J.B.-O. designed the signal adapting circuits and provided experimental facilities and data series; J.-M.F.-A., M.O.-L., and F.J.Q.L. programmed and assembled the estimation systems; all the authors performed experiments; J.-M.F.-A. wrote the paper (although all of the authors contributed with outlines and drafts to different sections); and A.M.-M. and F.J.B.-O. revised the manuscript.

Funding: This research was funded by the Spanish Ministry of Economy and Competitiveness under Project TEC2016-77632-C3-2-R.

Acknowledgments: This research was supported by the Spanish Ministry of Economy and Competitiveness under Project TEC2016-77632-C3-2-R.

Conflicts of Interest: The authors declare no conflict of interest. In addition, the founding sponsors had no role in the design of the study; in the collection, analyses, or interpretation of data; in the writing of the manuscript, or in the decision to publish the results.

References

1. GAP.5.7.1.3, GAPS Guidelines. In *Electric Power Quality and Reliability*; Global Asset Protection Services LLC: Hartford, CT, USA, 2015.
2. Moreno-Munoz, A.; Flores, J.M.; Oterino, D.; de la Rosa, J.J. Power line conditioner based on CA PWM Chopper. In Proceedings of the IEEE International Symposium on Industrial Electronics, Vigo, Spain, 4–7 June 2007; pp. 2454–2456. [CrossRef]
3. Moreno-Munoz, A.; Flores, J.M.; Oterino, D.; de la Rosa, J.J. Voltage regulator system based on a PWM AC chopper converter. In Proceedings of the 2011 IEEE International Symposium on Industrial Electronics (ISIE), Gdansk, Poland, 27–30 June 2011; pp. 468–473. [CrossRef]
4. Shainker, R.; Keebler, P.; Geist, T.; Fortenbery, B. *Enhanced Ride-Through for Industrial Power Supplies: Feasibility, Testing, Design and Assessment of Future Intelligent Embedded Solutions*; Electrical Power Research Institute: Palo Alto, CA, USA, 2005.
5. Guerrero, J.M. Microgrids: Connecting Renewable Energy Sources into the Smartgrid. In Proceedings of the Tutorial at 20th IEEE ISIE, Gdansk, Poland, 27–30 June 2011; pp. 2400–2566. [CrossRef]
6. Moreno-Munoz, A.; De la Rosa, J.J.G.; Flores-Arias, J.M.; Bellido-Outerino, F.J.; Gil-de-Castro, A. Energy efficiency criteria in uninterruptible power supply selection. *Appl. Energy* **2011**, *88*, 1312–1321. [CrossRef]
7. Moreno-Munoz, A.; Gonzalez, M.; Linan, M.; Gonzalez, J.J. Power quality in high-tech plants: A case study. In Proceedings of the IEEE Compatibility in Power Electronics, Gdynia, Poland, 1 June 2005; pp. 30–38. [CrossRef]

8.   Moreno-Muñoz, A. *Power Quality: Mitigation Technologies in a Distributed Environment*; Springer: London, UK, 2007; ISBN 978-1-84628-771-8.

9.   CSEP 2 Interoperability. 2013. Available online: http://www.csep.org/ (accessed on 16 September 2013).

10.   AENOR. *UNE-EN 50160:2011/A1:2015 Voltage Characteristics of Electricity Supplied by Public Electricity Networks*; The Spanish Association for Standardization and Certification: Madrid, Spain, 2015.

11.   IEEE. *IEEE Recommended Practice for Monitoring Electric Power Quality*; IEEE Std 1159-2009 (Rev. of Std 1159-1995); IEEE: Piscataway, NJ, USA, 2009; pp. c1–c81.

12.   Bollen, M.; Gordon, J.R.; Djokic, S.; Stockman, K.; Milanovic, J.; Neumann, R.; Ethier, G.; Zavoda, F. Voltage dip immunity of equipment and installations—Status and need for further work. In Proceedings of the China International Conference on Electricity Distribution (CICED), Nanjing, China, 13–16 September 2010; ISBN 978-1-4577-0066-8.

13.   Caruso, M.; Di Tommaso, A.O.; Miceli, R.; Galluzzo, G.R.; Romano, P.; Schettino, G.; Viola, F. Design and experimental characterization of a low-cost, real-time, wireless AC monitoring system based on ATmega 328P-PU microcontroller. In Proceedings of the 2015 AEIT International Annual Conference (AEIT), Naples, Italy, 14–16 October 2015; pp. 1–6. [CrossRef]

14.   Neto, I.M.B.; Lopes, A.I.; De, M.S.M.A.; Brito, M.M.D.A.; Silva, D.R.; Nogueira, M.B.; Rodrigues, M.C. THDI Measurement System of Home Energy Signal Based on IoT. In Proceedings of the 2018 Workshop on Metrology for Industry 4.0 and IoT, Brescia, Italy, 16–18 April 2018; pp. 180–185. [CrossRef]

15.   Koziy, K.; Gou, B.; Aslakson, J. A Low-Cost Power-Quality Meter with Series Arc-Fault Detection Capability for Smart Grid. *IEEE Trans. Power Deliv.* **2013**, *28*, 1584–1591. [CrossRef]

16.   Mujumdar, U.B.; Joshi, J.S. Microcontroller based true RMS current measurement under harmonic conditions. In Proceedings of the 2010 IEEE International Conference on Sustainable Energy Technologies (ICSET), Kandy, Sri Lanka, 6–9 December 2010; pp. 1–5. [CrossRef]

17.   Aurilio, G.; Gallo, D.; Landi, C.; Luiso, M.; Graditi, G. A low cost smart meter network for a smart utility. In Proceedings of the 2014 IEEE International Instrumentation and Measurement Technology Conference (I2MTC) Proceedings, Montevideo, Uruguay, 12–15 May 2014; pp. 380–385. [CrossRef]

18.   Gencer, Ö.; Öztürk, S.; Erfidan, T. A new approach to voltage sag detection based on wavelet transform. *Int. J. Electr. Power Energy Syst.* **2010**, *32*, 133–140. [CrossRef]

19.   International Electrotechnical Commission. *IEC 61000-4-30: 2015. Electromagnetic compatibility (EMC)—Part 4-30: Testing and Measurement Techniques—Power Quality Measurement Methods*; International Electrotechnical Commission: Geneva, Switzerland, 2015.

20.   Styvaktakis, E.; Bollen, M.H.J.; Gu, I.Y.H. Automatic classification of power system events using RMS voltage measurements. In Proceedings of the IEEE Power Engineering Society Summer Meeting, Chicago, IL, USA, 21–25 July 2002; pp. 824–829, ISBN 0-7803-7518-1. [CrossRef]

21.   Perez, E.; Barros, J. Voltage Event Detection and Characterization Methods: A Comparative Study. In Proceedings of the IEEE/PES Transmission & Distribution Conference and Exposition: Latin America, Caracas, Venezuela, 15–18 August 2006; pp. 1–6, ISBN 1-4244-0287-5. [CrossRef]

22.   Barros, J.; Perez, E. Limitations in the Use of R.M.S. Value in Power Quality Analysis. In Proceedings of the IEEE Instrumentation and Measurement Technology Conference, Sorrento, Italy, 24–27 April 2006; pp. 2261–2264.

23.   Deckmann, S.M.; Ferrira, A.A. About voltage sags and swells analysis. In Proceedings of the 10th International Conference on Harmonics and Quality of Power, Rio de Janeiro, Brazil, 6–9 October 2002; Volume 1, pp. 144–148, ISBN 0-7803-7671-4.

24.   Kai, D.; Cheng, K.W.E.; Xue, X.D.; Divakar, B.P.; Xu, C.D.; Che, Y.B.; Wang, D.H.; Dong, P. A Novel Detection Method for Voltage Sags. In Proceedings of the 2nd International Conference on Power Electronics Systems and Applications, (ICPESA '06), Hong Kong, China, 12–14 November 2006; pp. 250–255, ISBN 962-367-544-5. [CrossRef]

25.   Liu, S.; Li, P.; Huang, X. The Quasi-Synchronous Sampling and its Algorithms on Relay Protection. In Proceedings of the 2010 Asia-Pacific Power and Energy Engineering Conference (APPEEC), Yichang, China, 16–18 September 2010; pp. 1–4.

26.   Wang, X.; Zhang, X.; Zhang, L.; Li, M. True AC Root-mean-square Measurement Based on the Calculation of Cycle Double-end Sampling Values. In Proceedings of the 2010 WASE International Conference on Information Engineering (ICIE), Beidaihe, China, 14–15 August 2010; Volume 4, pp. 177–180.

27. Amaris, H.; Alvarez, C.; Alonso, M.; Florez, D.; Lobos, T.; Janik, P.; Rezmer, J.; Waclawek, Z. Application of advanced signal processing methods for accurate detection of voltage dips. In Proceedings of the 13th International Conference on Harmonics and Quality of Power, Wollongong, NSW, Australia, 28 September–1 October 2008; pp. 1–6.

28. Flores-Arias, J.M.; Bellido-Outeiriño, F.J.; Moreno-Munoz, A. A fast RMS meter for detecting sag events in household environments. In Proceedings of the 2014 IEEE International Conference on Consumer Electronics (ICCE), Las Vegas, NV, USA, 10–13 January 2014; pp. 321–322. [CrossRef]

29. Available online: http://www.lem.com/docs/products/lv%2025-p%20sp2%20e.pdf (accessed on 24 October 2015).

30. Available online: https://ww1.microchip.com/downloads/en/DeviceDoc/39564c.pdf (accessed on 19 March 2016).

31. Flores-Arias, J.M.; Outeiriño, F.J.B.; Palacios-Garcia, E.J.; Gil-de-Castro, A.; Moreno-Munoz, A.; Quero, M.A. Low cost de-energizing warning meter algorithm for sensitive loads. In Proceedings of the 2017 IEEE International Instrumentation and Measurement Technology Conference (I2MTC), Turin, Italy, 22–25 May 2017; pp. 1–4.

Article

Analysis of Dense-Mesh Distribution Network Operation Using Long-Term Monitoring Data

Michal Ptacek [1,*], Vaclav Vycital [1], Petr Toman [1] and Jan Vaculik [2]

[1] Department of Electrical Power Engineering, Brno University of Technology, Technicka 12, 61600 Brno, Czech Republic; vycital@feec.vutbr.cz (V.V.); toman@feec.vutbr.cz (P.T.)
[2] E.ON Distribuce, a.s., Hady 2, 61400 Brno, Czech Republic; jan.vaculik@eon.cz
* Correspondence: ptacekm@feec.vutbr.cz; Tel.: +420-541-146-209

Received: 22 August 2019; Accepted: 12 November 2019; Published: 14 November 2019

Abstract: The technical and economic aspects and the possibility of the mesh network topology offering many radial configurations lead to the fact that large municipal networks are generally under radial operation. However, it is very important to analyze the operation and control of the mesh networks, especially in terms of their safety and durability and in the frame of the smart grid concept, respectively. The article deals with the analysis of the operation of the dense-mesh municipal distribution network of E.ON Distribuce a.s. based on the long-term data from power quality monitors. It also shows a brief view of the current lack of data usability from monitors installed in distribution networks in the context of smart grid.

Keywords: dense-mesh topology; municipal distribution network; smart grid; power quality monitor; long term; operation analysis

---

1. Introduction

One of the main outcomes resulting from the transition towards new smart grids is expected the better observability and monitoring of the whole grid [1]. This observability is expected due to the bulk installation of smart meters, as well as the monitoring of transformer stations, ring main units, etc. [1,2]. Even though there is still lot of concern and research devoted to the processing of such a huge amount of recorded data [3] and its reliable communication back to distribution system operator database [4,5], the usefulness of measurement data availability is already obvious [1,2,6–11]. One of the fields that will benefit from the detailed measurement data availability would be distribution system modelling, planning, and operation optimization [1,6,7,11–14], which will be more accurate.

A commonly performed task as part of distribution system planning is the calculation of load flow. Although the solution of a non-linear power flow problem is quite well addressed [15–17], there is still some effort mainly devoted in computational and time efficiency of this numerically intensive task [18]. A part of load flow studies is also appropriate the modelling of connected loads, which can be as with constant impedance, constant current, or constant power, i.e., ZIP load model [19]. Basically, the historical data are very useful in the case of modelling verification, as well as the mapping of possible extreme states of the network. Thus, it is expected that the availability of more accurate data from bulk deployment of measurement will definitely lead to more accurate load models, power flow studies, load forecasting, prosumers tariffs setting, etc.

However, data processing shows that even the currently applied smart metering (SM) technology has a number of drawbacks. Current communication technologies have been shown to exhibit considerable delays and fail to achieve high reliability [4,5,20]. Another quite restricting bottleneck is the non-unified sorting of data that were measured from individual phases. It cannot be assumed that all measured data are correctly assigned to reference phases of power system [21].

The problem of non-unified phase allocation can be addressed by one of the following three methods. The easiest is by proper documentation of all changes to distribution system connection by maintenance personnel. Another one, although costly, is by using special per phase signal injection and receiver devices [22,23] that can match the non-unified phases with reference ones based on the phase with injected signal. The last method is by post-analysis of measured data and finding some type of correlation between series of measurement with non-unified phases and reference measurement [21,24–30].

All of the aforementioned aspects also apply to the case of monitoring of individual MV/LV distribution transformer stations (DTS), which are equipped with power quality (PQ) monitors (analyzers with PQM) by distribution system operators (DSO). Thus, this article is focused on preliminary analysis of bulk measurement of all DTSs in the distribution grid of Brno-stred, that is expected to be pilot project of transition of this extensive traditional urban distribution network towards the smart grid. The analysis is focused on finding the usability and suitability of long term measured data (half year 06/2016–02/2017) from the LV side of all 82 transformers supplying the whole LV distribution system (DS) of the Brno-stred dense mesh with more than 3800 customers. The analysis was primarily focused on finding statistically significant operating states that will be used and helpful for future per phase modelling of this LV network. As the problem with non-unified phases was also identified during the analysis, it was also addressed in this study. The analysis of DS Brno-stred is also unique due to the fact that the LV network is operated as a dense mesh urban network to increase the reliability [12]. As this type of operation might not be so common, not many publications were published dealing with dense mesh networks. Publications [6,9] state that, due to the transition to smart grids, the urban dense mesh networks might need more attention to identify their strong and weak points in the context of changes of these old traditional networks towards bidirectional power flow grids with a high penetration of renewable sources.

The per phase statistical analysis might be helpful for the estimation of load model parameters and forecasting future loading. For example, paper [31] used F-statistic to obtain the load model as a function of voltage and frequency while using data from phasor measurement units. Other researches [32] presented a simple histograms of three years historical data to analyze the correlation between power and voltage changes. Another papers [33–35] used significant statistical values, i.e., mean, median, standard deviation (SD), percentiles (5, 95, 99 etc.), of historical measured data for the estimation of load parameters as well as for evaluation of load forecasting. Gaussian mixed model is also a very powerful statistical approach for the estimation of loads. Its application on historical data was presented in [36–38]. In [39], the loads were modelled by normal and log normal distribution function with division for each of seven days within week. As the input to the load modelling the researchers used usually only summed 3-phase active power values, bulk frequency, phase to phase voltage, and current. Very few or almost none of the researchers dealt with modelling of reactive power loading as well as conducting analysis on per phase basis. Thus, the statistical analysis is done in this paper for the active and reactive power values for each phase separately in the view of dense mesh network (i.e., analysis for all transformers at the global level and also for most (T47) and the least (T61) loaded transformers for different time intervals during days).

2. PQ Monitor Measured Data Processing

The main function of the PQ monitor is to record voltage and current in a three-phase system (L1, L2, L3). The instantaneous values of the voltage and current in regular time intervals might be given by the sampling rate that is an integer multiple (e.g., 32) of the system frequency/period ($T_s = 20$ ms). In the context of monitoring power quality based on [40,41], in a three-phase system the rms values of voltage and current for each phase in measuring time interval ($T_\mathrm{m} = 200$ ms) are obtained by the following equations

$$U_{\mathrm{x,rms}} = \sqrt{\frac{1}{N}\sum_{n=1}^{N} u_x^2(n)}, \tag{1}$$

where $U_{\mathrm{x,rms}}$ is the rms voltage value in the phase x (i.e., 1, 2 or 3), N is the total number of measured samples, n is a serial number, and $u_x(n)$ is the data sequence of the instantaneous phase values of the sampled voltage waveform.

$$I_{\mathrm{x,rms}} = \sqrt{\frac{1}{N}\sum_{n=1}^{N} i_x^2(n)}, \tag{2}$$

where $I_{\mathrm{x,rms}}$ is the rms value of the current and $i_x(n)$ is the data sequence of the instantaneous phase values of the sampled current waveform.

The apparent power for each phase is evaluated from Equations (1) and (2) at a given moment, being measured over a given time interval (i.e., T_m)

$$S_\mathrm{x} = U_{\mathrm{x,rms}} \cdot I_{\mathrm{x,rms}} \tag{3}$$

Furthermore, the active power for each phase is defined as

$$P_\mathrm{x} = \frac{1}{N}\sum_{n=1}^{N} u_\mathrm{x}(n) \cdot i_\mathrm{x}(n) \tag{4}$$

The measurement of reactive energy, as implemented by the manufacturers, generally varies. The problem here might also be caused by the varied definition of reactive power. Usually, it is either the reactive component of power at the fundamental harmonic frequency, which is preferred by the standards, or it is defined by the sum of all reactive powers at the finite number of harmonic frequencies, or it can be calculated including the deformation power as

$$Q_\mathrm{x} = \sqrt{S_x^2 - P_x^2} \tag{5}$$

These active and reactive powers/energy flows are further sorted as per four quadrants (Q1–Q4) by [42], with the voltage at the measuring point being taken as the reference vector with zero phase angle. Reactive energy is always defined in association with active energy. Thus, the reactive energy is separately defined for each quadrant. Figure 1 shows the power flow quadrants.

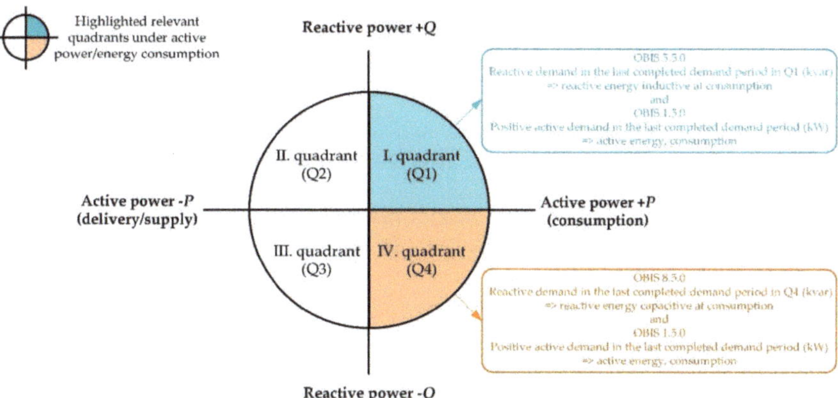

Figure 1. The power flow quadrants and relevant Object Identification System (OBIS) codes.

In general, PQ monitors measure and evaluate a lot of quantities. Thus, the unique marking of the measured quantities (e.g., aggregated active and reactive powers) is accomplished with the use of the DLMS standard [43], through Object Identification System (OBIS) codes, which are a part of the COSEM specifications and some of them are transferred to IEC standards, also see Figure 1.

From the power quality monitoring point of view based on [40], it is also necessary to consider standard aggregation interval for PQ monitors as N = 10 min. Thus, the time aggregation of the voltage can be also defined with respecting [41]

$$U_{x,agr} = \sqrt{\frac{1}{N}\sum_{n=1}^{N} U^2_{x,rms}(n)} \tag{6}$$

The aggregated value of the current is determined in the same way.

In the case of the short-term voltage drop event, Equation (1) is applied over measuring time interval $T_s/2$ = 10 ms. The obtained $U_{x,rms/2}$ for each phase is compared with the limits 90% of the nominal voltage according to [40]. Thus, it can be considered that the beginning and ending time of this event are recorded with the maximal inaccuracy 10 ms.

3. Analyzed Distribution Network and Reference Definition

3.1. Dense-Mesh Distribution Network Characteristics

Lots of low voltage (LV) municipal distribution systems in the Czech Republic were formerly often operated with dense-mesh topology. Figure 2 illustrates the general concept of the municipal dense-mesh network that was connected to the individual transformers and distribution transformer stations, respectively.

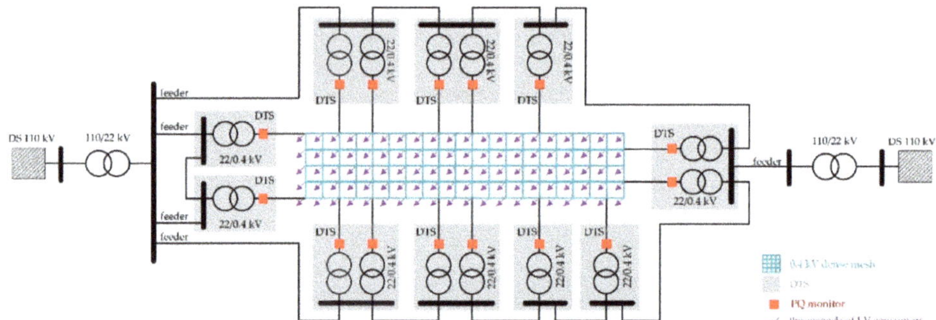

Figure 2. Illustration of the connection concept of the municipal low voltage (LV) dense-mesh distribution system to individual distribution transformation stations and the feeders.

From Figure 2, a high degree of complexity of dense-mesh connections is also visible. Therefore, distribution systems using this concept are usually only operated under test operation in the Czech Republic. The test operation is generally required due to a high number of failures, poor automation of these networks, and their insufficient monitoring. These problems in most cases led to the reconfiguration of the dense-mesh topology to a radial one. Almost all originally dense-mesh distribution systems are operated as with radial topology and they do not allow, technically, the automated reconfiguration back to the dense-mesh ones in general. However, there are still two distribution systems in the Czech Republic that operated as the dense-mesh. One of them, also the largest one, is the Brno-stred, which is analyzed in this paper.

The city planning and technical development of the city of Brno are closely connected and they correspond with its electric energy demands. Thus, the development has direct influence on the municipal distribution network structure and it also affects the composition of the connected power sources. Due to its size, Brno is divided into individual districts, with each having different distribution network topology, which takes the location of the district into account and also follows its main infrastructure (e.g., industry, transport, municipal council, hospitals, administrative, financial, and educational institutions). The Brno-stred district has a LV distribution network with a dense-mesh topology. Figure 3 shows the complexity of the power supply of this dense-mesh topology by individual feeders and transformers.

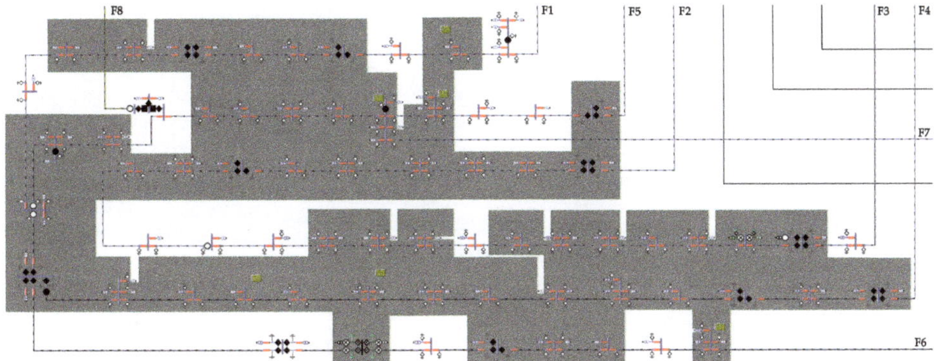

Figure 3. Detail connection of the 56 distribution transformation stations (grey highlighted) of the Brno-stred dense-mesh municipal network.

The Brno-stred distribution system (DS) comprises of 82 transformers 22/0.4 kV, each with an apparent power of 630 kVA with Dyn connection. The total installed power of the transformers is approximately 51.66 MVA. Transformers are located in 56 DTS and they provide power supply to more than 3800 customers connected in the LV 0.4 kV dense mesh. The DS supply is provided by eight 22 kV feeders, marked F1, F2, through to F8. PQ monitors are always installed in the secondary circuit of each transformer and they both have usual synchronization of the internal time by the oscillator frequency and synchronization of the internal time through the network voltage frequency. The analyzed data were anonymized because the municipal distribution network is a part of the critical electricity/energy infrastructure (e.g., transformers T1–T82 assigned randomly inconsistent with marking established by DSO, the manufacturer and model of PQ monitors are not mentioned). Table 1 shows numbers of transformers and DTSs, connected to the feeders F1–F6.

Table 1. Numbers of transformers and distribution transformer stations connected to feeders.

Number of Elements / Feeders	F1	F2	F3	F4	F5	F6
Number of T	11	14	13	19	13	10
Number of DTS	8	9	10	13	9	6

The feeders F1 to F6 are connected to a 110/22 kV (40 MVA) substation through the MV cable and they provide a permanent power supply of the Brno-stred dense-mesh network. Feeders F7 and F8 are connected in case one of F1–F6 is out of service due to a fault or maintenance.

It is very difficult to depict the technical LV scheme clearly in more detail due to the significant extensiveness and the complexity of LV dense-mesh topology. Moreover, it is not required for the initial analysis of power flows at a level of individual transformers. On the other hand, it is necessary to understand that just dense-mesh topology is very unique and specific one, e.g., see Figure 2, and it also has a fundamental impact on the power flows. Thus, achieved results are not directly comparable with the results for radial topology, because there are more complex interdependencies.

3.2. Input Data and Reference Definition

E.ON Distribuce a.s. provided data from PQ monitors comprising of rms voltage, rms current, and average powers for each phase measured with 5 min. intervals between 06/2016 and 02/2017. A three-phase summary value was also recorded in the case of measured powers. The provided data of the powers are not sorted in the individual record interval/measuring time interval, as is generally considered in Figure 1 and (1) through (5), but only one summed average active/reactive power value over all energy quadrants is available in each 5 min. interval (i.e., 5 min. measuring time interval). Therefore, if there were changes in the direction/character of active/reactive power during this 5 min. interval, the measured data are distorted by the significant effect of averaging. Generally, it must be realized that the algorithm itself implemented to measure active/reactive power might essentially have an impact on the attained results. The installed PQ monitors also consider reactive power to be a complement of active power to calculate the apparent power as (5). With regard to the power flow directions (see Figure 1), the data from the installed PQ monitors apply the so-called customer reference arrow system with a reference direction from the HV system to the LV system, see Figure 4.

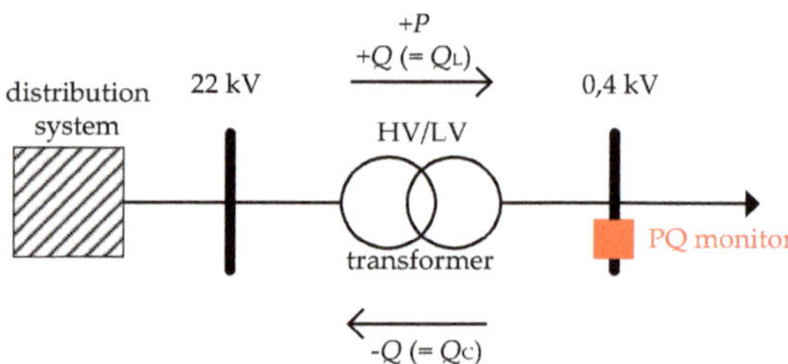

Figure 4. Definition of the reference power flow direction and the location of PQ monitors.

3.3. Conducted Analysis

The measurement data from all transformers from the Brno-stred dense mesh was statistically analyzed for finding statistically significant operating states. Although, the available measurement data was from the period from 06/2016 until 02/2017 for all 82 transformers, it was found out that quite comprehensive analysis can be done only for 56 transformers from time period 01/2017 due to occurrence of various flaws in the data. The complete data were only available for this 56 transformers T1–T14, T16–T54, and T60–T62, and only during 01/2017. Complete data represent 288 records/day/transformer, i.e., 8928 records of 5 min. aggregated values are available for each transformer, specifically 6336 records for the weekdays and 2592 records for the weekends. The various flaws occurring in the rest of the measurement data made the analysis of them impossible, and thus the rest of the data were excluded from the analysis. The possible flaws were as follows: (i) incomplete data, (ii) invalid time vector, and (iii) corrupt data file etc.

In the conducted statistical analysis, the following analyses were conducted, as described in Sections 4.1–4.4:

- Total three-phase and non-unified single phase active and reactive power loading of all 56 transformers during 01/2017.
- Non-unified phase voltage magnitudes and limits of all 56 transformers during 01/2017.
- The power loading of transformer with minimum T61 and maximum T47 power loading was further analyzed in more detail without unified phases for shorter time periods during different part of the weeks. In the analyses, the power loading of both transformers were analyzed with the help of statistical percentiles, minimum, maximum, mean, and standard deviation values.
- The unification of phase measurement of all transformers.

For the unification were used recorded voltage events by PQ monitors. By default, the PQ monitor enables the recording of the voltage events (i.e., the short-term voltage drop in any phase) based on [40]. With regard to the voltage, nine short-term voltage drops (two times in three phases, seven times in two phases) were recorded through PQ monitors for T1–T62 in the period 06/2016–02/2017. The short-term voltage drop records include the initial drop time, drop duration, and the minimum effective voltage and current values in each phase.

4. Results

4.1. Power Loading-Over All Transformers

The basic criterion that was applied by E.ON Distribuce a.s. on the dense-mesh network while it is under operation consists in keeping the loads of the individual transformers within exactly defined limits. Meeting the criterion then leads to ensuring the correct and reliable functioning of the network as a whole. This basic criterion results from the design of the complex topology and size of the network, as well as the fact that it was not possible in the past to implement extensive and efficient ("smart") monitoring of this network. The criterion is mainly based on the ability to ensure a certain extent of autonomous reliability of the network. Autonomous reliability consists in ensuring uninterrupted dense-mesh network operation, even if several transformers fail. The network concept ensures the ability of unaffected transformers to take over the load of the faulted transformers. The DSO requires an operating load of the individual transformers of about 25% in a steady state, or for the maximum power load to ideally not exceed 50% of the transformer rated power, for the proper functioning of the network as a whole.

Thus, the fulfilment of this criterion is assessed in the first part of the analysis, where maximum and minimum power loading of each of 56 transformers (i.e., T1–T14, T16–T54, and T60–T62) was identified. This analysis was only conducted for time period of 01/2017, where complete date for all 56 transformers were available (Section 3.3). The power loads were determined as the sum of the power loads of the individual phases. Figure 5 illustrates the reached maximum power load of individual transformers.

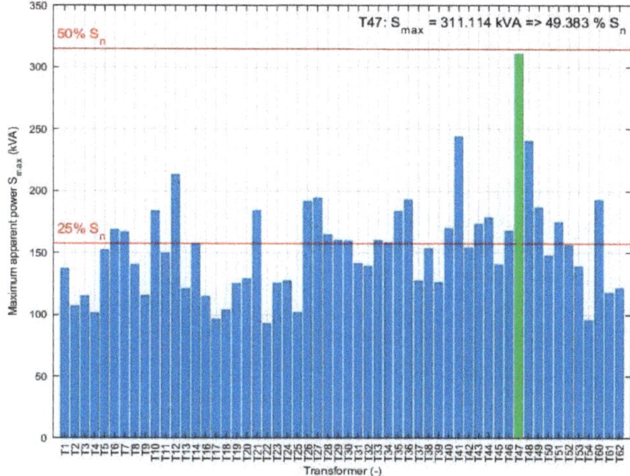

Figure 5. Comparison of the maximum power loads of individual transformers-01/2017.

Figure 5 shows that the maximum power load of approximately 311 kVA (more than 49% S_n) is reached by transformer T47. A total of 32 transformers have the maximum power load below 25% S_n and 23 transformers are in the range (25–40%) S_n. The operating criterion of this DS can be considered to be fulfilled, as the maximums are not higher than 50% S_n based on analyzed 5 min. data. However, it should be noted that the DSO does not exactly specify the technical requirements for performing the measurement based on which fulfillment of the criterion should be verified. Figure 6 shows the minimum power load to illustrate the opposite extreme loading of individual transformers.

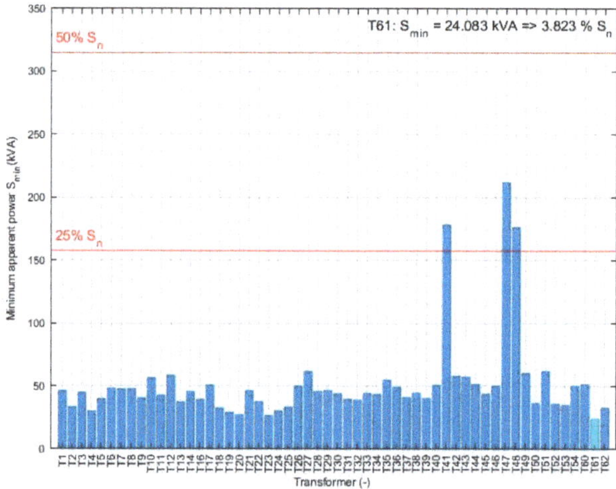

Figure 6. Comparison of the minimum power loads of the individual transformers-01/2017.

The minimum power load of approx. 24 kVA (i.e., approx. 3.8% S_n) is recorded for transformer T61. The values indicate that most transformers do not exceed the level of 50 kVA (approx. 8% S_n).

As the results above provide only a basic picture of the possible extreme operating conditions, the information needs to be understood in the context of averaging over the used 5 min. aggregation interval and further averaged over the three phases as only one value. Therefore, the analysis also includes the statistical evaluation of the magnitudes, together with power flow direction (see also reference direction in Figure 1) of the active and reactive powers in the secondary circuit of each transformer. Figure 6 shows the development of the measured cumulative frequencies of the active and the reactive powers of all analyzed transformers. The single-phase power load data were mixed together for all three phases and Figure 7 depicts cumulative frequencies over all three phases without distinction between separate phases to obtain the cumulative frequencies (single phase data are depicted, not summed powers for all three phases). For a better understanding of the occurrence of the active and the reactive power, it also shows the values separately for the Weekdays and the Weekend, and it is generally proceeded 3 × 8928 records/month/transformer.

Figure 7 shows that only the consumed active power, approximately 5–84 kW, was realized on transformers and the supplied active power was not realized (for the monitored period 01/2017). Approximately 75% of all active powers ranged up to 40 kW on Weekdays and up to 28 kW during Weekends. From the reactive power point of view, the consumption of approximately (0–71) kvar and the supply of approximately (−18–0) kvar were both realized. It also applies to the reactive power values that the ratio of values for the supplied and consumed reactive power is 40/60 for the Weekdays and 50/50 for the Weekend. Approx. 80% of all values of the supplied reactive power ranged from −18 kvar to −9 kvar (on the Weekdays and on the Weekend) and 90% of all values of the consumed reactive power ranged from 0 kvar to 20 kvar (on the Weekdays) and to approximately 10 kvar (on the Weekend).

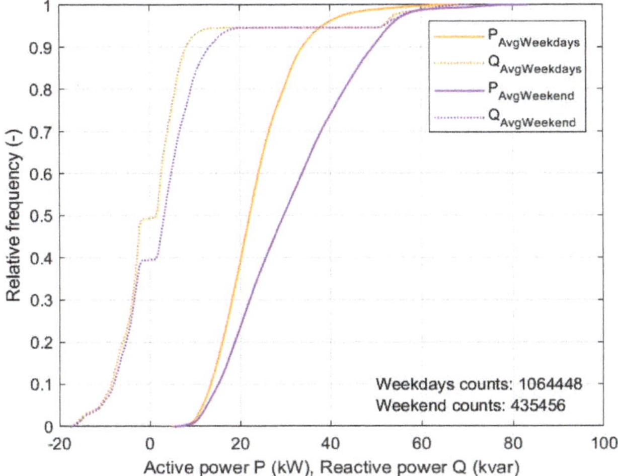

Figure 7. Comparison of the cumulative frequencies of the single-phase active and reactive power of 56 transformers-Weekdays vs. Weekend, 01/2017.

Although there were clearly identified the specific directions of the active and reactive power/energy flows at LV (i.e., consumption vs. supply), it is necessary to point out that all of the installed transformers (i.e., T1–T82) have Dyn connection, and thus the directions need not to explicitly correspond to the power flow directions at HV. With regard to the dense-mesh topology, theoretically plausible explanations for the single-phase supply of the active power may be e.g., (i) the unique configuration and non-symmetrical load character with a significant effect on the size and voltage angle of individual transformers, (ii) failure states on 22 kV side and blowing of one HV fuse, and (iii) operating condition while considering the existence of an unsuitable configuration, where power also flows over this dense-mesh network rather than through the HV network.

Generally, Figure 7 entails the loss of information regarding the possible concurrence of consumption and supply at the level of individual phases of all transformers, because the unified connection of phase measurement does not exist there.

4.2. Voltage Magnitude Distribution Analysis

The kind of complementary analysis that was carried out in this paper is voltage tolerance analysis. The measurement data were evaluated for compliance with voltage tolerance limits of +10% and −15% U_n for 100% of time [40] for 'all' 56 transformers in the Brno-stred dense mesh. The non-unified phase measurement might not be a problem here, as this information regarding voltage compliance can still be found in the available data. Figure 8 shows the histograms of the individual rms phase voltages of 56 transformers without considering the unification of their phase measurements. (i.e., the single-phase histogram was constructed over voltage measurement data from denoted phase one from all transformers together, etc.). The histogram data samples are divided into 50 bins.

The analysis is performed over samples of the individual phase voltages of all transformers, which amounts to 1,064,448 samples for the Weekdays (i.e., 6336 records/phase/transformer/month) and 435,456 samples for weekends (i.e., 2592 records/phase/transformer/month). The values show that all rms voltage values are within the required voltage tolerance [40] during the Weekdays and Weekend. Although the voltage tolerance should be assessed on 10 min. aggregation interval, the eventual reaggregation of 5 min. samples to average 10 min. values would only smoothen the voltage waveform and thus the voltage tolerance would be still met. The problem of non-unified phases between different measurements might lead to problems in identifying which particular phase is voltage tolerance compliant or non-compliant.

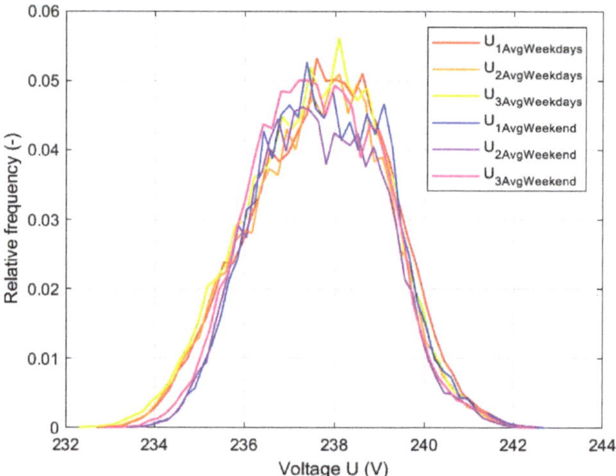

Figure 8. Comparison of relative frequencies of phase voltages of 56 transformers-Weekdays vs. Weekend, 01/2017.

4.3. Transformers with the Maximal T47 and Minimal Power Load T61

Although the unified measurement is not resolved, it is relevant to carry out the partial analysis of individual transformers per each individual phase for 01/2017. A detailed picture can thus be obtained while using percentiles (PCTL) of the occurrence frequencies of power values during the day. In the context of the information above, the article further provides a detailed assessment of the active and reactive power through percentiles/histograms, in particular it presents individual assessments for T47 and T61 differentiating the Weekdays vs. the Weekend.

4.3.1. T47 and T61 Not Sorted Active and Reactive Power to Quadrants

Firstly, Figure 9 shows an overview of the development of the relative frequencies of three-phase apparent power for T47 (with maximum power load) and at T61 (with minimum power load).

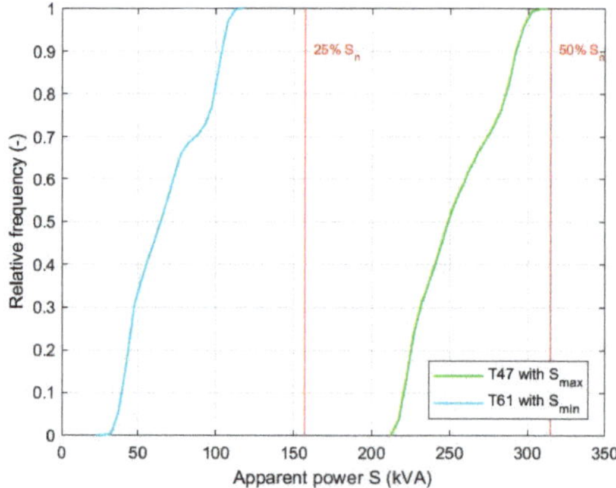

Figure 9. Comparison of relative frequencies of three-phase apparent power for T47 and T61-01/2017.

The histograms show that, for T47, all of the values ranged from 33–49% S_n, approximately 90% of all values were then below 46% of the nominal apparent transformer power. For T61, approximately 70% of all values were below 12% of the nominal apparent power of the transformer and the maximum power did not exceeded 20%. In general, if different energy flows/powers (consumption vs. supply) are simultaneously realized per individual phases, the average three-phase value is unsuitable and it entails a loss of information and lower energy/power is obtained. Thus, Tables 2 and 3 quantify the corresponding percentiles for Figures 10 and 11, which present an overall comparison of cumulative frequencies of the active and reactive power per phase for T47 and T61.

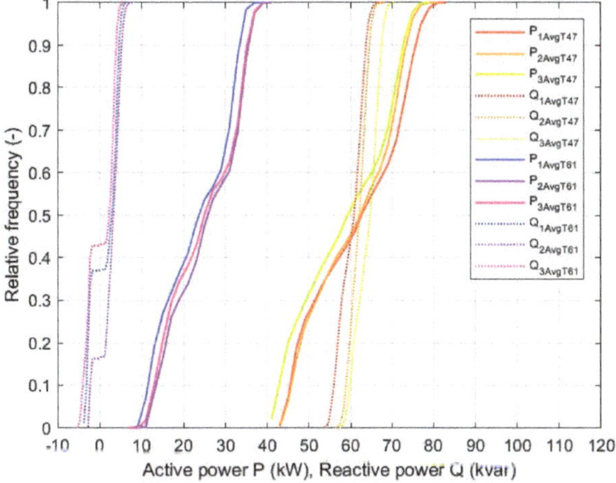

Figure 10. Cumulative frequencies of the single-phase active and reactive power for T47 and T61 during the Weekdays-01/2017.

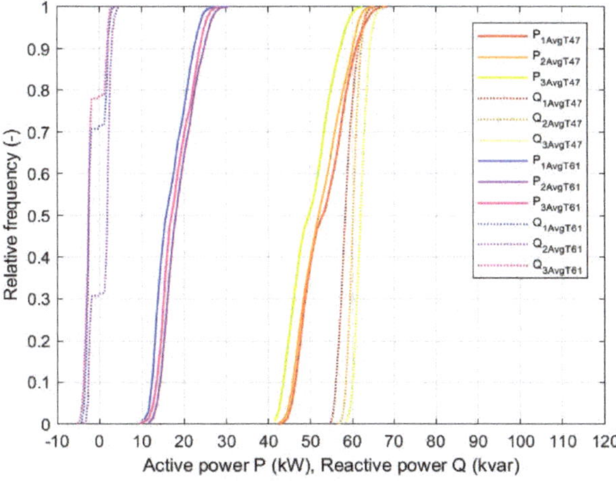

Figure 11. Cumulative frequencies of the single-phase average active and reactive power for T47 and T61 during the Weekend-01/2017.

Table 2. Percentiles of the single-phase active and reactive power for T47 during the Weekdays and the Weekend-01/2017.

Parts of the Week	PCTL	$P_{1AvgT47}$ (kW)	$P_{2AvgT47}$ (kW)	$P_{3AvgT47}$ (kW)	$Q_{1AvgT47}$ (kvar)	$Q_{2AvgT47}$ (kvar)	$Q_{3AvgT47}$ (kvar)
Weekend	50th	53.45	52.4	50.15	58.55	60.31	62.25
	90th	61.44	60.61	57.61	61.8	62.38	64.52
Weekdays	50th	63.6	63.2	60.45	61.05	62.33	64.94
	90th	76.68	74.37	73.74	63.93	64.74	67.75

Table 3. Percentiles of the single-phase active and reactive power for T61 during the Weekdays and the Weekend-01/2017.

Parts of the Week	PCTL	$P_{1AvgT61}$ (kW)	$P_{2AvgT61}$ (kW)	$P_{3AvgT61}$ (kW)	$Q_{1AvgT61}$ (kvar)	$Q_{2AvgT61}$ (kvar)	$Q_{3AvgT61}$ (kvar)
Weekend	50th	16.23	18.55	17.38	−2.19	2.15	−2.3
	90th	22.91	25.3	24.29	2.36	3.22	2.14
Weekdays	50th	24.56	26.64	25.93	3.22	3.58	2.51
	90th	34.33	36.39	36.1	4.98	5.61	4.24

The results in Table 2 prove following:

- in terms of the active power, phase L1 had the highest load and phase L3 the lowest, both during the Weekend and on the Weekdays. On average, the load on the Weekdays was approximately 20% higher than during the Weekend (for 90th PTCL);
- in terms of reactive power, phase L3 had the highest load and phase L1 the lowest, both during the Weekend and on the Weekdays. On average, the loading on the Weekdays was approximately 4% higher than during the Weekend (for 90th PTCL); and,
- the comparison of percentiles in individual phases indicates that the power distribution in the individual phases is relatively even, with 90th and 50th PTCLs differing in individual phases by only about 3–4 kW, for both the analyzed Weekdays and Weekend periods (i.e., for example the difference for the 90th PTCL on the Weekdays-the largest L1 ~61.44 kW, and the smallest L3 57.61 kW difference of about 4 kW). As for reactive power, the difference was similar in size to the active power of about 3–4 kvar. The reactive power was comparable in size to the active power.

The results in Table 3 prove the following:

- in terms of the consumed active power, phase L2 had the highest load and phase L1 the lowest, both during the Weekdays and the Weekend. On average, the load on the Weekdays was approximately 32% higher than during the Weekend (for 90th PTCL); and,
- if the transformer provides the consumption as well as the supply of the reactive power in the monitored period and the analyzed data are not divided according to the power character (i.e., the data series cumulates all of the consumed/supplied reactive powers), and then the evaluation of the reached reactive power through percentiles is not representative.

In general, the results in Table 3 also confirm that, in order to do the basic statistical evaluation of the transformer that provides the consumption of the active power and at the same time the consumption or the supply of the reactive power, it is firstly appropriate to sort the data based on the character of the reactive power.

4.3.2. T47 and T61 Sorted Active and Reactive Powers to Quadrants and Their Statistical Values

Sorting is separately performed for each phase in the context of quadrants Q1 and Q4, as in Figure 1. Thus, the following Table 4 provides this sorting by presenting a percentage of the numbers of single-phase reactive power records, with both the differentiation between the Weekdays vs. Weekend and the distribution considering four time-periods during the day A (22:00:00–03:59:59), B (4:00:00–07:59:59), C (08:00:00–15:59:59), and D (16:00:00–21:59:59).

Table 4. Percentages of the numbers of single-phase reactive power records for T61 sorted by–consumed Q ($Q > 0$), supplied Q ($Q < 0$), no Q ($Q = 0$), red text highlights dominant one per phase.

Parts of the Week	Time Interval	Number of Records	$Q_{1AvgT61}$ >0 (%)	<0 (%)	=0 (%)	$Q_{2AvgT61}$ >0 (%)	<0 (%)	=0 (%)	$Q_{3AvgT61}$ >0 (%)	<0 (%)	=0 (%)
Weekend	A	648	18.7	79.9	1.4	84.7	15.1	0.2	8	91.4	0.6
	B	432	33.8	64.6	1.6	77.8	21.1	1.2	33.1	65.5	1.4
	C	864	13.3	86.5	0.2	35.3	64.1	0.6	7.4	92.5	0.1
	D	648	54.5	44.6	0.9	91	8.5	0.5	45.1	53.7	1.2
Weekdays	A	1584	5.4	94.1	0.4	40.9	58.3	0.8	0.9	98.9	0.2
	B	1056	55.3	44.4	0.3	89	9.8	1.1	45.8	53.6	0.6
	C	2112	100	0	0	100	0	0	99.1	0.7	0.1
	D	1584	75.4	24.2	0.4	99.9	0.1	0	63.8	35.6	0.6

The sorting for T47 is irrelevant, because there is no change of the character of the reactive power (i.e., only Q consumption, 100% of the records fulfill condition $Q > 0$), therefore there are the results for T61 in Table 4, and these show:

- during the Weekend/A and Weekend/B, the number of records corresponding to the supplied reactive power predominates in phases L1 and L3, and the consumed reactive power predominates in phase L2;
- during the Weekend/D and Weekdays/B the number of records of the consumed reactive power significantly predominates in L2, while in L1 and L2 the balance of represented samples is leveled;
- during the Weekend/C and Weekdays/A the number of records of the supplied reactive power predominates in all phases; and,
- significant and predominating reactive power character (the consumption or the supply) is identifiable per phase for each specific time intervals during the day in individual parts of the week, e.g., solely the supplied reactive power is identified for T61 during the Weekdays/C period in phases L1 and L2, and it is also almost 100% in phase L3. A similar and also strong dependence is also observed in the Weekdays/D.

The data also generally include the values that stand for zero reactive power (see Table 4). However, these were not considered further in the statistical evaluation due to their percentage being insignificant.

The following Tables 5–7 present a summary statistical analysis by the percentiles. Specifically, for T61, Tables 6 and 7 present the percentile results performed on the sorted samples according to Table 4. The data were initially sorted according to the character of the reactive power into two groups for consumed Q (quadrant Q1) and supplied Q (quadrant Q4) to determine the percentiles. In this sorting, the corresponding active power samples P were also sorted for the corresponding Q samples. Over such sorted data, the percentiles were subsequently set separately for the active power (for given character of the reactive power) and for the reactive power.

Table 5. Percentiles of the single-phase active power and reactive power at the time of the reactive power consumption for T47-01/2017.

Parts of the Week	Time Interval	PCTL	$P_{1AvgT47}$ (kW)	$P_{2AvgT47}$ (kW)	$P_{3AvgT47}$ (kW)	$Q_{1AvgT47}$ (kvar)	$Q_{2AvgT47}$ (kvar)	$Q_{3AvgT47}$ (kvar)
Weekend	A	5th	46.70	47.35	43.60	56.85	59.02	60.38
		25th	48.41	49.09	45.45	57.76	59.94	61.24
		50th	50.09	50.76	47.01	58.41	60.59	62.06
		75th	51.96	52.43	48.95	59.38	61.37	63.10
		90th	54.41	54.98	50.99	60.61	62.13	64.22
	B	5th	45.05	44.45	42.15	55.61	57.42	58.94
		25th	46.29	45.62	43.42	56.33	58.62	59.91
		50th	47.36	46.50	44.34	56.93	59.25	60.72
		75th	48.30	47.33	45.50	57.65	59.82	61.40
		90th	49.27	48.15	46.60	58.16	60.31	62.14
	C	5th	46.70	45.56	44.82	55.60	58.15	60.20
		25th	52.03	49.86	49.09	57.30	59.23	61.61
		50th	57.09	54.65	53.34	58.63	60.15	62.35
		75th	60.31	59.34	56.53	59.80	61.01	63.20
		90th	62.66	61.21	59.15	60.94	61.74	64.11
	D	5th	50.06	48.84	47.27	56.16	58.39	60.84
		25th	56.15	55.41	52.74	59.08	60.24	62.46
		50th	58.13	57.47	54.41	60.64	61.47	63.63
		75th	60.26	59.99	56.45	62.10	62.40	64.59
		90th	62.86	61.77	58.43	63.34	63.13	65.76
Weekdays	A	5th	44.75	45.19	41.94	55.52	58.56	59.33
		25th	46.56	47.21	44.05	56.81	59.63	60.39
		50th	48.10	49.19	45.93	57.62	60.35	61.32
		75th	51.73	52.94	49.46	58.78	61.18	62.87
		90th	54.94	55.76	52.43	60.19	61.94	64.03
	B	5th	44.88	44.93	42.26	54.82	57.69	58.49
		25th	46.57	46.27	43.57	55.61	58.64	59.69
		50th	49.66	48.89	47.77	56.48	59.29	60.55
		75th	55.02	54.12	53.17	57.49	60.30	62.28
		90th	58.87	56.93	55.98	58.58	61.33	63.44
	C	5th	65.63	63.86	63.18	59.46	61.62	64.34
		25th	72.97	70.86	70.27	61.39	62.91	65.85
		50th	75.00	72.80	72.26	62.46	63.73	66.71
		75th	76.99	74.62	74.05	63.46	64.54	67.56
		90th	78.76	76.19	75.59	64.25	65.22	68.22
	D	5th	58.54	59.23	55.70	59.94	61.16	63.24
		25th	62.83	62.96	59.44	61.74	62.36	64.99
		50th	67.07	66.08	63.41	62.77	63.32	65.98
		75th	71.48	70.24	68.56	63.69	64.20	66.94
		90th	73.74	72.29	70.89	64.63	65.06	68.11

Table 6. Percentiles of the single-phase active power and reactive power at the time of the reactive power consumption for T61-01/2017.

Parts of the Week	Time Interval	PCTL	$P_{1AvgT61}$ (kW)	$P_{2AvgT61}$ (kW)	$P_{3AvgT61}$ (kW)	$Q_{1AvgT61}$ (kvar)	$Q_{2AvgT61}$ (kvar)	$Q_{3AvgT61}$ (kvar)
Weekend	A	5th	13.84	13.79	15.34	1.89	1.83	1.79
		25th	14.94	14.99	16.08	2.00	2.16	1.88
		50th	15.93	16.04	17.72	2.09	2.53	1.96
		75th	16.80	17.55	18.79	2.19	2.91	2.06
		90th	18.00	18.93	19.84	2.26	3.22	2.13
	B	5th	12.47	15.59	14.35	1.64	1.61	1.53
		25th	13.25	16.57	15.39	1.74	1.82	1.68
		50th	13.99	17.25	15.96	1.81	2.18	1.79
		75th	15.02	17.95	16.61	1.89	2.75	1.90
		90th	15.44	18.75	17.07	2.00	3.16	2.00
	C	5th	19.18	19.77	21.12	1.91	1.88	1.84
		25th	20.74	21.84	22.64	2.04	2.06	1.97
		50th	22.14	23.14	24.31	2.18	2.19	2.05
		75th	23.67	24.63	25.06	2.41	2.37	2.38
		90th	24.93	25.82	25.82	2.55	2.55	2.69
	D	5th	18.94	19.38	22.00	2.08	2.06	1.95
		25th	21.63	21.57	23.48	2.26	2.39	2.15
		50th	23.14	23.54	24.59	2.45	2.76	2.31
		75th	24.19	26.16	25.86	2.77	3.37	2.50
		90th	25.05	27.54	26.87	3.10	3.96	2.79
Weekdays	A	5th	15.55	13.96	16.89	1.97	1.92	1.95
		25th	17.30	16.13	18.61	2.10	2.10	2.04
		50th	18.63	18.35	19.29	2.26	2.28	2.13
		75th	19.92	20.64	23.02	2.40	2.54	2.24
		90th	20.68	22.00	23.65	2.55	2.97	2.42
	B	5th	14.44	16.33	16.91	1.81	1.61	1.78
		25th	18.29	17.44	19.86	2.84	1.82	2.09
		50th	20.60	21.38	22.11	3.58	2.86	2.55
		75th	23.17	25.00	24.16	4.07	3.56	3.13
		90th	25.11	26.43	25.53	4.60	3.96	3.53
	C	5th	27.59	28.27	27.08	3.54	3.58	2.84
		25th	31.63	33.22	32.97	4.11	4.30	3.34
		50th	33.08	34.93	34.73	4.50	4.80	3.69
		75th	34.39	36.19	35.93	4.91	5.27	4.08
		90th	35.39	37.41	37.16	5.32	5.72	4.45
	D	5th	22.34	23.01	24.79	2.36	2.47	2.22
		25th	25.13	25.56	30.15	2.65	3.09	2.84
		50th	30.20	29.95	33.78	3.65	4.09	3.65
		75th	32.74	35.05	35.56	4.42	5.47	4.25
		90th	34.00	36.44	36.70	5.07	6.21	4.75

Table 7. Percentiles of the single-phase active power and reactive power at the time of the reactive power supply for T61-01/2017.

Parts of the Week	Time Interval	PCTL	$P_{1AvgT61}$ (kW)	$P_{2AvgT61}$ (kW)	$P_{3AvgT61}$ (kW)	$Q_{1AvgT61}$ (kvar)	$Q_{2AvgT61}$ (kvar)	$Q_{3AvgT61}$ (kvar)
Weekend	A	5th	11.81	13.35	12.88	−3.28	−2.53	−4.30
		25th	13.14	14.82	14.33	−2.60	−2.32	−2.78
		50th	13.98	15.61	15.36	−2.25	−2.12	−2.33
		75th	14.90	16.30	16.64	−2.06	−1.99	−2.09
		90th	15.86	17.17	18.18	−1.97	−1.90	−1.95
	B	5th	12.35	14.85	13.31	−2.47	−2.04	−2.58
		25th	12.96	15.50	14.96	−2.20	−1.89	−2.28
		50th	13.48	16.21	15.56	−2.04	−1.78	−2.07
		75th	14.19	16.75	16.06	−1.88	−1.67	−1.85
		90th	14.90	17.70	16.55	−1.78	−1.61	−1.69
	C	5th	11.91	13.00	12.35	−3.65	−2.91	−3.68
		25th	14.96	15.42	15.67	−3.08	−2.45	−3.27
		50th	18.18	18.41	19.01	−2.76	−2.24	−2.91
		75th	20.68	20.49	22.23	−2.44	−2.09	−2.45
		90th	22.11	22.72	23.62	−2.21	−1.96	−2.17
	D	5th	13.76	14.40	14.47	−3.40	−2.39	−4.26
		25th	16.80	15.57	17.99	−2.71	−2.29	−3.18
		50th	18.15	16.58	20.06	−2.43	−2.17	−2.48
		75th	19.77	18.72	21.54	−2.25	−2.09	−2.27
		90th	20.86	19.15	22.71	−2.11	−1.95	−2.12
Weekdays	A	5th	10.74	12.30	11.80	−3.49	−2.39	−4.54
		25th	11.88	13.25	13.31	−3.08	−2.20	−4.10
		50th	12.91	13.99	14.65	−2.78	−2.07	−3.55
		75th	15.30	14.92	16.70	−2.44	−1.97	−2.88
		90th	17.51	16.31	18.98	−2.24	−1.86	−2.41
	B	5th	11.93	12.98	14.03	−2.66	−2.20	−3.57
		25th	13.39	15.37	15.47	−2.21	−1.94	−2.43
		50th	13.99	16.34	16.21	−2.02	−1.78	−2.17
		75th	14.56	16.71	17.04	−1.88	−1.66	−1.99
		90th	15.22	16.98	18.09	−1.79	−1.52	−1.83
	C	5th	0.00	0.00	23.09	0.00	0.00	−3.12
		25th	0.00	0.00	25.08	0.00	0.00	−2.86
		50th	0.00	0.00	26.50	0.00	0.00	−2.60
		75th	0.00	0.00	28.07	0.00	0.00	−2.57
		90th	0.00	0.00	32.20	0.00	0.00	−2.52
	D	5th	19.74	23.14	21.35	−2.77	−2.37	−2.82
		25th	21.47	23.14	23.51	−2.59	−2.37	−2.57
		50th	22.71	23.14	24.85	−2.49	−2.37	−2.42
		75th	23.82	23.14	26.10	−2.38	−2.37	−2.29
		90th	24.70	23.14	26.90	−2.27	−2.37	−2.19

For transformer T47, based on the comparison of 90th PTCL results from Table 2 (only the summary Weekdays and Weekend percentiles are considered) and Table 5 (considers both the summary percentiles and the individual time intervals A, B, C, D during day) can be observed with following dependencies:

- the summary Weekend PCTLs of the single-phase active and reactive power achieved are comparable to those achieved during the Weekend/C and Weekend/D with differences of approximately 1–3%;
- a similar dependence is also observed at the summary Weekdays PCTLs vs. Weekdays/C and Weekdays/D, although these differ by approximately 1–4%;
- for other time intervals A and B it is possible to observed more significant differences against the summary PCTLs

  ○ the summary Weekend PCTLs of the active power are approx. 10% higher than those for the Weekend/A, for the reactive power they are approx. 1–2% higher,

- ○ the summary Weekend PCTLs of the active power are approx. 20% higher than those for the Weekend/B, for the reactive power they are approx. 4–5% higher,
- ○ the summary Weekdays PCTLs of the active power are approx. 25–29% higher than those for the Weekdays/A, for the reactive power they are approx. 5% higher, and
- ○ the summary Weekdays PCTLs of the active power are approx. 24% higher than those for the Weekdays/B, for the reactive power they are approx. 5–8% higher.

Finding the interdependencies for T61 is very complicated, not only for the confrontation summary percentiles (in Table 3) with the results in Tables 6 and 7, but also generally from the individual percentiles in case of the change of the character of reactive power flow. Thus, it would first be necessary to define the specific purpose for which the analysis is to be prioritized (e.g., reactive power flows, voltage magnitude, unbalance, etc.) and it should be performed over the phase unified data of all transformers. On the other hand, the search for basic interdependencies can also be done while using other statistical indicators. Therefore, the following part of the analysis evaluates the values of the maximal and minimal power, the mean and the standard deviation in more detail over the sorted data. Tables 8–10 show the obtained results.

Table 8. Statistical results for T47 at the time of the reactive power consumption including the highlighted total maximal and minimal powers and the highest absolute mean values per phase.

Parts of the Week	Time Interval	Quantity	$P_{1AvgT47}$ (kW)	$P_{2AvgT47}$ (kW)	$P_{3AvgT47}$ (kW)	$Q_{1AvgT47}$ (kvar)	$Q_{2AvgT47}$ (kvar)	$Q_{3AvgT47}$ (kvar)
Weekend	A	Max	60.89	58.75	54.31	63.13	63.97	66.04
		Min	44.79	45.68	41.77	55.67	57.63	59.56
		Mean	50.54	51.02	47.34	58.68	60.67	62.26
		SD	2.93	2.59	2.53	1.41	1.08	1.33
	B	Max	54.48	50.09	49.66	59.39	61.24	63.49
		Min	43.99	42.65	41.46	54.52	56.55	58.15
		Mean	47.39	46.51	44.51	56.98	59.17	60.69
		SD	1.59	1.26	1.59	0.90	0.92	1.08
	C	Max	67.22	63.91	62.63	63.58	64.59	66.40
		Min	42.97	42.17	42.68	54.85	57.00	59.03
		Mean	56.16	54.36	52.94	58.60	60.14	62.38
		SD	5.38	5.37	4.75	1.79	1.24	1.27
	D	Max	68.30	65.14	61.80	65.84	64.71	67.95
		Min	48.13	47.07	45.45	54.99	57.51	59.68
		Mean	57.96	57.17	54.17	60.48	61.25	63.55
		SD	4.00	3.92	3.33	2.28	1.54	1.60
Weekdays	A	Max	61.75	60.63	57.70	63.44	63.70	66.24
		Min	43.22	43.25	40.32	53.92	56.91	58.07
		Mean	49.26	50.10	46.85	57.90	60.42	61.63
		SD	3.73	3.73	3.68	1.65	1.14	1.64
	B	Max	64.31	61.97	62.69	61.27	64.29	66.74
		Min	43.94	43.54	40.92	53.60	56.43	57.35
		Mean	51.09	50.20	48.56	56.66	59.54	60.98
		SD	5.18	4.48	5.24	1.36	1.32	1.77
	C	Max	83.12	80.15	79.53	67.63	67.38	70.21
		Min	58.61	57.61	56.73	56.77	59.36	61.91
		Mean	74.45	72.20	71.63	62.35	63.71	66.64
		SD	3.94	3.75	3.72	1.58	1.21	1.28
	D	Max	80.01	77.56	75.99	67.23	68.40	70.97
		Min	53.80	54.65	52.00	58.22	59.89	61.61
		Mean	67.05	66.44	63.86	62.71	63.30	65.99
		SD	5.24	4.51	5.26	1.51	1.32	1.58

Table 9. Statistical results for T61 at the time of the reactive power consumption, including the highlighted total maximal and minimal powers and the highest absolute mean values per phase.

Parts of the Week	Time Interval	Quantity	$P_{1AvgT61}$ (kW)	$P_{2AvgT61}$ (kW)	$P_{3AvgT61}$ (kW)	$Q_{1AvgT61}$ (kvar)	$Q_{2AvgT61}$ (kvar)	$Q_{3AvgT61}$ (kvar)
Weekend	A	Max	19.52	23.56	21.94	2.52	4.63	2.21
		Min	13.53	12.65	15.04	1.78	1.53	1.76
		Mean	15.98	16.38	17.68	2.10	2.57	1.97
		SD	1.38	1.88	1.75	0.14	0.51	0.12
	B	Max	16.27	20.52	18.18	2.37	3.74	2.30
		Min	12.11	12.37	13.59	1.55	1.39	1.45
		Mean	14.12	17.26	15.94	1.82	2.31	1.80
		SD	1.02	1.13	0.88	0.13	0.57	0.17
	C	Max	26.67	27.87	28.04	3.11	3.45	3.31
		Min	18.09	17.27	20.74	1.82	1.74	1.75
		Mean	22.24	23.12	24.01	2.25	2.23	2.18
		SD	1.91	2.01	1.70	0.27	0.25	0.34
	D	Max	27.61	30.17	28.29	3.93	4.91	3.47
		Min	16.95	16.70	18.12	1.85	1.87	1.58
		Mean	22.82	23.78	24.61	2.55	2.93	2.35
		SD	2.00	2.82	1.76	0.38	0.69	0.29
Weekdays	A	Max	21.51	25.02	24.02	3.05	4.70	2.44
		Min	14.73	12.59	16.87	1.87	1.66	1.95
		Mean	18.58	18.31	20.40	2.27	2.40	2.16
		SD	1.63	2.78	2.49	0.23	0.47	0.15
	B	Max	28.41	29.26	31.06	5.67	5.17	5.63
		Min	13.42	13.51	15.84	1.58	1.41	1.55
		Mean	20.66	21.32	22.02	3.45	2.77	2.66
		SD	3.40	3.93	2.85	0.92	0.93	0.67
	C	Max	38.21	41.54	39.78	6.55	7.81	5.84
		Min	7.61	8.21	7.86	1.68	1.08	1.54
		Mean	32.69	34.40	34.04	4.53	4.80	3.73
		SD	2.57	2.86	3.10	0.63	0.74	0.56
	D	Max	36.87	39.72	39.61	6.77	7.90	5.83
		Min	18.89	20.37	21.94	2.10	1.99	1.84
		Mean	29.21	30.20	32.57	3.66	4.29	3.59
		SD	4.14	5.00	3.97	1.03	1.36	0.89

The results in Table 8 show that T47 has the maximal consumed active and reactive power during the Weekdays approx. 15–17 kW and 2–3 kvar higher than during the Weekend. Furthermore, transformer T47 reaches approximately the same average values of the minimal consumed active and reactive power during the Weekdays and the Weekend.

The results in Table 9 show that the maximal consumed active and reactive power for transformer T61 during the Weekdays is approx. 11 kW and 2.5–3 kvar higher than during the Weekend. The minimal consumed active power during the Weekdays is approx. 4–6 kW lower than during the Weekend and the minimal consumed reactive power is similar for the Weekdays and Weekend.

The results in Table 10 show that the maximal consumed active power for transformer T61 has differences of approx. 3–6 kW and this transformer reaches similar values of the maximal supplied reactive power during the Weekdays and Weekend. This similarity is also observed for the minimal consumed active and supplied reactive power, respectively.

Table 10. Statistical results for T61 at the time of the reactive power supply including the highlighted total maximal and minimal powers and the highest absolute mean values per phase.

Parts of the Week	Time Interval	Quantity	$P_{1AvgT61}$ (kW)	$P_{2AvgT61}$ (kW)	$P_{3AvgT61}$ (kW)	$Q_{1AvgT61}$ (kvar)	$Q_{2AvgT61}$ (kvar)	$Q_{3AvgT61}$ (kvar)
Weekend	A	Max	19.03	18.35	21.97	−1.74	−1.74	−1.71
		Min	10.17	12.91	12.05	−3.82	−2.76	−5.07
		Mean	14.08	15.56	15.62	−2.38	−2.16	−2.61
		SD	1.45	1.26	1.88	0.43	0.22	0.75
	B	Max	16.01	18.58	17.44	−1.51	−1.49	−1.49
		Min	10.34	14.19	11.31	−2.71	−2.42	−3.22
		Mean	13.60	16.26	15.42	−2.06	−1.79	−2.08
		SD	0.91	0.95	1.02	0.23	0.16	0.30
	C	Max	24.45	24.53	26.30	−1.81	−1.63	−1.70
		Min	10.12	11.63	9.52	−4.19	−3.53	−4.15
		Mean	17.73	18.10	18.75	−2.79	−2.30	−2.88
		SD	3.39	3.25	3.85	0.47	0.31	0.51
	D	Max	24.59	20.83	26.39	−1.90	−1.81	−1.71
		Min	12.45	14.13	13.05	−3.69	−2.57	−4.70
		Mean	18.12	16.98	19.68	−2.53	−2.18	−2.77
		SD	2.25	1.80	2.67	0.41	0.16	0.70
Weekdays	A	Max	21.11	20.32	23.91	−1.87	−1.56	−1.91
		Min	9.60	11.40	10.26	−3.98	−2.72	−5.19
		Mean	13.68	14.22	15.17	−2.78	−2.08	−3.48
		SD	2.39	1.41	2.54	0.42	0.18	0.72
	B	Max	17.30	17.92	20.12	−1.48	−1.36	−1.50
		Min	10.03	12.22	11.28	−3.69	−2.44	−4.90
		Mean	13.95	15.80	16.26	−2.08	−1.80	−2.29
		SD	1.08	1.38	1.42	0.31	0.22	0.53
	C	Max	0.00	0.00	32.37	0.00	0.00	−2.42
		Min	0.00	0.00	22.88	0.00	0.00	−3.13
		Mean	0.00	0.00	27.18	0.00	0.00	−2.71
		SD	0.00	0.00	2.96	0.00	0.00	0.21
	D	Max	27.41	23.14	28.98	−2.07	−2.37	−1.84
		Min	17.78	23.14	18.61	−2.98	−2.37	−3.73
		Mean	22.62	23.14	24.73	−2.49	−2.37	−2.45
		SD	1.69	0.00	1.82	0.16	0.00	0.22

The general conclusions can be drawn when comparing the results of Tables 8–10:

- the highest mean values for T47 are usually achieved in the Weekdays/C. The minimal differences of approx. 11% are in the comparison with the Weekdays/D, for other interval the differences are roughly 30–35%;
- the highest mean values achieved for T47 during the Weekend are in the Weekend/D but differences with other Weekend intervals are under 18% for the active power and under 6% the reactive power;
- the mean values of the active power for T47 are approx. 14–25% higher in the intervals Weekdays/C and Weekdays/D than those for Weekend/C and Weekend/D, for the reactive power they are similar in all intervals and the most of differences are lower than 6%; and,
- similar dependencies can be found for T61 as for T47, when the values are usually higher during the Weekdays than during the Weekend. Specifically, the active power values during the Weekdays are usually 11–33% higher than those during the Weekend and the reactive power differs in the range 1–54%. Although high percentage differences between corresponding intervals were achieved, it should be noted that, in terms of absolute values, these are comparable to differences for the transformer T47 when these are not higher than 5% and 1% of the nominal power of transformer, e.g., for the active power (10.5 kW/per phase) and the reactive power (2.5 kvar/per phase).

4.4. Unification of Phase Allocation

The authors have demonstrated earlier in [21] the use of the phase identification method on the experimental one-year worth of voltage data from smart meters that were installed in DS with radial topology. However, the sensitivity of this method is highly dependable on the amount of long-term data and it is better suited for radial topology of distribution network. Therefore, in the case of DS Brno-stred with an extensive dense-mesh topology, a different post processing approach of phase identification from the one that was published in [21] was used. The approach is basically based on the use of recorded voltage events/drops by PQ monitors for T1–T62. Table 11 quantifies these voltage drops in the transformer T1, which is considered as a reference one from the phase identification point of view.

Table 11. Short-term voltage drops in the reference transformer T1-06/2016–02/2017.

Number of the Events	U_{1T1}/U_n (–)	U_{2T1}/U_n (–)	U_{3T1}/U_n (–)	Affected Phases	Status
1	0.878	0.884	0.883	L1-L2-L3	not used
2	0.992	0.759	0.775	L2-L3	used for L1
3	0.995	0.753	0.754	L2-L3	used for L1
4	0.701	0.671	1.016	L1-L2	used for L3
5	0.876	1.014	0.876	L1-L3	used for L2
6	0.725	0.733	0.989	L1-L2	used for L3
7	1.001	0.754	0.683	L2-L3	used for L1
8	0.844	1.002	0.820	L1-L3	used for L2
9	0.878	0.885	0.882	L1-L2-L3	not used

Generally, the short-term voltage drop records include the initial drop time, drop duration, and the minimum effective voltage and current values in each phase. Table 11 shows the three-phase voltage drops that occurred twice and the two-phase voltage drops that occurred seven times. If the short-term voltage drops in one (L1, L2 or L3) or two phases (L1-L2, L1-L3, or L2-L3) are available, it is theoretically possible to use them to unify the measurement phases with reference one in a whole monitored system. In the case of a single-phase voltage drop, this significant drop can also be observed in each one specific phase of transformers. On the other hand, in the case of two-phase voltage drop, the object of observation is the phase without voltage drop. Thus, Table 12 brings the results of the phase identification with the reference T1 while using the mentioned two-phase voltage drop events (i.e., events 2–8 in Table 11) that were recorded by individual PQ monitors.

The results indicate good usability of the short-voltage drop events for the phase identification/unification of the measurement, when the phases were assigned to transformers T1–T59. However, in the case of three transformers (T2, T3, and T42), a reverse phase sequence was assigned. The rest of transformers was assigned a correct phase sequence. The phases of transformers T60–T62 remained unidentified because at the time of the voltage events recorded at T1, the PQ monitors at these transformers have not been installed yet. The results also show the ability to identify the individual phases of the measurement in 100% of the cases when PQ monitors recorded the short-voltage drop events. Specifically, 37 transformers have the same phase measurement position as the reference one. However, it is necessary to verify the results by comparing them with the actual PQ monitors HW connection directly at DTS. As this verification has not been done yet, this article further presents only a detailed analysis over the non-unified data.

Table 12. Results of the phases identification of the measurement–reference T1.

TR No.	Identified Phases			Status	TR No.	Identified Phases			Status
1	L1	L2	L3	reference	32	L1	L2	L3	correct
2	L3	L2	L1	reverse	33	L1	L2	L3	correct
3	L3	L2	L1	reverse	34	L2	L1	L3	correct
4	L3	L1	L2	correct	35	L2	L3	L1	correct
5	L1	L2	L3	correct	36	L1	L2	L3	correct
6	L1	L2	L3	correct	37	L1	L2	L3	correct
7	L1	L2	L3	correct	38	L2	L3	L1	correct
8	L1	L2	L3	correct	39	L1	L2	L3	correct
9	L3	L1	L2	correct	40	L3	L1	L2	correct
10	L2	L1	L3	correct	41	L1	L2	L3	correct
11	L1	L3	L2	correct	42	L3	L2	L1	reverse
12	L1	L2	L3	correct	43	L1	L2	L3	correct
13	L3	L1	L2	correct	44	L1	L2	L3	correct
14	L1	L2	L3	correct	45	L1	L2	L3	correct
15	L1	L2	L3	correct	46	L3	L1	L2	correct
16	L1	L2	L3	correct	47	L1	L2	L3	correct
17	L1	L2	L3	correct	48	L1	L2	L3	correct
18	L1	L2	L3	correct	49	L1	L2	L3	correct
19	L2	L3	L1	correct	50	L1	L2	L3	correct
20	L2	L1	L3	correct	51	L1	L2	L3	correct
21	L2	L3	L1	correct	52	L2	L3	L1	correct
22	L1	L2	L3	correct	53	L1	L2	L3	correct
23	L2	L1	L3	correct	54	L1	L2	L3	correct
24	L2	L3	L1	correct	55	L1	L2	L3	correct
25	L1	L2	L3	correct	56	L1	L2	L3	correct
26	L2	L3	L1	correct	57	L1	L2	L3	correct
27	L1	L2	L3	correct	58	L1	L2	L3	correct
28	L3	L1	L2	correct	59	L1	L2	L3	correct
29	L1	L2	L3	correct	60	NaN	NaN	NaN	unidentified
30	L2	L1	L3	correct	61	NaN	NaN	NaN	unidentified
31	L1	L2	L3	correct	62	NaN	NaN	NaN	unidentified

5. Discussion and Conclusions

The mathematical modeling of a distribution system requires both detailed technical parameterization of individual network elements and the quantification of the power flows over the time. Especially in the case of a complex dense-mesh topology, detailed mapping of LV customer behavior is needed for the purpose of the model verification and its credibility. Even though the following conclusions are mainly the beneficial inputs for this specific model, from the global point of view, it also contributes to finding the interdependencies of power flows over time in more detail.

From the analysis of all transformers the results show:

- the higher power load is achieved on the Weekdays than on the Weekend, specifically 75% of all single-phase active powers range up to 40 kW on the Weekdays and 28 kW on the Weekend;
- the ratio of the supplied and the consumed reactive power is 40/60 on the Weekdays and 50/50 on the Weekend;
- the transformers with a higher power load do not show the changes of the reactive power character. Specifically, there is only the consumed reactive power and no supplied one; and,
- in general, for the transformers with a low power load the supplied reactive power is inversely proportional to the consumed active power.

In the case of the analyses of limit states based on detailed results for the transformer with the maximal power load (T47) and the minimal power load (T61), the following can be observed.

For T47 with maximal power load and with no change of the reactive power character the results show:

- it operates 35% of time with tan *phi* < 1 (on the Weekdays). It also operates 100% of time with tan *phi* > 1 at the level of phase L2 and L3, and it operates at L1 with tan *phi* > 1 approx. for 95% of time (on the Weekend);
- the active and reactive power during the Weekdays is approximately 20% and 4% higher than during the Weekend;
- the active and reactive powers in the individual phases are relatively even;
- in more detail, these summary weekdays and weekend power flows are comparable to those that were achieved during time interval C (08:00:00–15:59:59) and D (16:00:00–21:59:59) with differences of approximately 1–3% for the Weekdays and approximately 1–4% for the Weekend at the same parts of the week; and,
- the summary power flows are higher than during interval A (22:00:00–03:59:59) and B (4:00:00–07:59:59). Specifically, for the active and reactive power up to 29% and 8% during the Weekdays and 10% and 5% during the Weekend.

For T61 with minimal power load and with the change of the reactive power character the results shown:

- it always operates with tan *phi* < 1 at the level of individual phases, when the minimum active power is not less than the maximum reactive power value in a given interval;
- over the Weekend it has an approx. twofold increase in frequencies of the supplied reactive power in individual phases than during the Weekdays; and,
- the active power during the Weekdays is usually 11–33% higher than during the Weekend and for the reactive power the differences are in range 1–54%, but in terms of the absolute values these are comparable to differences for the transformer T47.

Furthermore, a novel approach that is presented in this article is the demonstration of the phase unification within the measurements of the short-term voltage drops that were recorded by PQ monitors instead of the most often used process demanding HW or SW techniques. Unlike these techniques, the shown approach is not both computationally demanding, because no complex evaluation of the continuous voltage trends are observed and also financially demanding because it used installed PQ monitors and no additional HW is required to be installed within the distribution system. Although the results show the ability to identify phases in 100% of the cases, this approach has still to be verified via DSO in the real distribution system and, as a result of the application of the short-term voltage drops, there is also a risk that no voltage events will generally occur. The analysis of statistically sufficient sample of data (at least 1 year) is also necessary in order to relevantly assess the DS behaviour based on long-term measured values from DTS.

Moreover, the results show that it is not necessary to assess the voltage magnitudes while using the measuring over the unified phase data. On the other hand, the unified data can provide better information in which particular phase the voltage tolerance is compliant/non-compliant, thus the complex dependences can be generally observed in the context of a whole distribution system.

The results also have a practical contribution directly to the DSO. The main operating criterion of the analyzed municipal distribution system has not been complexly evaluated prior to the installation of PQ monitors. Thus, the results confirm the fulfilment of the established criterion and the power load of all transformers is below the required upper limit of 50% of S_n. It should be noted that the power load has always to be calculated from the rms/trms voltage and current values. Unlike other incorrect approaches, this takes the influence of the process of data aggregation and the change of the reactive power character in the measurement interval into account as well.

Author Contributions: Conceptualization, M.P. and V.V.; Data curation, M.P., V.V. and J.V.; Formal analysis, M.P. and V.V.; Investigation, M.P.; Methodology, M.P. and V.V.; Project administration, M.P.; Resources, M.P. and V.V.; Supervision, M.P., P.T. and J.V.; Visualization, M.P.; Writing—original draft, M.P., V.V. and P.T.

Funding: Authors gratefully acknowledge financial support from the Technology Agency of the Czech Republic (project No. TK02030039, TK02030013 a TJ02000006).

Acknowledgments: This research work has been carried out in the Centre for Research and Utilization of Renewable Energy (CVVOZE).

Conflicts of Interest: The authors declare no conflict of interest.

References

1. Dileep, G. A survey on smart grid technologies and applications. *Renew. Energy* **2020**, *146*, 2589–2625. [CrossRef]
2. Pau, M.; Patti, E.; Barbierato, L.; Estebsari, A.; Pons, E.; Ponci, F.; Monti, A. A cloud-based smart metering infrastructure for distribution grid services and automation. *Sustain. Energy Grids Netw.* **2018**, *15*, 14–25. [CrossRef]
3. Zhichao, L.; Yuping, Z. Research on Distribution Network Operation and Control Technology Based on Big Data Analysis. In Proceedings of the 2018 China International Conference on Electricity Distribution (CICED), Tianjin, China, 17–19 September 2018; pp. 1158–1161.
4. Guo, Y.; Lin, T.; Liang, K.; Chen, G. Network Quality Monitoring for Typical Power Services. In Proceedings of the 2019 IEEE 3rd Information Technology, Networking, Electronic and Automation Control Conference (ITNEC), Chengdu, China, 15–17 March 2019; pp. 1437–1441.
5. Zhao, Z.; Zheng, R.R.; Xing, N.Z. The research on 35kV power station service data flow model in power grid network. In Proceedings of the 2014 International Conference on Information Science, Electronics and Electrical Engineering, Sapporo, Hokkaido, Japan, 26–28 April 2014; pp. 24–27.
6. Rudin, C., Waltz, D.; Anderson, R N.; Boulanger, A.; Salleb-Aouissi, A.; Chow, M.; Dutta, H.; Gross, P.N.; Huang, B.; Ierome, S.; et al. Machine Learning for the New York City Power Grid. *IEEE Trans. Pattern Anal. Mach. Intell.* **2012**, *34*, 328–345. [CrossRef] [PubMed]
7. Martinez-Pabon, M.; Eveleigh, T.; Tanju, B. Smart Meter Data Analytics for Optimal Customer Selection in Demand Response Programs. *Energy Procedia* **2017**, *107*, 49–59. [CrossRef]
8. Wan Yen, S.; Morris, S.; Ezra, M.A.G.; Jun Huat, T. Effect of smart meter data collection frequency in an early detection of shorter-duration voltage anomalies in smart grids. *Int. J. Electr. Power Energy Syst.* **2019**, *109*, 1–8. [CrossRef]
9. Liao, Y.; Weng, Y.; Liu, G.; Rajagopal, R. Urban MV and LV Distribution Grid Topology Estimation via Group Lasso. *IEEE Trans. Power Syst.* **2019**, *34*, 12–27. [CrossRef]
10. Elphick, S.; Ciufo, P.; Drury, G.; Smith, V.; Perera, S.; Gosbell, V. Large Scale Proactive Power-Quality Monitoring: An Example from Australia. *IEEE Trans. Power Deliv.* **2017**, *32*, 881–889. [CrossRef]
11. Mertens, E.A.; Dias, L.F.S.; Fernandes, F.A.; Bonatto, B.D.; Abreu, J.P.G.; Arango, H. Evaluation and trends of power quality indices in distribution system. In Proceedings of the 2007 9th International Conference on Electrical Power Quality and Utilisation, Barcelona, Spain, 9–11 October 2007; pp. 1–6.
12. Xuemei, Y.; Jinliang, J. Study on power supply of ring network of 10KV distribution network. In Proceedings of the 2008 China International Conference on Electricity Distribution, Guangzhou, China, 10–13 December 2008; pp. 1–5.
13. Tao, X.; Renmu, H.; Peng, W.; Dongjie, X. Applications of data mining technique for power system transient stability prediction. In Proceedings of the 2004 IEEE International Conference on Electric Utility Deregulation, Restructuring and Power Technologies. Proceedings, Hong Kong, China, 5–8 April 2004; pp. 389–392.
14. Schäfer, F.; Menke, J.; Braun, M. Contingency Analysis of Power Systems with Artificial Neural Networks. In Proceedings of the 2018 IEEE International Conference on Communications, Control, and Computing Technologies for Smart Grids (SmartGridComm), Aalborg, Denmark, 29–31 October 2018; pp. 1–6.
15. Tinney, W.F.; Hart, C.E. Power Flow Solution by Newton's Method. *IEEE Trans.Power Appar. Syst.* **1967**, *PAS-86*, 1449–1460. [CrossRef]
16. Stott, B.; Alsac, O. Fast Decoupled Load Flow. *IEEE Trans. Power Appar. Syst.* **1974**, *PAS-93*, 859–869. [CrossRef]

17. Stott, B. Decoupled Newton Load Flow. *IEEE Trans. Power Appar. Syst.* **1972**, *PAS-91*, 1955–1959. [CrossRef]
18. Rao, S.; Feng, Y.; Tylavsky, D.J.; Subramanian, M.K. The Holomorphic Embedding Method Applied to the Power-Flow Problem. *IEEE Trans. Power Syst.* **2016**, *31*, 3816–3828. [CrossRef]
19. Bokhari, A.; Alkan, A.; Dogan, R.; Diaz-Aguiló, M.; De Leon, F.; Czarkowski, D.; Zabar, Z.; Birenbaum, L.; Noel, A.; Uosef, RE. Experimental Determination of the ZIP Coefficients for Modern Residential, Commercial, and Industrial Loads. *IEEE Trans. Power Deliv.* **2014**, *29*, 1372–1381. [CrossRef]
20. Hala, T.; Drapela, J. On stabilization of voltage in LV distribution system employing MV/LV OLTC transformer with control based on Smart Metering. In Proceedings of the 2019 20th International Scientific Conference on Electric Power Engineering (EPE), Ostrava, CR, VŠB-Technical University of Ostrava, Ostrava, Czech Republic, 15–17 May 2019; pp. 1–6.
21. Vycital, V.; Ptacek, M.; Toman, P.; Topolanek, D.; Drapela, J.; Zamphiropolos, J. Phase identification in smart metering pilot project Komorany. In *CIRED 2019, CIRED—Open Access Proceedings Journal, Proceedings of the 25th International Conference and Exhibition on Electricity Distribution, Madrid, Spain, 3–6 June 2019*; AIM: El Segundo, CA, USA, 2019.
22. Chen, C.S.; Ku, T.T.; Lin, C.H. Design of phase identification system to support three-phase loading balance of distribution feeders. In Proceedings of the 2011 IEEE Industrial and Commercial Power Systems Technical Conference, Baltimore, MD, USA, 1–5 May 2011; pp. 1–8.
23. Shen, Z.; Jaksic, M.; Mattavelli, P.; Boroyevich, D.; Verhulst, J.; Belkhayat, M. Three-phase AC system impedance measurement unit (IMU) using chirp signal injection. In Proceedings of the 2013 Twenty-Eighth Annual IEEE Applied Power Electronics Conference and Exposition (APEC), Long Beach, CA, USA, 17–21 March 2013; pp. 2666–2673.
24. Pappu, S.J.; Bhatt, N.; Pasumarthy, R.; Rajeswaran, A. Identifying topology of low voltage distribution networks based on smart meter data. *IEEE Trans. Smart Grid* **2018**, *9*, 5113–5122. [CrossRef]
25. Dilek, M.; Broadwater, R.P.; Sequin, R. Phase prediction in distribution systems. In Proceedings of the 2002 IEEE Power Engineering Society Winter Meeting, New York, NY, USA, 27–31 January 2002; pp. 985–990.
26. Arya, V.; Seetharam, D.; Kalyanaraman, S.; Dontas, K.; Pavlovski, C.; Hoy, S.; Kalagnanam, J.R. Phase identification in smart grids. In Proceedings of the 2011 IEEE International Conference on Smart Grid Communications (SmartGridComm), Brussels, Belgium, 17–20 October 2011; pp. 25–30.
27. Pezeshki, H.; Wolfs, P. Correlation based method for phase identification in a three phase LV distribution network. In Proceedings of the 2012 22nd Australasian Universities Power Engineering Conference (AUPEC), Bali, Indonesia, 26–29 September 2012; pp. 1–7.
28. Short, T.A. Advanced metering for phase identification, transformer identification, and secondary modeling. *IEEE Trans. Smart Grid* **2013**, *4*, 651–658. [CrossRef]
29. Wen, M.H.; Arghandeh, R.; von Meier, A.; Poolla, K.; Li, VO. Phase identification in distribution networks with micro-synchrophasors. In Proceedings of the 2015 IEEE Power & Energy Society General Meeting, Denver, CO, USA, 26–30 July 2015; pp. 1–5.
30. Seal, B.K.; McGranaghan, M.F. Automatic identification of service phase for electric utility customers. In Proceedings of the 2011 IEEE Power and Energy Society General Meeting, Detroit, MI, USA, 24–28 July 2011; pp. 1–3.
31. Lv, J.; Pawlak, M.; Annakkage, U.D.; Bagen, B. Statistical testing for load models using measured data. *Electr. Power Syst. Res.* **2018**, *163*, 66–72. [CrossRef]
32. Srinivasan, K.; Lafond, C. Statistical analysis of load behavior parameters at four major loads. *IEEE Trans. Power Syst.* **1995**, *10*, 387–392. [CrossRef]
33. Loewenstern, Y.; Katzir, L.; Shmilovitz, D. Statistical analysis of power systems and application to load forecasting. In Proceedings of the 2014 IEEE 28th Convention of Electrical & Electronics Engineers in Israel (IEEEI), Eilat, Israel, 3–5 December 2014; pp. 1–5.
34. Ahmed, M.A.; Reddy, K.S.; Prakash, J.; Rai, A.; Singh, B.P. Statistical analysis of load and its frequency response for load forecasting in a medium voltage distribution system. In Proceedings of the 2017 International Conference on Energy, Communication, Data Analytics and Soft Computing (ICECDS), Chennai, India, 1–2 August 2017; pp. 1146–1149.
35. Seppala, A. Statistical distribution of customer load profiles. In Proceedings of the 1995 International Conference on Energy Management and Power Delivery EMPD '95, Singapore, 21–23 November 1995; pp. 696–701.

36. Dong, Z.; Wang, Y.; Zhao, J. Variational Bayesian inference for the probabilistic model of power load. *IET Gener. Transm. Distrib.* **2014**, *8*, 1860–1868. [CrossRef]

37. Singh, R.; Pal, B.C.; Jabr, R.A. Statistical Representation of Distribution System Loads Using Gaussian Mixture Model. *IEEE Trans. Power Syst.* **2010**, *25*, 29–37. [CrossRef]

38. Ganjavi, A.; Christopher, E.; Johnson, C.M.; Clare, J. A study on probability of distribution loads based on expectation maximization algorithm. In Proceedings of the 2017 IEEE Power & Energy Society Innovative Smart Grid Technologies Conference (ISGT), Washington, DC, USA, 23–26 April 2017; pp. 1–5.

39. Ahmad, A.; Azmira, I.; Ahmad, S.; Anwar Apandi, N.I. Statistical distributions of load profiling data. In Proceedings of the 2012 IEEE International Power Engineering and Optimization Conference, Melaka, Malaysia, 6–7 June 2012; pp. 199–203.

40. European standard EN 50160 ed.3, Voltage Characteristics of Electricity Supplied by Public Distribution Systems. 2003. Available online: https://www.orgalim.eu/position-papers/en-50160-voltage-characteristics-electricity-supplied-public-distribution-system (accessed on 14 March 2003).

41. International Electrotechnical Commission. Electromagnetic compatibility (EMC)-Part 4-30: Testing and measurement techniques-Power quality measurement methods. Available online: https://ci.nii.ac.jp/naid/10026582038/ (accessed on 10 October 2019).

42. IEC 62053-23, *Electricity Metering Equipment (a.c.)—Particular Requirements—Part 23: Staticmeters for Reactive Energy (Classes 2 and 3)*; International Electrotechnical Commission: Geneva, Switzerland, 2018.

43. IEC 62056-6-1, *Electricity Metering Data Exchange—The DLMS/COSEM Suite—Part 6-1: Object Identification System (OBIS)*; International Electrotechnical Commission: Geneva, Switzerland, 2017.

Article

An Extended Kalman Filter Approach for Accurate Instantaneous Dynamic Phasor Estimation

Matilde de Apráiz, Ramón I. Diego and Julio Barros *

Department Computer Science and Electronics, University of Cantabria, 39005 Santander, Spain;
matilde.deapraiz@unican.es (M.d.A.); ramon.diego@unican.es (R.I.D.)
* Correspondence: julio.barros@unican.es; Tel.: +34-942-201-355

Received: 4 October 2018; Accepted: 23 October 2018; Published: 26 October 2018

Abstract: This paper proposes the application of a non-linear Extended Kalman Filter (EKF) for accurate instantaneous dynamic phasor estimation. An EKF-based algorithm is proposed to better adapt to the dynamic measurement requirements and to provide real-time tracking of the fundamental harmonic components and power system frequencies. This method is evaluated using dynamic compliance tests defined in the IEEE C37.118.1-2011 synchrophasor measurement standard, providing promising results in phasor and frequency estimation, compliant with the accuracy required in the case of off-nominal frequency, amplitude and phase angle modulations, frequency ramps, and step changes in magnitude and phase angle. An important additional feature of the method is its capability for real-time detection of transient disturbances in voltage or current waveforms using the residual of the filter, which enables flagging of the estimation for suitable processing.

Keywords: power system measurements; dynamic phasor estimation; Kalman filters; phasor measurement; power quality

---

1. Introduction

Accurate estimation of synchrophasors of voltage and current waveforms is a key aspect of the correct operation of modern electric power systems. IEEE standard C37.118.1-2011 defines the measurement of synchrophasors under all operation conditions, specifying methods for evaluating these measurements and requirements for compliance with the standard, under both steady-state and dynamic conditions [1,2].

The traditional methods for synchrophasor estimation are based on a steady-state model that assumes phasor parameters are constant during the measurement window, but in general, power system voltage and currents are not steady-state sinusoidal waveforms, particularly during disturbances. Harmonic and other non-harmonic frequency components, step changes in magnitude and phase angle, oscillations, fast transients, and high-frequency components can be produced during electromagnetic and electromechanical transient disturbances. The estimation of phasors under these dynamic conditions may produce erroneous results as is reported in [3].

Discrete Fourier Transform (DFT)-based phasor estimation algorithms have traditionally been used because of their good performance in steady-state conditions, and they are used in many commercial phasor measurement units (PMUs), but they fail for off-nominal frequencies and under transient and dynamic conditions. Under these conditions, time-varying amplitude and phase angle models have been proposed to improve the accuracy of phasor estimation and for compliance with synchrophasor measurement standards [4–12].

Dynamic phasor estimation algorithms have been proposed [4–6], where the phasor is approximated with a Taylor polynomial expansion around the reference time. A Kalman filter was used [7,8] to estimate a dynamic phasor and its derivatives. A state-space model was used where the dynamic phasor was approximated by a kth Taylor polynomial expansion. The state-space model uses the phasor and its derivatives as the state variables and a state transition matrix obtained from the derivatives of each Taylor truncated dynamic phasor, applying the Kalman filter for estimation.

A Taylor-Kalman-Fourier filter for instantaneous estimation of oscillating phasors was proposed in [9]. The signal model was extended to the full set of harmonics, using an extended transition matrix where all harmonics are included, obtaining filters that were more robust to noise and able to estimate dynamic harmonics, providing phasor estimates free from harmonic infiltration.

A modified Taylor-Kalman filter for instantaneous phasor estimation based on an improved dynamic model was proposed in [10]. Other extensions of the Taylor-Kalman filter model have been introduced [11,12]. In [11], an output smoother was applied to improve the estimation accuracy of the method when large oscillations in amplitude and phase angle in harmonics and interharmonics affect the input waveforms. Reference [12] introduced a Taylor Extended Kalman Filter formulation that separately considers the amplitude and phase angle of the Taylor expansion. A new dynamic model was then defined that enables the reduction in the state space dimension considering different expansion orders of magnitude and phase angle, while including harmonics in the model.

A different approach from those based on the use of the synchrophasor Taylor series expansion model was proposed in [13]. An adaptive algorithm was used for phasor estimation, eliminating the effects of various transient disturbances on voltage and current waveforms. The algorithm pre-analyzes the waveforms in the observation window to detect and localize a discontinuity. The data detected before or after the singular point, excluding the transient, are used for phasor estimation.

In this paper, we propose the use of a non-linear adaptive Extended Kalman Filter (EKF) to better adapt to the dynamic requirements of power system signals. The fundamental component, the most relevant harmonic components, the dominant subsynchronous interharmonic component, and the power system frequency were used in a 13-state EKF to model the system and to estimate fundamental and harmonic phasors and the power system frequency. This method enables on-line tracking of the fundamental, harmonics, and subsynchronous interharmonic phasors and the detection of transient conditions in the input waveform using the residual of the Kalman filter.

The paper is organized as follows: Section 2 presents the main characteristics of the EKF and the signal model used in our proposal. Section 3 outlines the performance of the algorithm under off-nominal frequencies, frequency ramps, power oscillations, and step changes in magnitude and phase angle dynamic compliance tests defined in IEEE C37.118.1-2011. The obtained results are also analyzed in this section. Section 4 outlines the transient detection capability of the method and Section 5 presents the conclusions.

2. Kalman Filtering and Signal Model

2.1. Extended Kalman Filter

The Extended Kalman Filter (EKF) is a modified version of the linear Kalman filter that is applied to systems with non-linear processes and measurement equations. In each step of the recursive algorithm, the non-linear equations are linearized using a first order Taylor series to form a linear process, and then the linear Kalman filter model is applied.

The EKF defines the state transition function and the relationship between the states and the measurements by means of the nonlinear functions f and h, respectively:

$$x_{k+1} = f[x_k, k] + w_k$$

$$z_k = h[x_k, k] + v_k$$

where x_k and z_k are the state vector and the measurement at instant k, respectively; and w_k and v_k are the uncertainties introduced by the state transition and the measurement noise, both with zero mean and covariances Q_k and R_k, respectively. The nonlinear functions f and h are linearized using a first-order Taylor series as follows:

$$x_{k+1} = \Phi_k x_k + w_k$$

$$z_k = H_k x_k + v_k$$

with

$$\Phi_k = \frac{\partial f_i[x_k, k]}{\partial x_j}$$

$$H_k = \frac{\partial h_i[x_k, k]}{\partial x_j}$$

where f_i and h_i are the ith elements of functions f and h, respectively, and Φ_k and H_k are the state transition and the measurement matrices, respectively.

The Kalman filter is a two-step prediction-correction process. Starting with an initial estimate of the process x'_k, and its error covariance matrix P'_k, the measurement at instant k, z_k, is used to improve the estimation. A linear combination of the estimate and the measurement is chosen according to Equation (1):

$$x_k = x'_k + K_k(z_k - H_k x'_k) = x'_k + K_k \varepsilon_k \tag{1}$$

where x_k is the estimation update at instant k and K_k is the filter coefficient.

ε_k is the residual, defined as:

$$\varepsilon_k = z_k - H_k x'_k \tag{2}$$

which represents the difference between the measurement z_k and the estimation x'_k at instant k.

The state transition matrix Φ_k is used to project the filter ahead using the measurement at instant $k + 1$. Kalman filter equations can be found in [14].

2.2. Signal Model

Considering the input signal as a single-phase sinusoidal with superimposed harmonic and interharmonic components:

$$x(k) = \sum_{i}^{n} A_i \sin(2\pi i f k T_S + \theta_i) + A_{ih} \sin(2\pi f_{ih} k T_S + \theta_{ih})$$

where i is the harmonic order; A_i and θ_i are the amplitude and phase angle of each harmonic component respectively; A_{ih}, θ_{ih}, and f_{ih} are the amplitude, phase angle, and frequency of a dominant interharmonic component, respectively; f is the fundamental frequency; and T_s is the sampling period.

The successful application of a Kalman filter depends on the accuracy of the model used, which requires previous knowledge of the system. The lack of relevant components in the signal model may produce erroneous results in the estimation of the components of interest [11,12]. In order to estimate the dominant frequency components in a specific power system network, a DFT analysis was applied to the input signal to select the best possible state vector, which is the DFT output bins-harmonics and interharmonics-with the highest magnitudes, as well as their initialization values and error covariances.

Other proposals to obtain these dominant components in the signal to improve the accuracy of the model can be found in the literature. A compressive sensing approach was proposed in [15] to identify the most relevant components of the signal and to model them in the phasor estimation procedure. A frequency-domain decomposition of the residuals of a linear Kalman filter using the Short-Time Fourier Transform was used in [16] to obtain the dominant subsynchronous component of the signal to be included in the model of the system to achieve an accurate phasor estimation.

Fundamental and odd harmonic components from the third to the ninth are normally dominant in low-voltage supply systems and were proven in [17] as an efficient solution to model the signal. The use of a higher order model improves the accuracy of the estimation but also increases the computational complexity, so a trade-off is required between accuracy and model complexity. The presence of a dominant sub-synchronous interharmonic component close to the fundamental frequency may affect the estimation of the fundamental phasor [16]. Including this frequency component in the model of the system is necessary to obtain an accurate phasor estimate.

Another necessary and important aspect in the successful application of Kalman filters is the adequate selection of the filter parameters, the initial estimation of the state vector, and the covariance matrices, based on previous knowledge of the system to be modeled. The process and the measurement noise covariance matrices, Q_k and R_k respectively, act as tuning parameters to balance the dynamic response of the filter against the sensitivity to noise. Convergence rate and noise rejection of the filter requires a trade off against each other. The degree of stability improves as the process covariance increases and/or the noise covariance decreases [18], the filter response depending more on the ratio of these two matrices than on their specific values [19,20].

The state vector selected at instant k, x_k, is:

$$x_k = [x_1 x_2 \cdots x_{11} x_{12} x_{13}]_k \tag{3}$$

where x_1, x_3, ..., x_9 are the in-phase components of the fundamental and the dominant harmonic components; x_2, x_4, ..., x_{10} are the corresponding in-quadrature components; x_{11}, x_{12} are the in-phase and in-quadrature components of the dominant sub-synchronous interharmonic component at instant k; and x_{13} is the power system frequency. The magnitude and phase angle of each frequency component are computed using the instantaneous rectangular components considered in the model of the signal. The inclusion of the power system frequency in the state vector enables the filter parameters and the phasor to be updated with the new estimation of the frequency in each step of the recursive process, improving the accuracy of the estimation algorithm, and avoiding the need for a separate frequency estimation algorithm, as required in other previously proposed Kalman filter phasor estimation methods [16,21].

The on-line estimation of fundamental phasors and power system frequency with each new sample of the input signal allows increasing the reporting rate without decreasing the frequency resolution, which would be the case for the application DFT-based synchrophasor estimation methods.

3. Performance Analysis

The performance of the proposed phasor estimation method was evaluated using the dynamic compliance tests defined in the IEEE synchrophasor measurement standard and its amendment [1,2], for both P-class and M-class compliance. In each test, the estimated phasor at instant k is compared with the theoretical value provided as an input, computing the Total Vector Error (TVE), Frequency Error (FE), and Rate of Change of Frequency (ROCOF) Error (RFE) performance indices as measurement of the difference between the ideal values and their estimates as defined in [1].

This section presents the results obtained with the proposed method under the different dynamic tests defined in IEEE Standard C37.118.1-2011: frequency deviations, amplitude and phase angle modulations, frequency ramps, and step changes in magnitude and phase angle.

As previously mentioned in Section 2, a previous analysis of the signal in a specific power system network is required to select the dominant frequency components to be included in the state vector. To this end, the daily mean value of the fundamental and odd harmonic components from the 3rd to the 11th order and the mean square values of voltage supply measured in the specific power system network were selected as the initialization values of the state vector and its covariance matrix, respectively. We selected 50 Hz as the initial value of the power system frequency and a Q_k/R_k ratio of 10. This ratio was proven to be an efficient solution in the low-voltage distribution network in [17]. Other power system networks require a similar previous analysis to select the most adequate state vector and the initialization parameters of the Kalman filter. The sampling rate used in the simulations was 6.4 kS/s (128 samples/cycle in a 50 Hz system).

3.1. Frequency Deviation

In these tests, a constant frequency was is applied to the nominal frequency in the range of ± 5 Hz for M-class or ± 2 Hz for P-class compliance, with constant phase angle in steps of 0.1 Hz, as defined in [1], computing the maximum magnitudes of the three indices: TVE, FE and RFE. The maximum acceptable values of these performance indices are 1%, 0.005 Hz, and 0.01 Hz/s, respectively, for M-class compliance. According to the standard amendment [2], the values are 1%, 0.005 Hz, and 0.4 Hz/s for P-class compliance, respectively.

The test signal used in this test was [1]:

$$x(t) = X_m \cos(2\pi(f_0 + f_{dev})t + \phi)$$

where X_m is the magnitude, φ is the phase angle of the fundamental component, f_0 is the nominal power system frequency, and f_{dev} is the frequency deviation.

As an example of the performance of our method, Figure 1 shows the actual and the estimated magnitude and frequency in the case of a 230 V magnitude and 52 Hz voltage waveform. After a short initialization time, both the estimated magnitude and frequency quickly converged to their real values. In this case, the maximum TVE, FE, and RFE were 3.4×10^{-5}%, 5.48×10^{-3} mHz, and 2.34×10^{-5} Hz/s respectively.

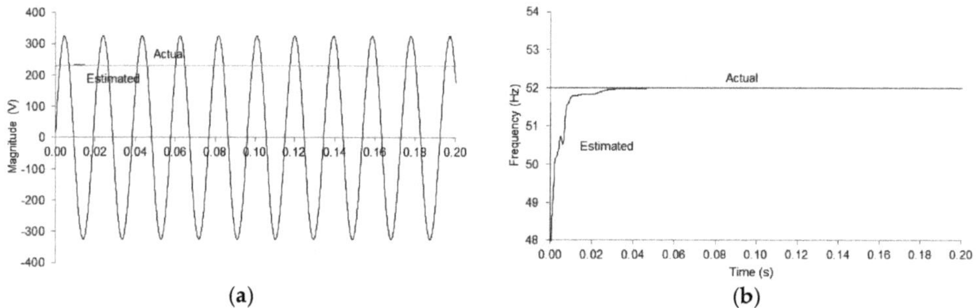

(**a**) (**b**)

Figure 1. (**a**) Estimated phasor magnitude and (**b**) frequency in the case of a 230 V magnitude and 52 Hz voltage waveform.

Figure 2 shows the maximum values of TVE, FE, and RFE during the five second record for each of the frequency deviation test signals. The maximum values of the three performance indices were all below the minimum requirements of the synchrophasor measurement standard for both P-class and M-class compliance.

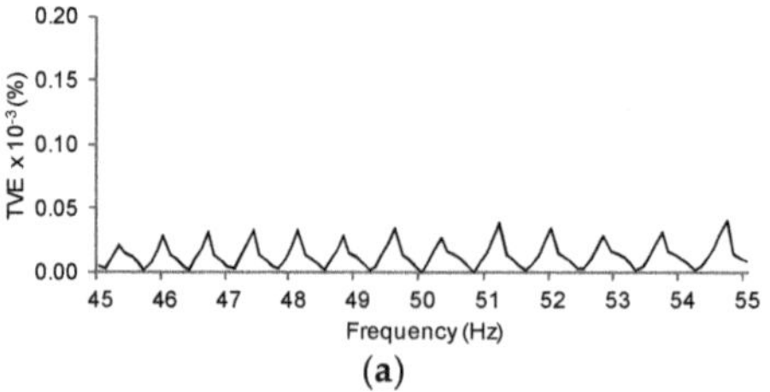

(**a**)

Figure 2. *Cont.*

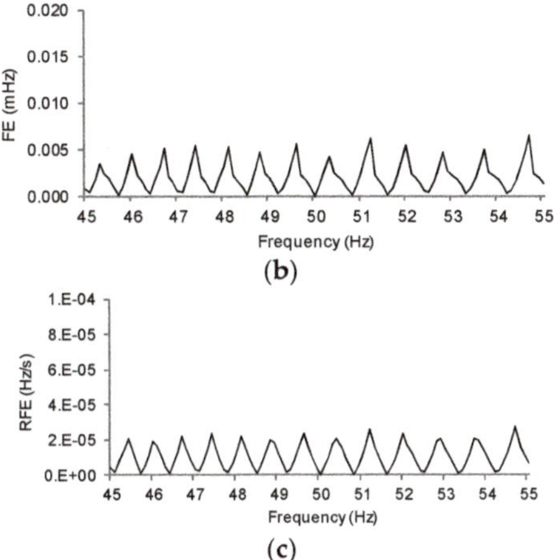

Figure 2. Maximum values of (**a**) Total Vector Error (TVE), (**b**) Frequency Error (FE), and (**c**) Rate of Change Frequency Error (RFE) for constant frequency deviation tests.

3.2. Amplitude and Phase Angle Modulation

These tests were performed by modulating the input signal with sinusoidal signals simultaneously applied to signal magnitude and phase angle, which were used to estimate the bandwidth of the estimation algorithm. These signals are mathematically represented for a single phase as [1]:

$$x(t) = X_m[1 + k_x \cos(wt)] \cos(w_0 t + k_a \cos(wt - \pi))$$

where X_m is the amplitude, w_0 is the nominal frequency, w is the modulating frequency, and k_x and k_a are the amplitude and phase angle modulation factors, respectively.

The IEEE synchrophasor measurement standard allows a maximum 3% TVE for modulating frequencies from 0.1 Hz to 5 Hz, and the maximum values of FE and RFE should be lower than 0.06 Hz and 3 Hz/s for P class compliance, respectively, and reporting rates higher than 20, and 0.3 Hz and 30 Hz/s, respectively, for M-class compliance at the same reporting rate. According to the standard, TVE, FE, and RFE indices should be measured over at least two full cycles of the signal in the different modulation tests.

Figure 3 shows the estimated amplitude and frequency in the case of a 230 V magnitude and 50 Hz voltage waveform with a 10% magnitude, and 0.1 Hz amplitude modulation starting from 1 s of the record. As can be seen in the figure, the estimated amplitude accurately tracked the actual voltage fluctuation with minimum frequency error.

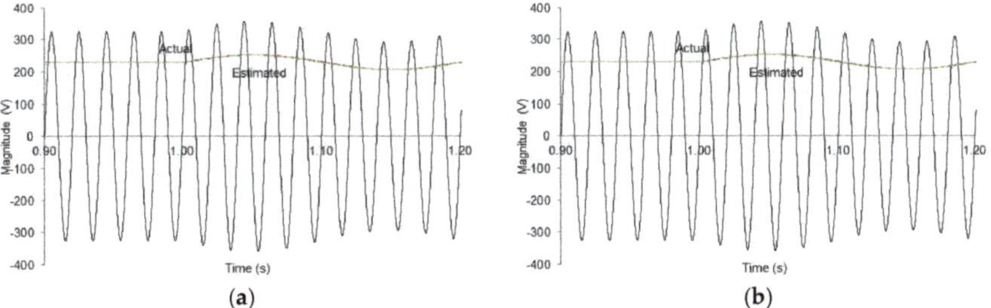

Figure 3. Performance of the method during a 10% magnitude and 0.1 Hz amplitude modulation test: (**a**) magnitude and (**b**) frequency.

Figure 4a,b show the maximum TVE, FE, and RFE values obtained for the different amplitude and phase modulation test signals for modulation frequencies from 0.1 Hz to 5 Hz, varying in steps of 0.1 Hz and 10% amplitude and 10% phase angle modulation. From the results reported, the maximum 3% TVE criteria was satisfied under all conditions. The same applies for the measurements of maximum values of FE and RFE indices.

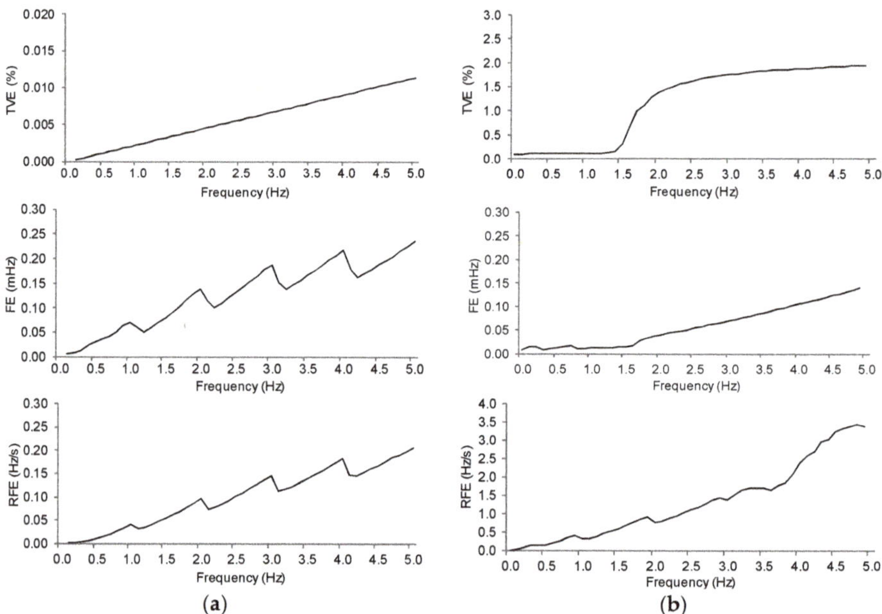

Figure 4. Effect of modulation frequency on maximum TVE, FE, and RFE for (**a**) amplitude modulation and (**b**) phase modulation tests.

3.3. Frequency Ramp

According to the synchrophasor measurement standard, the performance of a measurement method should also be evaluated during frequency ramp tests. Mathematically, the input signal can be represented by:

$$x(t) = X_m \cos[w_0 t + \pi R_f t^2]$$

where X_m is the amplitude, w_0 is the nominal frequency, and R_f ($=df/dt$) is the frequency ramp rate in Hz/s.

According to [2], the ramp rate and the frequency range in frequency ramp tests should be ±1.0 Hz/s and ±2 Hz for P-class compliance and ±1.0 Hz/s and ±5 Hz for M-class compliance, respectively. The measurement error limit in TVE is 1% in both cases, whereas the limits are 0.01 Hz and 0.4 Hz/s (P-class), and 0.01 Hz and 0.2 Hz/s (M-class) for FE and RFE, respectively. A transition time of ±2/F_s, where F_s is the reporting rate, is allowed for TVE, FE, and RFE compliance before and after the sudden change in ROCOF occurs.

Figure 5 shows the time evolution of the frequency and TVE computed with the 13-state EKF method proposed during a positive and a negative ramp test of 1 Hz/s in the range of ±2 Hz (P-class). All results reported were obtained considering the previously mentioned exclusion intervals, with F_s = 50. With the exception of the beginning of the frequency ramp, the method presented an accurate convergence during the test. The maximum values of TVE were 0.074% and 0.071% for the positive and negative ramp test, respectively.

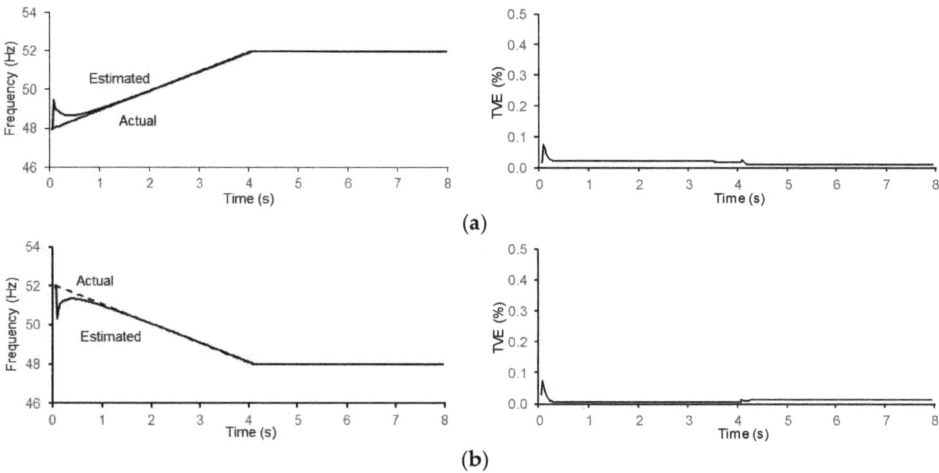

Figure 5. Time evolution of frequency and TVE in (**a**) positive and (**b**) negative frequency ramp test of 1 Hz/s in the ±2 Hz frequency range.

3.4. Step Changes in Magnitude and Phase Angle

In these tests, the test signals suddenly transitioned in the magnitude/phase angle between two steady states. The test signals were used to determine response time, delay time, and overshoot in the measurement. In our proposal, once the step change was applied, the 13-state EKF re-initialized to improve its transient response to the step change. Other proposals to improve the transient response of a Kalman filter have been previously reported [22–24]

Figure 6a,b show the estimated magnitude, the residual, and TVE during a +10% step change in magnitude ($k_x = 0.1$) and the estimated phase angle and TVE during a +10° phase angle step change ($k_a = 0.1$). The method was able to accurately track these transient conditions, with the residual showing a sudden transition in its magnitude at the beginning of these changes, and the TVE always remained below the maximum 1% limit for both P-class and M-class compliance, and for step changes in magnitude and phase angle.

Table 1 presents the response time, the delay time, and the overshoots/undershoots for TVE, FE, and RFE for ±10% and ±10° step changes in magnitude and phase angle, respectively, for P-class compliance and a reporting rate $F_s = 50$. The maximum response time, delay time, and overshoot are defined in Tables 9 and 10 of [2]. In the case of TVE and FE, the response time with the 13-state EKF was zero because the magnitude of both indices was always below the limits defined for P-class and M-class compliance. The maximum response time for RFE ($6/f_0$, i.e., 0.12 s for a 50 Hz system, for the step change in magnitude and phase angle in P-class, and $14/F_s$, i.e., 0.28 s for $F_s = 50$, for both step change tests in M-class) was not surpassed in any of the tests.

Table 1. Step changes in magnitude and phase angle test results.

Range (Hz)	TVE T_{res} (s)	FE T_{res} (s)	RFE T_{res} (s)	TVE Delay Time (s)	Overshoot (%)
+10%	0.0	0.0	0.00015	0.00031	0.524
−10%	0.0	0.0	0.00015	0.00031	0.729
+10°	0.0	0.0	0.00031	0.00015	0.035
−10°	0.0	0.0	0.00031	0.00015	0.028

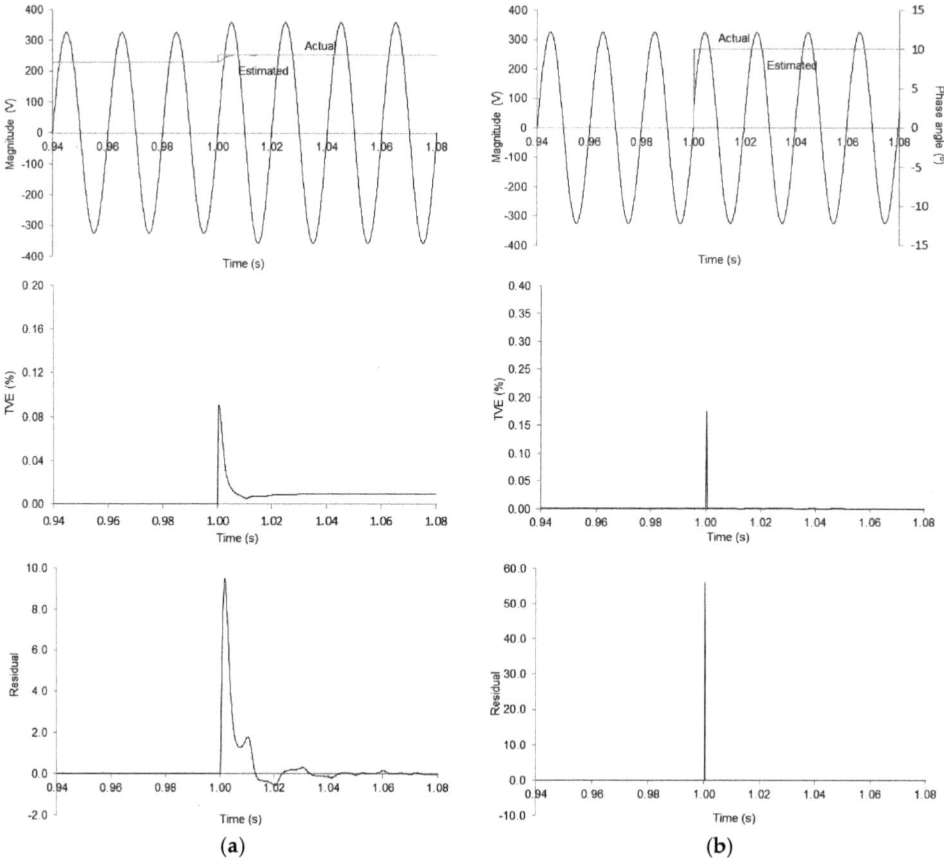

Figure 6. (**a**) Estimated magnitude, residual and TVE during a 10% step change in amplitude, and (**b**) estimated phase angle and TVE during a 10° step change in phase angle.

The results obtained show the high responsiveness and good estimation of step inputs of our method, with short response times and small overshoot.

4. Transient Detection

This section outlines an important additional feature of the method proposed: the real-time detection of sudden transients in the input signal using the residuals of the Kalman filter. In a Kalman filter, the residual ε_k, defined in Equation (2), reflects the difference between the actual measurement at instant k, z_k, and the predicted measurement $H_k x'_k$. This difference was computed in each step of the adaptive algorithm. A zero magnitude of the residual ε_k implies total agreement between the measurement and the prediction, whereas a sudden change in magnitude appears when a sudden mismatch occurs between the two, as was the case in the beginning of a transient condition in the signal under analysis. Large sudden changes in a signal produce larger magnitudes in the residuals ε_k.

The transient detection capability of the residuals depends on both the magnitude of the transient and the detection threshold selected. The detection of a transient condition enables the flagging of the estimation for its adequate processing and represents an important contribution of our method.

As an example of the capability of our method for step transient detection in the input signal, Figure 6a,b show the evolution of the residual ε_k during the +10% step change in magnitude and +10°

step change in phase angle tests, respectively. A sudden change in magnitude of the residual ε_k can be observed at the beginning of the two step changes.

Reference [25] demonstrated the performance of their method in the detection of other important transient conditions in voltage or current waveforms in power systems, such as voltage dips and short interruptions, high-frequency transients, time varying harmonics and power swings, as well as the procedure for the most adequate selection of the transient detection threshold.

5. Conclusions

In this paper, we presented the application of an Extended Kalman Filter for instantaneous dynamic phasor and frequency estimation, which includes fundamental and harmonic components, compliant with the dynamic requirements of the IEEE synchrophasor measurement standards. The estimates are instantaneous and updated with each new sample of the signal under analysis. The filter parameters are adapted on-line with each new estimate of the power system frequency. The algorithm is robust against off-nominal frequencies and power swings, has accurate convergence during the frequency ramp tests, and shows high responsiveness and good estimation to step inputs, with short response times and small overshoot. Simultaneous with the estimation of the phasor magnitude and phase angle and frequency, the method can detect a step change in the input signal using the residual of the filter in real-time, which enables the flagging of the estimation for suitable processing.

Author Contributions: Conceptualization, J.B. and R.I.D.; methodology, J.B., R.I.D., and M.d.A.; software, M.d.A.; validation, M.d.A. and R.I.D.; formal analysis, J.B. and R.I.D.; investigation, J.B. and R.I.D.; writing—original draft preparation, J.B.; writing—review and editing, J.B.; project administration, J.B. and R.I.D.

Funding: This work received financial support from the Spanish Ministry of Economía y Competitividad, Plan Estatal de Investigación Científica y Técnica y de Innovación 2013–2016, through the grant ENE2014-54039-R.

References

1. IEEE Std C37.118.1-2011. *IEEE Standard for Synchrophasor Measurements for Power Systems*; IEEE, Inc.: New York, NY, USA, 2011. [CrossRef]
2. IEEE Std C37.118.1a-2014. *IEEE Standard for Synchrophasor Measurements for Power Systems*; Amendment 1: Modification of Selected Performance Requirements; IEEE, Inc.: New York, NY, USA, 2014. [CrossRef]
3. Phadke, A.G.; Kasztenny, B. Synchronized phasor and frequency measurement under transient conditions. *IEEE Trans. Power Deliv.* **2009**, *24*, 89–95. [CrossRef]
4. de la O Serna, J.A. Dynamic phasor estimates for power systems oscillations. *IEEE Trans. Instrum. Meas.* **2007**, *56*, 1648–1657. [CrossRef]
5. Platas-Garza, M.A.; de la O Serna, J.A. Dynamic phasor and frequency estimates through maximally flat differentiators. *IEEE Trans. Instrum. Meas.* **2010**, *59*, 1803–1811. [CrossRef]
6. Mai, R.; He, Z.; Fu, L.; Kirby, B.; Bo, Z. A dynamic synchrophasor estimation algorithm for online application. *IEEE Trans. Power Deliv.* **2010**, *25*, 570–578. [CrossRef]
7. de la O Serna, J.A.; Rodriguez, J. Instantaneous dynamic phasor estimates with Kalman filter. In Proceedings of the 2010 IEEE PES General Meeting, Minneapolis, MN, USA, 25–29 July 2010. [CrossRef]
8. de la O Serna, J.A.; Rodriguez, J. Instantaneous oscillating phasor estimates with Taylor^K—Kalman filters. *IEEE Trans. Power Syst.* **2011**, *26*, 2336–2344. [CrossRef]
9. de la O Serna, J.A.; Rodriguez-Maldonado, J. Taylor–Kalman–Fourier filters for instantaneous oscillating phasor and harmonic estimates. *IEEE Trans. Instrum. Meas.* **2012**, *61*, 941–951. [CrossRef]
10. Junqui, L.; Fei, N.; Junjie, T.; Ponci, F.; Monti, A. A modified Taylor-Kalman filter for instantaneous dynamic phasor estimation. In Proceedings of the IEEE PES ISGT Europe 2012, Berlin, Germany, 14–17 October 2012. [CrossRef]
11. Fontanelli, D.; Macii, D.; Petri, D. Dynamic synchrophasor estimation using smoothed Kalman filtering. In Proceedings of the 2016 IEEE I^{2}MTC, Taipei, Taiwan, 23–26 May 2016. [CrossRef]

12. Ferrero, R.; Pegoraro, P.A.; Toscani, S. Dynamic fundamental and harmonic synchrophasor estimation by extended Kalman filter. In Proceedings of the IEEE International Workshop on Applied Measurements for Power Systems, Aachen, Germany, 28–30 September 2016. [CrossRef]

13. Ren, J.; Kezunovic, M. An adaptive phasor estimator for power system waveform containing transients. *IEEE Trans. Power Deliv.* **2012**, *27*, 735–745. [CrossRef]

14. Girgis, A.A.; Daniel Hwang, T.L. Optimal estimation of voltage phasors and frequency deviation using linear and non-linear Kalman filtering. *IEEE Trans. Power App. Syst.* **1984**, *PAS-103*, 2943–2949. [CrossRef]

15. Bertocco, M.; Frigo, G.; Narduzzi, C.; Muscas, C.; Pegoraro, P.A. Compressive sensing of a Taylor-Fourier multifrequency model for synchrophasor estimation. *IEEE Trans. Instrum. Meas.* **2015**, *64*, 3274–3283. [CrossRef]

16. Chakir, M.; Kamwa, I.; Le Huy, H. Extended C37.118.1 PMU algorithms for joint tracking of fundamental and harmonic phasors in stressed power systems and microgrids. *IEEE Trans. Power Deliv.* **2014**, *29*, 1465–1480. [CrossRef]

17. Pérez, E.; Barros, J. An extended Kalman filtering approach for detection and analysis of voltage dips in power systems. *Electr. Power Syst. Res.* **2008**, *78*, 618–625. [CrossRef]

18. Bitmead, R.R.; Chung Tsoi, A.H.; Parker, J.P. A Kalman filtering approach to short-time Fourier analysis. *IEEE Trans. Acoust. Speech Signal Process.* **1986**, *ASSP-34*, 1493–1501. [CrossRef]

19. Kennedy, K.; Lightbody, G.; Yacamini, R. Power system harmonic analysis using the Kalman filter. In Proceedings of the IEEE Power Engineering Society General Meeting, Toronto, ON, Canada, 13–17 July 2003; Volume 2, pp. 752–757.

20. Bittanti, S.; Savaresi, S.M. On the parametrization and design of an extended Kalman filter frequency tracker. *IEEE Trans. Autom. Control* **2000**, *45*, 1718–1724. [CrossRef]

21. Kamwa, I.; Samantaray, S.R.; Joos, G. Compliance analysis of PMU algorithms and devices for wide-area stabilizing control of large power systems. *IEEE Trans. Power Syst.* **2013**, *28*, 1766–1778. [CrossRef]

22. Liu, S. An adaptive Kalman filter for dynamic estimation of harmonic signal. In Proceedings of the IEEE 8th International Conference on Harmonics and Quality of Power, Athens, Greece, 14–16 October 1998; Volume 2, pp. 636–640.

23. Yu, K.K.C.; Watson, N.R.; Arrillaga, J. An adaptive Kalman filter for dynamic harmonic state estimation and harmonic injection tracking. *IEEE Trans Power Deliv.* **2005**, *20*, 1577–1584. [CrossRef]

24. Rosendo, J.A.; Gómez, A. Self tunning of Kalman filters for harmonic computation. *IEEE Trans. Power Deliv.* **2006**, *21*, 501–503.

25. Apráiz, M.; Diego, R.I.; Barros, J. Transient detection in phasor measurement units with Kalman filtering. In Proceedings of the 18th International Conference in Harmonics and Quality of Power, Ljubljana, Slovenia, 13–16 May 2018. [CrossRef]

Article

Implementation of Processing Functions for Autonomous Power Quality Measurement Equipment: A Performance Evaluation of CPU and FPGA-Based Embedded System

María-Ángeles Cifredo-Chacón [1,*], Fernando Perez-Peña [2], Ángel Quirós-Olozábal [3] and Juan-José González-de-la-Rosa [4]

[1] Microelectronic Circuit Design Group, Escuela Superior de Ingeniería, University of Cádiz, Avda. de la Universidad 10, E-11519 Puerto Real-Cádiz, Spain
[2] Applied Robotics Lab, Escuela Superior de Ingeniería, University of Cádiz, Avda. de la Universidad 10, E-11519 Puerto Real-Cádiz, Spain; fernandoperez.pena@uca.es
[3] Microelectronic Circuit Design Group, Escuela Superior de Ingeniería, University of Cádiz, Avda. de la Universidad 10, E-11519 Puerto Real-Cádiz, Spain; angel.quiros@uca.es
[4] Computational Instrumentation and Industrial Electronics Group, Escuela Politécnica Superior, University of Cádiz, Avda. Ramón Puyol S/N, E-11202 Algeciras-Cádiz, Spain; juanjose.delarosa@uca.es
* Correspondence: mangeles.cifredo@uca.es; Tel.: +34-956-483-315

Received: 6 February 2019; Accepted: 3 March 2019; Published: 9 March 2019

Abstract: Motivated by the effects of deregulation over power quality and the subsequent need of new types of measurements, this paper assesses different implementations of an estimate for the spectral kurtosis, considered as a low-level harmonic detection. Performance of a processor-based system is compared with a field programmable gate array (FPGA)-based solution, in order to evaluate the accuracy of this processing function for implementation in autonomous measurement equipment. The fourth-order spectrum, with applications in different fields, needs advanced digital signal processing, making it necessary to compare implementation alternatives. In order to obtain reproducible results, the implementations have been developed using common design and programming tools. Several characteristics of the implementations are compared, showing that the increasing complexity and reduced cost of the current FPGA models make the implementation of complex mathematical functions feasible. We show that FPGAs improve the processing capability of the best processor using an operating frequency 33 times lower. This fact strongly supports its implementation in hand-held instruments.

Keywords: reconfigurable computing; FPGA; power quality; spectral kurtosis; digital signal processing; embedded system

---

1. Introduction

During the last fifteen years, we have been progressing to a large deployment of power electronics with the goal of conducting non-linear loads in a frame of distributed energy resources at generation, transmission, and distribution stages. This fact, along with utilities deregulation, elicits the need for a new concept of power quality (PQ) that should affect standards and measurement equipment aligned with an effective operation system [1,2].

Indeed, one of the new challenges of PQ analyzers resides in the fact of registering new PQ events that have been arising in the smart grid (SG) and dramatically affect sensitive electronics. As a consequence, there is a special need for PQ monitoring analysis techniques that account for new disturbance assessment techniques. These not only detect the presence of anomalous power, but also allow adopting protection measures, which eventually enable the correct on/off load or mitigation strategies. In particular, the focus of new class C instruments is on the quasi steady condition, more specifically on waveform distortion (e.g., harmonic, inter-harmonic) [2–4].

Depending on the situation being considered (e.g., type of disturbances, measurement campaign), different equipment can be used. In the frame of the current electrical network, harmonics are gaining attention, mainly due to their proliferation as a consequence of the non-linear loads in distributed energy resources (DERs). However, traditional harmonic and spectrum analyzers are mainly based on fast Fourier transform (FFT) (e.g., power spectrum) and may not register low-level (but persistent) harmonics or any other electrical disturbance. Certainly, a conventional disturbance monitor is useful for initial tests at the point under study, but it is difficult to extract the characteristics of a disturbance from the information available. Therefore, it is recommended to add new capabilities for a more detailed or specific analysis of a power quality issue [4,5].

The selection of the most appropriate implementation for a given digital signal processing (DSP) algorithm intending to build autonomous PQ measurement equipment is always an important decision, in order to meet the design objectives.

In general terms, a processor-based implementation gives the greatest design flexibility with the lowest design effort, but provides the lowest hardware exploitation and energy efficiency. On the other hand, a specifically manufactured device obtains the best hardware exploitation and energy efficiency, but requires a very high design effort with the lowest design flexibility.

Reconfigurable devices, and particularly field programmable gate arrays (FPGAs), have been considered since their introduction as a trade-off solution for the implementation of DSP algorithms. Consequently, many of the improvements in the technological capabilities and design tools have been targeted for this task, extending the use of FPGAs to application fields previously reserved to processors because of their required flexibility, or to specific hardware because of their high demanding performance.

These improvements have meant that typical DSP functions, like digital filtering or FFT, have an easy and straightforward implementation using FPGAs, with the provision of many adjustable hardware cores that can be used in a similar way to software functions in a processor program. Thus, the implementation of a complex DSP function in a FPGA can be considered as an affordable design task in many situations, enabling its inclusion in a real-time digital processing system as small and autonomous measurement equipment.

The spectral kurtosis (SK) is an example of a complex DSP function, of interest to the PQ field and whose implementation in a FPGA can be advantageous over a processor-based alternative. This function is a fourth order statistic estimator that can be considered as the frequency domain version of the statistical variance named kurtosis. It is a useful tool to indicate the presence of transients in a signal and their locations in the frequency domain. It also provides additional information with respect to second order quantities given by the power spectrum density. This information can be used to discriminate between constant amplitude harmonics, time-varying amplitude harmonics, and noise (see Appendix A for more detailed information).

The SK has been used in several applications: (1) for detecting and removing radio frequency interference in radio astronomy data [6], (2) for bearing fault detection in asynchronous machines [7], and (3) for termite detection [8], and PQ analysis can be improved by the use of this function in the diagnosis of presence of underlying electrical perturbations [9].

The SK is calculated in all these contributions by means of a processor-based system using a specialized high-level language such as MATLAB^{TM}, Octave, SciPy Python's library, etc., and involves the use of a general-purpose desktop computer. A laptop computer may be required depending on the location of the system under test.

This implementation is the most accurate when the objective is to demonstrate the capability of the algorithm to perform the desired analysis or detection task. However, when real-time operation capability, energy efficiency, equipment size, and many other practical characteristics of autonomous

PQ measurement equipment are taken into account, an alternative implementation using FPGAs must be considered. The viability of this solution for the implementation of the SK has been shown in Reference [10], but it makes no assessment of different implementation alternatives.

Comparative studies between FPGA and processor-based implementation of different algorithms can be found in many previous works. Specifically, it is quite common to find comparative implementations using FPGAs, common desktop processors, and graphics processing units (GPUs) [11–16]. These studies are oriented to show the advantages of using the so-called General Purpose Computing on GPUs' paradigm. This paradigm was first introduced by NVIDIA Corporation in 2007 with its Compute Unified Device Architecture, an application programming interface that allows the use of a GPU for general purpose computing, exploiting the highly parallelized GPU hardware resources to implement non-graphic applications. These studies show that this is a very attractive alternative (probably the best one) if we are considering the use of a conventional computer equipped with a high-end graphics card. Although it could be a suitable implementation for easily accessible systems with unlimited power, it is not an option for autonomous measuring equipment, where power consumption and size are key aspects.

In this paper we present a performance evaluation of different implementations of the SK algorithm. Processor-based and FPGA-based implementations are compared, focusing on the ultimate goal of obtaining autonomous PQ measuring equipment.

2. Materials and Methods

In order to establish a consistent comparison between the different implementations of the SK, explained in details in Appendix A, the same basic flowchart is followed to evaluate Equation (A2). This flowchart is represented in Figure 1.

The comparison only takes into account the evaluation of the SK, and it is assumed that data are available at the rate needed by the computational core (processor or FPGA). The relationship between data sampling frequency and computation will be analyzed in section Results.

The values of M, N, and input data width have been selected in order to obtain results with statistical validity, sufficient frequency, and amplitude resolution. Thus, values of $M = 1000$, $N = 2048$, and 16 bits of data width have been chosen.

The FFT function has been represented in the flowchart as a single step because it is available as a software function or as a hardware core.

The flowchart also includes two nested loops to evaluate the summation functions of the Equation (A2). The outer one was repeated M times (index i), one for each data frame, while the inner one is repeated $N/2$ times (index m), one for each frequency step (although data frames are N points length, the negative and symmetric spectrum of FFT has not be considered, as usual). From the summation of the functions evaluated for each frequency step, a final loop was executed $N/2$ times to evaluate the SK.

Since input data are acquired by means of an analog to digital converter (ADC), an integer format was applied to the input data. However, the mathematical treatment greatly increased the range of the intermediate values. Consequently, a conversion to floating point was performed, but it was not explicitly shown in the flowchart. Output data were obtained also in floating point format.

All the operations in the flowchart are represented sequentially, without taking into account that some of them can be executed concurrently.

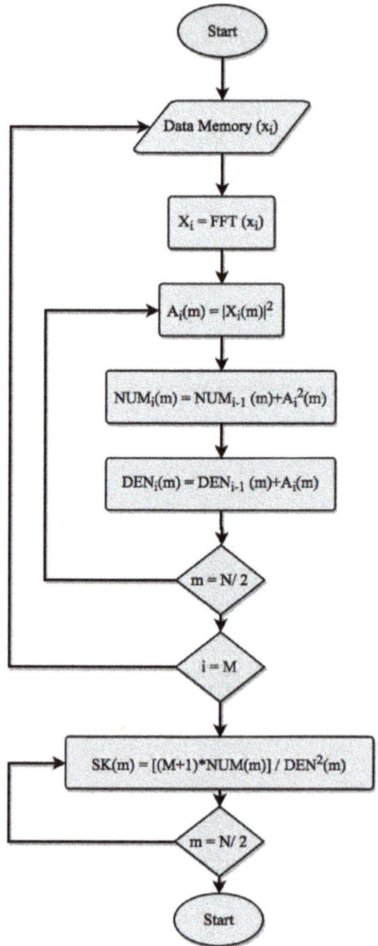

Figure 1. Spectral kurtosis algorithm flowchart.

2.1. FPGA Implementation

The FPGA version of the SK operator is built by means of the hardware description language VHDL. Only the synthesis subset of VHDL'93 is used to design the circuit [17]. The hardware description was synthesized using the Xilinx Synthesis Tool, XST, which was included in Integrated Synthesis Environment, ISE 14.2 software suite provided by Xilinx^{TM}, which is the manufacturer of the chosen devices.

As it is intended to make a comparison from a practical point of view, some Intellectual Property (IP) cores from Xilinx were used in order to facilitate and speed up the design. This fact not only decreases the FPGA design cycle, but also ensures accurate results, since IP cores were verified and optimized.

The computation of the FFT was accomplished with one of these IP cores, and it was performed using integer format to avoid losing resolution [18]. The output of this core was converted to floating-point format (as has been previously indicated) and mathematically processed. Both format conversion and mathematical process were implemented using floating point arithmetic IP cores [19]. All IP cores were parameterized, requiring a previous setup process to adapt them to the SK algorithm design, allowing the specification of different design strategies and objectives. This feature is one of the main advantages of a FPGA implementation.

Although the processing was designed and implemented using Xilinx software tools and devices, the design was described in VHDL, and the logical resources and IP cores used in the implementation are commonly available Therefore, it can be adapted to other similar devices from others companies.

The circuit is divided into two main modules:

- A datapath that computes the SK and is built from IP cores.
- A finite state machine that controls the whole circuit operation.

Making use of the previously mentioned adaptability of the IP cores, two versions of the SK FPGA implementation were designed and tested in order to achieve the best results in either performance or area. Additionally, both designs were implemented using two different FPGA series from Xilinx: Spartan6 and Artix-7, which are the lowest end series of the manufacturer. Particularly, the devices used are XC6LS16 and XC7A100T, housed in Digilent's Nexys3 and Nexys4 evaluation boards, respectively.

To obtain the best results in terms of performance, the IP cores have to be setup for minimal latency, implying the use of a pipelined structure for the FFT [18] that enables the use of a streaming input/output, i.e., a new set of data which can be read at the input while the previous one is still being processed.

Focusing on the use of resources, to obtain the best results in terms of using them, the IP cores have to be setup for maximal latency, implying the use of a conventional radix-2 structure for the FFT that needs the use of a burst input/output, i.e., it is necessary to completely process a set of data before reading a new one.

For IP cores other than FFT, the latency can also be adjusted, but a pipeline structure is always used [19].

By combining the two-implementation setup with the two target devices, we obtained four different experiments, Table 1.

Table 1. Field programmable gate arrays (FPGAs) experiments.

Experiment	FPGA Device	FFT-Core Architecture/Latency
FPGA 1 FPGA 2	XC7A100T	Radix-2 (Burst I/O)/Maximal Pipelined (Streaming I/O)/Minimal
FPGA 3 FPGA 4	XC6LX16	Radix-2 (Burst I/O)/Maximal Pipelined (Streaming I/O)/Minimal

In order to test the implementations, some additional circuitry was added to the SK algorithm circuit, as shown in Figure 2.

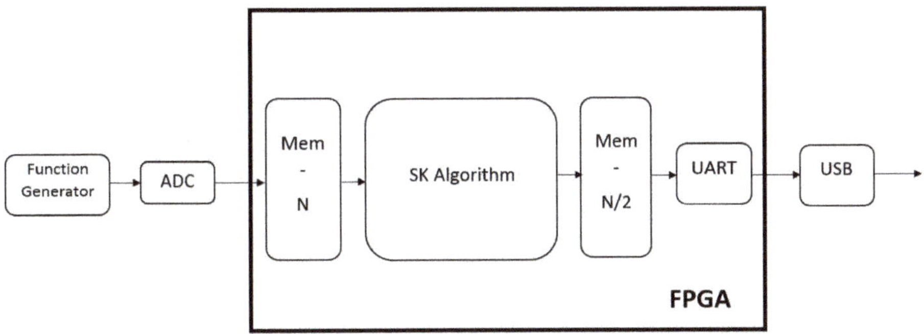

Figure 2. Experimental setup for FPGA implementations.

The FPGA takes its input from an external ADC that is connected to a function generator during the test process. The output of the SK evaluator is sent using an UART through an USB port in order to be analyzed and presented in a computer. The USB circuitry is part of the evaluation board.

To synchronize the ADC and the SK evaluator, an internal dual port memory stores the samples until a set is complete (N samples have been acquired). Then the SK evaluator computes the sample set without interrupting the sampling process.

From the M (1000) sets of N (2048) samples of input signal, the SK evaluator generates $N/2$ (1024) values corresponding to the SK value for each frequency step. These values are stored in an internal memory to synchronize the SK evaluator with the UART operation.

The SK evaluator receives the data in integer format and produces its results in floating point format, specifically in single-precision IEEE-754 format.

2.2. Processor-Based Implementation

The processor-based implementation is written in the Python programming language for the following reasons: (1) its scientific and numeric capabilities, which assure that the mathematical functions needed to build Equation (A2), such as FFT or floating-point arithmetic are available; (2) its portability, which guarantees that the code can be executed on platforms built around processors with different architectures; and (3) accessibility, which makes the results easily reproducible.

As the comparison is oriented to the implementation of autonomous measurement equipment, the processors used to execute and test the algorithm were selected from those that are typically used to build embedded systems. Specifically, an ARM Cortex-A7, an Intel Atom N270, and an Intel i7-3517U were selected. The first one is representative of the type of processors that can be found in systems like a Raspberry Pi, while the devices from Intel are in the low and high end of the processors used in industrial embedded computers from companies like Wordsworth. Table 2 contains a summary of the characteristics of the processor-based implementations.

Table 2. Processor-based experiments.

Experiment	Processor	Operating Frequency	RAM
Processor 1	Intel Atom N270	1.6 GHz	1 GB
Processor 2	ARM Cortex-A7	900 MHz	1 GB
Processor 3	Intel i7-3517U	1.9 GHz	1 GB

The experimental setup for the processor-based SK implementations was simpler than the corresponding setup for the FPGA implementations. This is because we were only interested in the evaluation of the SK algorithm. Therefore, the input test signal could be synthesized by the software instead of being acquired, and consequently, the ADC was not needed. Additionally, the results can also be represented using the same software that performs the SK algorithm.

2.3. Preliminary Test

Before any performance evaluation, all the implementations were tested to verify that their behavior was correct. Several types of signals were tested and the resulting SK in each case was the expected.

As a sample of this preliminary test, Figures 3 and 4 respectively represent the SK obtained from experiments FPGA 1 and Processor 1. In both cases the input signal was a 100 Hz sine wave (the amplitude was not significant because the SK shows the variability of the harmonic content).

In both figures, the expected single spectral line (SK = 1) corresponding to the harmonic content at 100 Hz, can be seen. Also, the absence of harmonic content (SK = 2) for the rest of frequencies was visible.

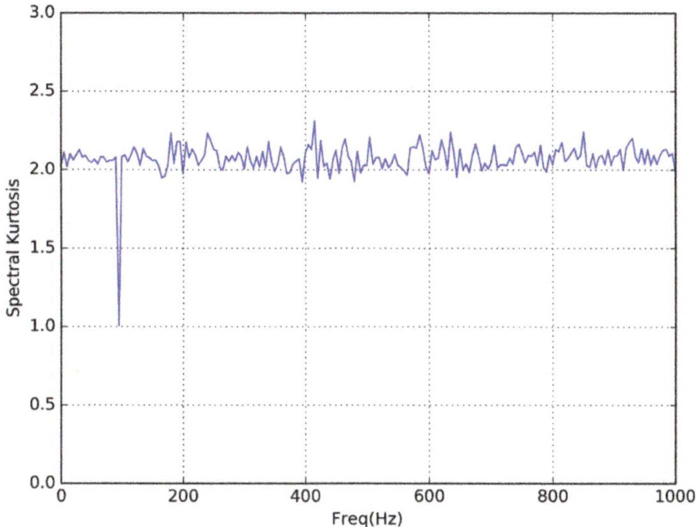

Figure 3. Spectral kurtosis (SK) from experiment FPGA 1. Sine wave signal, 100 Hz.

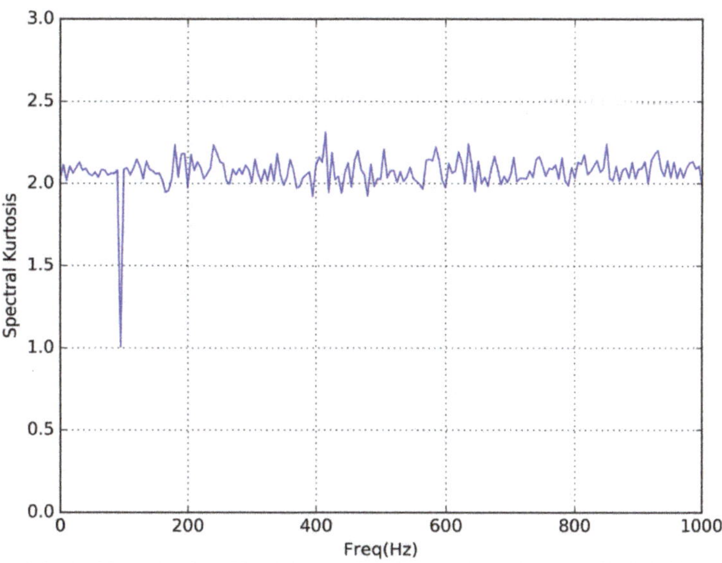

Figure 4. SK from experiment Processor 1. Sine wave signal, 100 Hz.

As can be seen in the figures, the results obtained from the two types of implementation are exactly the same.

2.4. Performance Test

The main objective of this test was to compare the different implementations in terms of processing time. To evaluate this characteristic, input data was fed to the SK processing unit at the maximum possible rate. For the FPGA-based circuits, this condition was accomplished by passing the ADC and directly feeding the digital input. For the processor-based implementations, datasets were synthesized by software and stored in memory before the SK code was launched. This simulates the practical situation where data was available to the SK processor on demand, enabling the evaluation of the SK processing time.

One of the main differences that can be anticipated between FPGA and processor-based implementations is that processing time is constant for FPGA circuits but not for processor-based implementations. This is because the operating system is running onto the processor to arrange and distribute the processor time among different processes. Therefore, since the total processing time for the SK evaluator is essentially due to the repetition of the main loop in Figure 1 M times, the best way to compare such different implementations is to divide the total time employed to complete the whole algorithm by M, obtaining the averaged time needed to perform a single iteration.

For FPGA implementations, the measurement of this time was performed through the handshake signals of the different IP cores. These signals allow a direct monitoring of the computational process and are easily extracted from the FPGA.

For processor-based implementations, time measurement instructions have been included in the code to measure the elapsed time dedicated to solve the SK equation.

3. Results

3.1. Implementation Results for FPGAs

Although the main objective of the study is to compare the processing time, it is also of interest to show the cost in terms of used resources of the different FPGA implementations.

In general terms, to optimize timing results it is necessary to parallelize the implementation and, consequently, use a greater amount of hardware resources. The results represented in Table 3 confirm this general assertion and quantifies it, showing that the time optimal implementation uses around a 60% more hardware resources than the area optimal circuit. This implies that the fastest version cannot be implemented in the smaller device.

Table 3. FPGA implementation results.

FPGA Resource	XC7A100T				XC6LX16			
	Time Optimal		Area Optimal		Time Optimal		Area Optimal	
	Used	Percentage	Used	Percentage	Used	Percentage	Used	Percentage
Slices	2937	18.5%	1763	11.1%			1696	74%
Block RAM	13	9.6%	9	6.7%	It does not fit		17	53%
DSP Slices	34	14.2%	21	8.7%			31	97%
Frequency (MHz)	225		225		N/A		163	

The results of Table 3 are divided into three different categories: Slices (Xilinx's denomination for general purpose logic unit), block RAM, and DSP blocks. This is because the exhaustion of any of the three categories implies the impossibility of implementing the circuit, even if there are free resources in the other categories.

Table 3 also includes the maximum operating frequency for each implementation, a feature that will be very important to determine the time behavior of the circuits.

3.2. Preliminary Time Evaluation for FPGA-Based Implementations

One of the advantages of FPGA-based implementations is that their time behavior is mostly predictable. Their latency and maximum operating frequency can be evaluated from their structure, and this fact is exploited by the design tools that can offer very accurate time information.

While in this paper all the experimental results were measured, time predictions from design tools were used to elaborate extrapolations of the circuit behavior for different operating frequencies and implementation conditions. This is the reason this analysis was included.

Taking into account the flowchart of Figure 1, two main parts can be distinguished in the evaluation of the SK. The first one performs the FFT and obtains the summations of Equation (A2), while the second one computes the product and division of this equation.

The first part is clearly more complex, and implies M iterations that run $N/2$ samples each, while the second part executes one single cycle that runs $N/2$ samples. Taking into account that M has to be large to give statistical validity to the results, it has to be concluded that the total execution time was expected to be almost proportional to M and completely dependent on the execution time of a single iteration of the first part of the flowchart.

For an FPGA-based implementation the FFT operation can be divided into three stages: data reading, processing, and result extraction. The first and the last stages use N clock cycles each, while the processing stage has a variable duration depending on the implementation parameters.

In any case, the evaluation of the squared terms and summations can be executed concurrently with FFT because this can be done as soon as data become available at the FFT module output, when result extraction stage begins. Therefore, the time needed to perform this evaluation was not added to the time required by the FFT, and only the longest one was used to determine the total execution time of one iteration.

The evaluation of the square terms and summations, as well as product and division, were always implemented by means of pipeline technique. Consequently, for each frame of data they used $N/2$ clock cycles.

For a pipelined structure the stages previously described for the FFT can be executed concurrently. Thus, at the same time a new set of data was being read, the results of the previous set were extracted, and different sets of data were processed simultaneously. The number of clock cycles to perform one FFT was therefore equal to N.

Nevertheless, for a radix-2 structure, the FFT stages were not concurrent and the number of clock cycles needed to complete the operation was greater than N. We have established experimentally a relationship between N and the number of clock cycles needed to complete the FFT. This relationship was obtained for $512 \leq N \leq 4096$.

Table 4 shows the number of clock cycles needed to perform the different operations used in the SK evaluation for the different implementations as a function of N.

Table 4. Number of clock cycles needed to perform each operation.

Operation	Any Device	XC7A100T	XC6LX16
	Time Optimal	Area Optimal	Area Optimal
FFT	N	$8.27 \times N - 997$	$8.27 \times N - 981$
Squared and summations	$N/2$	$N/2$	$N/2$
Product and division	$N/2$	$N/2$	$N/2$

The results in Table 4 are valid for the intermediate iterations of the main loop in Figure 1, which was repeated M times. For the first and the last iteration the number of clock cycles can be greater because of the latency of the different blocks.

The latency for the evaluation of squared terms and summations had no effect because it was small (54 cycles max) compared to $N/2$, which was the number of cycles of the result extraction of the FFT that was not used (corresponding to the negative frequency spectrum).

Latency for the product and the division had little effect because this operation was performed only once at the end of the overall process.

Finally, the latency for the FFT was significant for the time optimal implementation only, because for the radix-2 implementation there was not concurrency. It was experimentally evaluated that the latency for the time optimal implementation of FFT (L_{FFT}) can be obtained from Equation (1).

$$L_{FFT} = 2 \times N + 113 \tag{1}$$

Taking into account all of this information, the total processing time (T_{SK}) expressed in number of clock cycles can be evaluated using Equation (2) for an area optimal implementation, and using Equation (3) for a time optimal implementation, where $L_{P\&D}$ is the latency of the product and division operation, which is 40 cycles max.

$$T_{SK} = M \times (8.27 \times N - 997) + N/2 + L_{P\&D} \tag{2}$$

$$T_{SK} = M \times N + L_{FFT} + \frac{N}{2} + L_{P\&D} = M \times N + (2 \times N + 113) + N/2 + L_{P\&D} \tag{3}$$

For the usual values of M and N, these equations can be approximated by Equations (4) and (5), respectively.

$$T_{SK} \approx M \times (8.27 \times N - 997) \tag{4}$$

$$T_{SK} \approx (M + 2.5) \times N \tag{5}$$

3.3. Time Results

As it has been previously indicated, time results are presented in terms of the average time per iteration, which can be obtained dividing the total SK calculation time by M. Assuming that the input data buffer is big enough, this quantity provides a good representation of the data processing capacity of the different implementations. Figures 5–7 show clearly the convenience of using this average time to compare different implementations. They are graphical representations of the time taken by each iteration of the main loop of the flowchart in Figure 1. The horizontal dotted line represents the average time per iteration for each experiment.

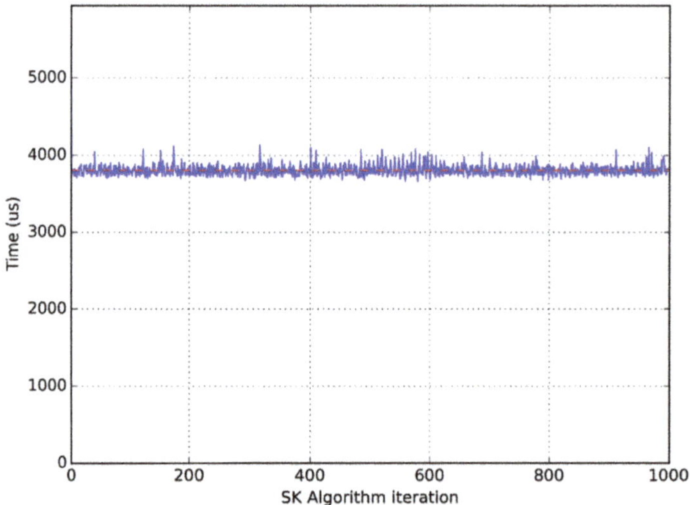

Figure 5. Elapsed time to calculate each iteration. ARM^{TM} Cortex A7.

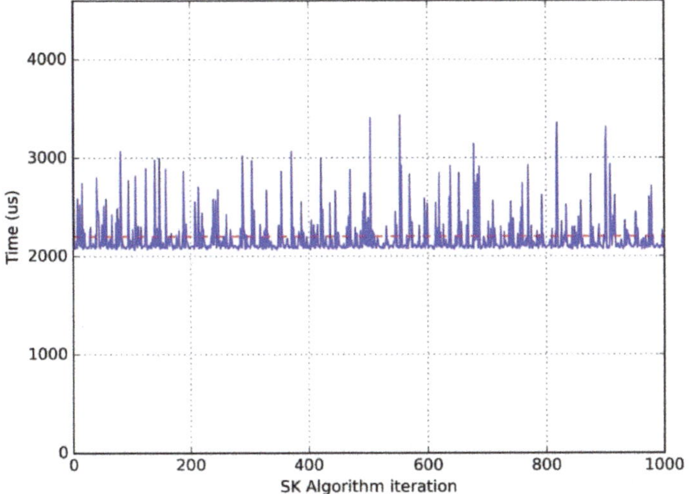

Figure 6. Elapsed time to calculate each iteration. Atom N270.

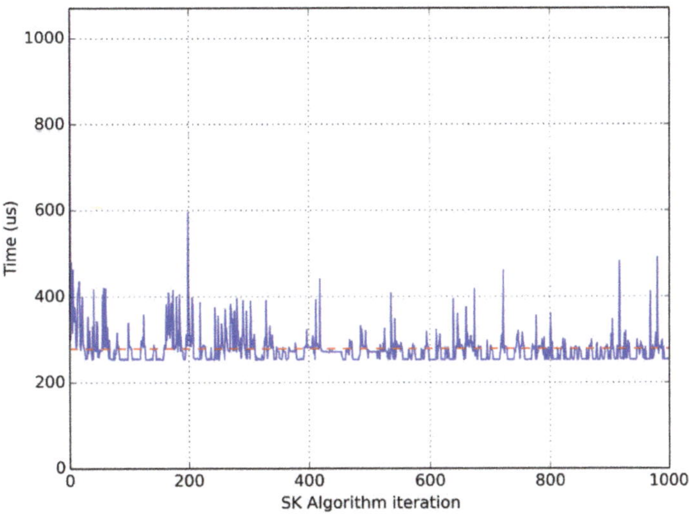

Figure 7. Elapsed time to calculate each iteration. i7-3517U.

It is clear that the time used to perform the same computation in the same processor greatly varies from one iteration to another, which advises the use of an average function to compare different implementations.

Also taking into account that the evaluated implementations are targeted to obtain autonomous PQ measurement equipment, it has to be assumed that input data to be processed are continuously acquired with a fixed sampling frequency.

The minimum value for this frequency is fixed by the characteristics of the signal to be acquired and it will be specific for each application. Consequently, it is not possible in a general study to select a particular value for this parameter, but it can be determined for each implementation which is the maximum value that can be reached. This value being a figure of merit that enables the comparison between different implementations.

The evaluation of the maximum sampling frequency that is compatible with a continuous data processing for each implementation, can be calculated by dividing the previously defined average time per iteration ($T_{cy,avg}$) by the number of samples that has to be acquired in this time (N).

Table 5 summarizes the operating frequency (f_{Op}) regarding to all the implementations, the average time per iteration ($T_{It,avg}$), which is graphically represented in Figure 8, and the maximum data sampling frequency ($f_{S,max}$) that can be reached assuming a continuous data processing. In the last column of the table, a new parameter was included: the frequency exploitation factor (FEP), defined by de expression FEP $= f_{S,max}/f_{Op}$.

Table 5. Time results summary.

Implementation	f_{Op} (MHz)	$T_{It,avg}$ (µs)	$f_{S,max}$ (MHz)	FEP
XC7A100T Area optimal	200	79.71	25.6	0.128
XC7A100T Time optimal	200	10.27	199	0.995
XC6LX16 Area optimal	150	106.4	19.2	0.128
Atom N270	1600	2229	0.918	5.74×10^{-4}
ARM Cortex-A7	900	3800	0.538	5.99×10^{-4}

FEP = frequency exploitation factor.

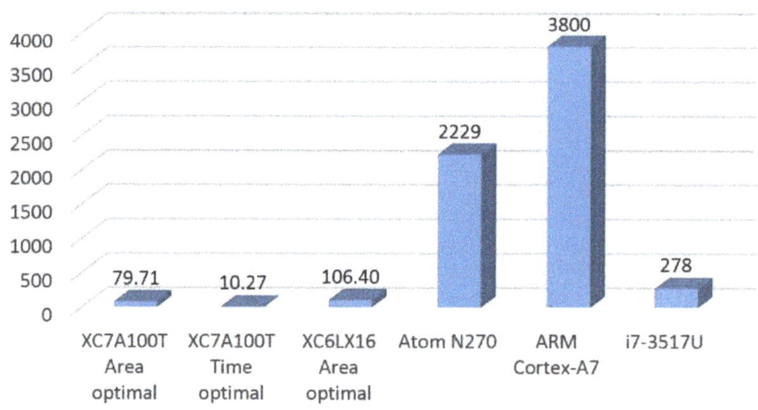

Figure 8. Average time per iteration.

For instance, this factor shows that the worst FPGA implementation can reach the same processing capability as the best processor implementation with an operating frequency 33 times lower. It also shows that different FPGAs with the same implementation options are equivalent if they operate at the same frequency, revealing very similar hardware architectures.

4. Discussion

The main conclusion of the work is that an FPGA can be between 34 and 1740 times more frequency efficient than a processor for a typical SK function implementation. This implies that a lowest-end FPGA implementing a minimal resource use hardware structure (XC6LX16 area optimal) and with an operating frequency as relatively low as 60 MHz can reach a higher sampling frequency for a continuous processing than a high-end processor (i7-3517U) running at 1.9 GHz.

Performance results are not only better for FPGAs but also much more predictable and stable. This reduces the risk that the initial choice of an implementation does not fulfill the specifications when materialized.

Additionally, FPGA-based implementations can be tailored to different requirements, enabling the optimization of the solution to achieve diverse objectives. In this regard, it was shown that a time optimal implementation can reach a sampling frequency about eight times greater at the expense of an increase of 60% in the hardware resources used.

Despite this overwhelming result, processor-based implementations are still a reasonable option for low sampling frequencies, in the range of several hundred KHz for low-end processors, and up to several MHz for high-end processors, because of their ease of use and availability.

The experimental outcomes show that the FPGA technology processes the spectral kurtosis in a shorter time than general-purpose processors, even working at a lower frequency. This feature is accomplished even when low-cost models are used.

Since FPGAs and processors are made using CMOS technology, it can be assumed that most of the power consumption is dynamic and consequently, proportional to operation frequency.

The fact that two of the platforms used in this paper are laptop-based implies that we cannot include a quantitative evaluation of the power consumption, since a laptop incorporates many loads unrelated to processing. However, considering the proportional relationship between power consumption and working frequency, it can be concluded that FPGAs provide the best results in terms of power consumption. This is an additional advantage of using FPGA technology to develop autonomous PQ measurement equipment.

Author Contributions: Conceptualization, M.-Á.C.-C.; methodology, M.-Á.C.-C., F.P.-P. and Á.Q.-O.; software, M.-Á.C.-C., F.P.-P. and Á.Q.-O.; validation, M.-Á.C.-C., F.P.-P. and Á.Q.-O.; formal analysis, Á.Q.-O. and J.-J.G.-d.-l.-R.; data curation M.-Á.C.-C., F.P.-P. and Á.Q.-O.; writing—original draft preparation, M.-Á.C.-C.; writing—review and editing, F.P.-P. and Á.Q.-O.; visualization, M.-Á.C.-C. and F.P.-P.; funding acquisition, J.-J.G.-d.-l.-R.

Funding: This research was funded by the University of Cádiz and by the Spanish Ministry of Economy and Competitiveness through the project TEC2016-77632-C3-3-R.

Acknowledgments: This research is supported by the Spanish Ministry of Economy, Industry and Competitiveness and the EU (AEI/FEDER/UE) in the workflow of the State Plan of Excellency and Challenges for Research, via the project TEC2016-77632-C3-3-R-Control and Management of Isolable NanoGrids: Smart Instruments for Solar Forecasting and Energy Monitoring (COMING-SISEM), which involves the development of new measurement techniques applied to monitoring the PQ in micro-grids.

Conflicts of Interest: The authors declare no conflict of interest. The funders had no role in the design of the study; in the collection, analyses, or interpretation of data; in the writing of the manuscript, or in the decision to publish the results.

Appendix A

The SK measures the variability of each frequency component of a signal that is assumed to be periodic [20]. To perform this measurement, the signal is sampled for a period of time, the acquisition window and its spectrum are obtained from that data frame using FFT. With the spectra obtained repeating several times this process (acquisition plus FFT), a statistical kurtosis operator is computed for each frequency component to evaluate if (and how) the spectrum of the signal has changed through the time comprised by the successive acquisition windows.

In statistics, the kurtosis operator quantifies the shape of a data distribution, assigning a particular and characteristic value to a Gaussian distribution, a lower value to flatter data distributions and greater values to more peaked data distributions. The value assigned to the Gaussian distribution, which can be considered as a reference, changes from one approach to another, but this is not significant in order to discriminate the different distributions.

Consequently, in the frequency domain, the SK is a representation of the statistical kurtosis associated to each spectral component of a signal through several acquisition windows. If for a given frequency there are not harmonic content, the FFT values for that frequency will match a Gaussian distribution basically due to noise, and consequently the SK will take the reference value for that frequency.

For a frequency with a constant harmonic content the distribution is ideally uniform and its SK will be lower than the reference value, specifically the minimum value of the SK. This is very useful to detect very low-level harmonics in a noisy environment [10], as usual in PQ measurements.

Finally, a frequency with a non-permanent harmonic content have a more peaked distribution and its SK will be higher than the reference value. In this way a small transient can be detected and its frequency located.

To formulate the SK, it is assumed that M data frames of signal x are sampled as input, each frame containing N points. For data frame number i, the FFT is computed and its module is represented by $|X_i|$. With these definitions the SK can be evaluated using Equation (A1).

$$SK(m) = \frac{M}{M-1}\left[\frac{(M+1)\sum_{i=1}^{M}|X_i(m)|^4}{\left(\sum_{i=1}^{M}|X_i(m)|^2\right)^2} - 2\right] \tag{A1}$$

where $SK(m)$ represents the SK value for frequency index m and $|X_i(m)|$ represents the module of the FFT for that frequency index and for the data frame number i.

It has to be considered that the SK is a statistic indicator, thus its value is significant for large values of M, the number of data frames used to evaluate the estimator.

With this definition and consideration, a frequency without harmonic content will be assigned a SK value of 0 (the previously considered as a reference value), a frequency with a permanent harmonic content will be assigned a minimum value of -1, and a frequency with a variable amplitude harmonic content will be assigned a value larger than 0.

Although this definition of SK is the most convenient from a mathematical point of view because it is normalized and referenced to 0, it is possible to simplify Equation (A1) without losing any of its significance and usefulness. This simplification is shown in Equation (A2).

$$SK(m) = \frac{(M+1)\sum_{i=1}^{M}|X_i(m)|^4}{\left(\sum_{i=1}^{M}|X_i(m)|^2\right)^2} \tag{A2}$$

With this definition, a frequency without harmonic content will be assigned a SK value of 2 (which is the new reference value), a frequency with a permanent harmonic content will be assigned a minimum value of $(M+1)/M$, that approximates to 1 for large values of M, and a frequency with a variable amplitude harmonic content will be assigned a value larger than 2.

From a practical point of view, the task of distinguishing the different situations is equally difficult, while the evaluation is slightly simpler. Consequently, the SK will be evaluated using Equation (A2).

References

1. Arrillaga, J.; Bollen, M.H.J.; Watson, N.R. Power quality following deregulation. *Proc. IEEE* **2000**, *88*, 246–261. [CrossRef]
2. Arrillaga, J.; Watson, N.R. Power Quality. In *Power System Restructuring and Deregulation*; Lai, L.L., Ed.; John Wiley & Sons, Ltd.: Hoboken, NJ, USA, 2001; pp. 330–352.
3. Florencias-Oliveros, O.; Agüera-Pérez, A.; González-de-la-Rosa, J.J.; Palomares-Salas, J.C.; Sierra-Fernández, J.M.; Jiménez-Montero, A. Cluster analysis for Power Quality monitoring. In Proceedings of the 11th IEEE International Conference on Compatibility, Power Electronics and Power Engineering (CPE-POWERENG), Cadiz, Spain, 4–6 April 2017; pp. 626–631.
4. González-de-la-Rosa, J.J.; Sierra-Fernández, J.M.; Palomares-Salas, J.C.; Agüera-Pérez, A.; Jiménez-Montero, A. An Application of the Spectral Kurtosis to Separate Hybrid Power Quality Events. *Energies* **2015**, *8*, 9777–9793. [CrossRef]
5. Bingham, R.P. Measurement instruments for power quality monitoring. In Proceedings of the 2008 IEEE/PES Transmission and Distribution Conference and Exposition, Chicago, IL, USA, 21–24 April 2008; pp. 1–3.

6.   Abidin, Z.Z.; Ibrahim, Z.A.; Rosli, Z.; Hassan, S.R. Methods and applications of Radio Frequency Interference surveys for radio astronomy in Malaysia. In Proceedings of the 2011 IEEE International Conference on Space Science and Communication (IconSpace), Penang, Malaysia, 12–13 July 2011; pp. 178–181.

7.   Sawalhi, N.; Randall, R.B. The application of spectral kurtosis to bearing diagnostics. In Proceedings of the 2004 Annual Conference of the Australian Acoustical Society, Gold Coast, Australia, 3–5 November 2004; Volume 1, pp. 393–398.

8.   de la Rosa, J.J.G.; Muñoz, A.M. Higher-order cumulants and spectral kurtosis for early detection of subterranean termites. *Mech. Syst. Signal Process.* **2008**, *22*, 279–294. [CrossRef]

9.   de la Rosa, J.J.G.; Sierra-Fernández, J.M.; Agüera-Pérez, A.; Salas, J.C.P.; Moreno-Muñoz, A. An application of the spectral kurtosis to characterize power quality events. *J. Electr. Power Energy Syst.* **2013**, *49*, 386–398. [CrossRef]

10. Quirós-Olozábal, A.; González-de-la-Rosa, J.J.; Cifredo-Chacón, M.A.; Sierra-Fernández, J.M. A novel FPGA-based system for real-time calculation of the Spectral Kurtosis: A prospective application to harmonic detection. *Measurement* **2016**, *86*, 101–113. [CrossRef]

11. Tian, X.; Benkrid, K. Mersenne Twister Random Number Generation on FPGA, CPU and GPU. In Proceedings of the NASA/ESA Conference on Adaptive Hardware and Systems, San Francisco, CA, USA, 29 July–1 August 2009; pp. 460–464.

12. Kalarot, R.; Morris, J. Comparison of FPGA and GPU implementations of Real-time Stereo Vision. In Proceedings of the IEEE Computer Society Conference on Computer Vision and Pattern Recognition, San Francisco, CA, USA, 13–18 June 2010; pp. 9–15.

13. Zou, D.; Dou, Y.; Xia, F. Optimization schemes and performance evaluation of Smith–Waterman algorithm on CPU, GPU and FPGA. In Proceedings of the International Conference on Field-Programmable Technology, New Delhi, India, 12–14 December 2011; pp. 1–6.

14. Duan, B.; Wang, W.; Li, X.; Zhang, C.; Zhang, P.; Sun, N. Floating-point Mixed-radix FFT Core Generation for FPGA and Comparison with GPU and CPU. *Concurr. Comput. Pract. Exp.* **2012**, *24*, 1625–1644.

15. Drieseberg, J.; Siemers, C. C to Cellular Automata and Execution on CPU, GPU and FPGA. In Proceedings of the International Conference on High Performance Computing and Simulation, Madrid, Spain, 2–6 July 2012; pp. 216–222.

16. Russo, L.M.; Pedrino, E.C.; Kato, E.; Roda, V.O. Image Convolution Processing: A GPU versus FPGA Comparison. In Proceedings of the VIII Southern Conference on Programmable Logic, Bento Gonçalves, Brazil, 20–23 March 2012; pp. 1–6.

17. *1076.6-2004—IEEE Standard for VHDL Register Transfer Level (RTL) Synthesis*; IEEE Standard Association: Piscataway, NJ, USA, 2004.

18. Xilinx, Inc. *LogiCORE IP Fast Fourier Transform v7.1*; Xilinx, Inc.: San Jose, CA, USA, 2011.

19. Xilinx, Inc. *LogiCORE IP Floating-Point Operator v5.0*; Xilinx, Inc.: San Jose, CA, USA, 2011.

20. Vrabie, V.; Granjon, P.; Serviere, C. Spectral kurtosis: From definition to application. In Proceedings of the IEEE International Workshop on Nonlinear Signal and Image Processing, Grado-Trieste, Italy, 8–11 June 2003; Volume 1, pp. 1–5.

Article

Power Quality Disturbance Monitoring and Classification Based on Improved PCA and Convolution Neural Network for Wind-Grid Distribution Systems

Yue Shen, Muhammad Abubakar, Hui Liu * and Fida Hussain

School of Electrical and Information Engineering, Jiangsu University, Zhenjiang 212013, China; shen@ujs.edu.cn (Y.S.); mabubakarqazi@gmail.com (M.A.); fida.hussain07@yahoo.com (F.H.)
* Correspondence: amity@ujs.edu.cn; Tel.: +86-138-1245-9812

Received: 10 March 2019; Accepted: 28 March 2019; Published: 3 April 2019

Abstract: The excessive use of power semiconductor devices in a grid utility increases the malfunction of the control system, produces power quality disturbances (PQDs) and reduces the electrical component life. The present work proposes a novel algorithm based on Improved Principal Component Analysis (IPCA) and 1-Dimensional Convolution Neural Network (1-D-CNN) for detection and classification of PQDs. Firstly, IPCA is used to extract the statistical features of PQDs such as Root Mean Square, Skewness, Range, Kurtosis, Crest Factor, Form Factor. IPCA is decomposed into four levels. The principal component (PC) is obtained by IPCA, and it contains a maximum amount of original data as compare to PCA. 1-D-CNN is also used to extract features such as mean, energy, standard deviation, Shannon entropy, and log-energy entropy. The statistical analysis is employed for optimal feature selection. Secondly, these improved features of the PQDs are fed to the 1-D-CNN-based classifier to gain maximum classification accuracy. The proposed IPCA-1-D-CNN is utilized for classification of 12 types of synthetic and simulated single and multiple PQDs. The simulated PQDs are generated from a modified IEEE bus system with wind energy penetration in the balanced distribution system. Finally, the proposed IPCA-1-D-CNN algorithm has been tested with noise (50 dB to 20 dB) and noiseless environment. The obtained results are compared with SVM and other existing techniques. The comparative results show that the proposed method gives significantly higher classification accuracy.

Keywords: power quality disturbance; convolution neural network; improved principal component analysis; wind-grid distribution

1. Introduction

Power quality (PQ) is becoming the primary concern as serious issues affecting sustainable energy, energy security, and the environmen tend to arise. Distributed generation (DG) based on renewable energy sources and conventional grid is a concern as it uses modern power electronics devices for control, heavy non-linear loads, microprocessor and computer solutions [1–3]. Real-time commercial PQ analyzer solutions are available. Some of the principal manufacturers of PQ analyzers are Fluke (Everett, WA, USA) Yokogawa (Tokyo, Japan), and FLIR (Wilsonville, OR, USA). PQ analyzer solutions with basic functionality are expensive yet they cannot analyze the complex and extensive data [4]. Non-stationary PQ disturbances occur due to fluctuations and loads, which change the capability of the signals. The sudden change in frequency, magnitude, current and phase angle can cause PQ disturbances. Automatic classification and detection of PQ disturbances with appropriate methods have solved this issue [5–8].

Generally, the process of identification of PQ disturbances consists of feature extraction, feature selection, and classification [9]. Voltage sag, swell, notch, spike, harmonics, flickers, and interruption are the primary types of PQ disturbances [10]. For feature extraction, many signal processing techniques are employed to detect the power system disturbances. Fourier transform (FT), Short-Time Fourier Transform (STFT), Discrete Fourier Transform (DFT), Fast Fourier Transform (FFT), Wavelet Transform (WT) are fundamental signal processing techniques for feature extraction of PQ disturbances. FT and STFT have been applied only to stationary signals, STFT is an extension of DFT which does not apply to non-stationary signals due to the fixed window size used. The limitation of STFT is explained by using a variable window size for low and high frequencies in WT. In WT, the selection of suitable mother wavelet function and the sampling rate are the key concern to carry out the frequency components. The extended version of WT is discrete WT (DWT). Furthermore, DWT is upgraded into the discrete wavelet packet transform [11–16].

The S-transform (ST) is also a useful tool to analyze PQ disturbances; it is a combination of WT and STFT, and it is an essential technique for time-frequency localization [17]. The major drawback of the S-transform is a redundant representation of the time-frequency domain [18]. Likewise, many PQ analyzer techniques are examined such as Kalman Filter (KF), Discrete Orthogonal ST (DOST), Curvelet Transform (CT), Hilbert Transform (HT), Hilbert-Huang Transform (HHT), Singular Value Decomposition (SVD), Wigner distribution function (WDF), Gabor transform(GT), Singular spectrum analysis (SSA), Empirical mode decomposition (EMD) and hybrid transform-based methods [19–28].

Principal component analysis (PCA) is a fundamental multivariate statistical approach to find a set of projection vectors that can maximize the variance of data [29]. Data reduction and interpretation is the key objective of PCA. IPCA is an improved version of PCA that converges faster and results in the less number of eigenvalues than PCA. IPCA is similar to PCA, except for the way it selects the eigenvectors from the feature matrix. These eigenvectors can be used for feature extraction, reduction, interpretation and identification patterns of data. PCA is already applied in many applications such as electrocardiogram (ECG) signals, fault extraction, face recognition and hyperspectral images [30–33]. PCA has been found a more suitable decomposition technique for automatic detection and feature extraction for PQ disturbances, particularly transient disturbances. However, the proposed feature extraction method provided more efficient results due to its precise data handling capability. IPCA is more suitable for detection and feature extraction of PQ disturbances, but IPCA alone cannot categorize the various type of PQ disturbances [34]. IPCA can only detect the disturbances in normal signals, but it can not fulfill the task of classification without using a suitable classifier [35,36].

Furthermore, feature selection and classification is an essential part of power quality classification [37]. In previous studies, pre-processed input and statistically analyzed data were directly applied as an input to the classifier. Many neural network (NN) classifiers such as an artificial neural network (ANN), probabilistic neural network (PNN), have already been investigated for the classification of PQ disturbances [38–41]. Support vector machine (SVM), fast extreme learning machine (FELM) and nearest centroid neighbor (NCN) are more dominant techniques to the traditional NN regarding classification accuracy and computational cost [42–44]. In deep learning, CNN is found the most accepted one and successfully applied to the hyperspectral image, large-scale audio face recognition and image classification [45–48]. CNN is one of the leading machine learning classifiers for images that use 2-dimensional (2-D) image data. CNN was first introduced by Fukushima [49], improved by LeCun [50], and simplified by Ciresan [51]. In this paper, the 1-D-CNN classifier has been proposed for the classification of PQ disturbances. A five layer of architecture is introduced in the current study that is simple and effective for the optimal feature extraction and classification of PQ disturbances. Several simulations validate the outstanding performance of this proposed method as compared to the SVM [52].

Many hybrid approaches are proposed for optimal feature extraction and classification of PQ disturbances. Hybrid feature extraction techniques FT, STFT, WT, HT, and Time to Time Transform (TTT) and different classification methods are proposed to get the accurate results for the different type of PQ disturbances [37]. Hybrid feature extraction and selection techniques WT, FFT, HHT and Genetic Algorithms (GA), Particle Swarm Optimization (PSO), Statistical Analysis (SA) are proposed for optimal feature extraction and selection [53].

In this study, the IPCA and 1-D-CNN-based hybrid approaches are proposed for data reduction and feature extraction of PQ disturbance signals. PCA is also utilized for the feature extraction and compared with the proposed feature extraction technique. In the case of IPCA, the PQ disturbance signal is normalized to form a covariance matrix and then it is further processed to decompose the signal. IPCA is employed to extract the statistical features such as root mean square, skewness, range, kurtosis, crest factor, form factor, and four levels of decomposition are considered in this study. 1-D-CNN is also used to extract features such as mean, energy, standard deviation, Shannon entropy, and, log-energy entropy. Finally, the optimally selected features are fed to the proposed 1-D-CNN-based classifier and SVM for the classification of PQ disturbances. This proposed method is also applied to wind penetration model based on a modified IEEE 13-bus system.

This paper is organized as follows: in Section 2, the proposed methodology based on IPCA and 1-D-CNN for feature extraction and classification is described. Section 3 explains the feature extraction based on IPCA, 1-D-CNN and statistical analysis. The proposed algorithm is covered in Section 4. Section 5 discusses the experiments. Section 6 is about results and discussion. Conclusions are mentioned in Section 7.

2. Proposed Method for Feature Extraction and Classification

2.1. Improved Principal Component Analysis

Principal Component Analysis (PCA) is a very famous technique to extract, compress, simplify and analyze data sets. PCA is widely used in many signal processing, image processing, and data acquisition applications. PCA is a linear combination of the random variables (original data) and extracts the best vector space (eigenvector) which helps to reduce the dimensions of original data, and ultimately reduces the computational time of feature extraction and classification of PQDs [29]. These eigenvectors are also called principal components (PC) of the dataset. There are two ways to present PCA, Firstly, the decomposition of the covariance matrix that is based on eigenvalues. Secondly, the composition of the singular value of the data matrix [54].

The information contained in the raw data sets is divided into two parts: (1) PCA produces an eigenvector that contains overlapping information. (2) the output variables of eigenvector (different information) are revealed through a variance, which in general is not considered, but it has a substantial impact. This vital defect needs to be improved in conventional PCA.

This considerable weakness can be improved by taking the mean of each class. Moreover, the average of each class is a linear combination of the raw data within. Smooth feature reduction process and less train time are the advantages of IPCA. The implementation of IPCA is consists of three steps. The covariance matrix, eigenvalue calculation and, the data projection. The flow steps of this process are also explained in Figure 1. The details of the algorithm are as follows.

Improved Principal Component Analysis Algorithm

(1) Formation of the Covariance matrix

Assume that Y_n is the 1-D training vector array of PQDs is defined as $Y_n = [y_1, y_2, \ldots, y_n]^T \in \Re^N$ of length N. In the case of IPCA, the typical approach is described as normalized the raw data:

$$y'_{kl} = y_{kl} / \left(\frac{1}{n} \sum_{k=1}^{n} y_{kl} \right) \quad (k = 1, 2, \ldots, n \text{ and } l = 1, 2, \ldots, m) \tag{1}$$

where "y'_{kl}" is the normalization value of "y_{kl}". "n" stands for sample size and "m" for dimension number. The advantage of this method is that there is no difference in the correlation matrix and no information is lost. The covariance matrix can be obtained as:

$$C = \frac{1}{n} \sum_{i=1}^{n} [(y_{ik} - \overline{y}_k) - (y_{il} - \overline{y}_l)] \tag{2}$$

(2) Eigenvalue Calculation

The eigenvalue calculation is to decompose the covariance matrix "C". It can be diagonalized as:

$$C = \sum_{x=1}^{N} \lambda_x A_x A_x^T = A \Lambda A^T \tag{3}$$

where "λ_x" is the eigenvalue of the matrix "C". A_x is the sub-eigenvalues of the vector, $[A_1, A_2, \ldots, A_X]$ is the orthogonal basis. "x" is the rank of the matrix and "Λ" is diagonal a matrix $(\lambda_1, \lambda_2, \ldots, \lambda_x)$ and, we have $\lambda_1 > \lambda_2 >, \ldots, \lambda_x$.

(3) Calculation of Principal Components

After normalization and set up of covariance matrix C to calculate the eigenvalues and eigenvectors and the original data vector "M_n" is transferred to the uncorrelated vector "z_n" can be formed as:

$$z_n = A^T M_n = \begin{bmatrix} a_{11} & a_{12} & \cdots & a_{1m} \\ a_{21} & a_{22} & & a_{2m} \\ \vdots & & \ddots & \vdots \\ a_{m1} & a_{m2} & \cdots & a_{mm} \end{bmatrix}_{M \times M} \tag{4}$$

Cumulative contribution rate η_n contains the information proportion of the first n principal components. Moreover, the threshold value η_n is set to 85% to collect enough information. Contribution rate η_k and cumulated contribution rate η_n can be calculated from the following equations:

$$\eta_k = \left(\frac{\lambda_k}{\sum_{i=1}^{m} \lambda_i} \right) \times 100 ; \ (k = 1, 2, \ldots, m) \tag{5}$$

$$\eta_n = \sum_{k=1}^{n} \frac{\lambda_k}{\sum_{k=1}^{m} \lambda_k} > 85\%, (i = 1, 2, \ldots, q) \tag{6}$$

After determining the exact value of "n", the principal component of "R_k" samples is obtained as:

$$R_k = \begin{bmatrix} R_{k1} \\ R_{k2} \\ \vdots \\ R_{kn} \end{bmatrix} = \begin{bmatrix} \delta_{11} & \delta_{21} & \cdots & \delta_{q1} \\ \delta_{12} & \delta_{22} & & \delta_{q2} \\ \vdots & & \ddots & \vdots \\ \delta_{1n} & \delta_{2n} & \cdots & \delta_{qn} \end{bmatrix} y_k^T \tag{7}$$

where, $(i = 1, 2, \ldots, q)$

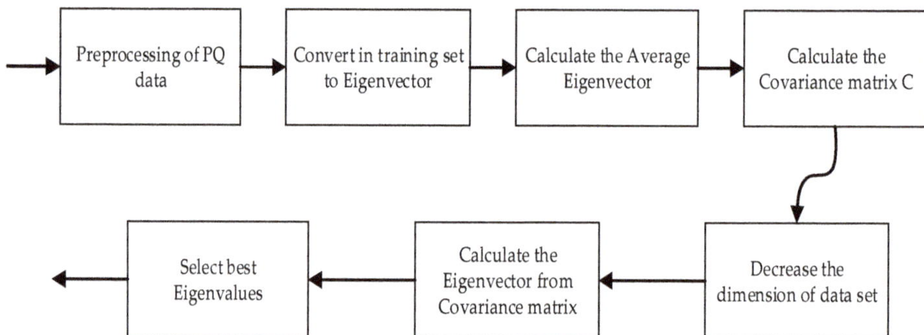

Figure 1. Flow steps for feature extraction using IPCA.

2.2. Convolutional Neural Network (CNN)

CNN is inspired from a feed-forward neural network that consists of the combination of convolution layer, maximum pooling layer and a fully connected layer of the neural network layer that provides a simple model for the mammalian visual context [55]. CNN recently outperformed some other conventional methods on many vision related tasks [56], including image classification [57], object detection [58] and face recognition [59]. CNN has been demonstrated to provide even better classification performance than the traditional SVM classifiers [52] and the conventional deep neural networks (DNNs) [57] in a visually related area. In this paper, we have found that the modified architecture of CNN can be effectively employed to classify the PQ disturbances. A typical CNN architecture is shown in Figure 2.

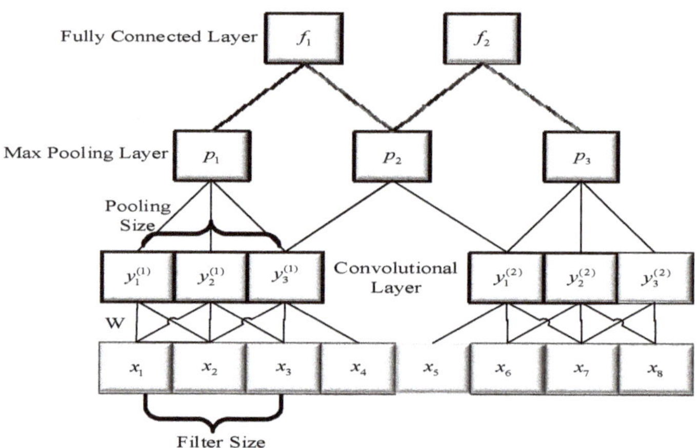

Figure 2. Architecture of the convolution neural network.

2.2.1. Architecture of 1-D-CNN

CNN primarily features an extraction and classification technique. The multilayer architecture makes it more complicated. This complex architecture can be reviewed for many applications [60]. Each layer extracts the different level of feature and has to learn a large number of features. Traditionally, the CNN architecture is implemented in two dimensions in the various applications (i.e., width and height). Many numbers of kernels and configurations are required for convolution operation. On the other hand, PQ disturbances classification works on a single dimension (1-D) domain because the dimension of the signal is in 1-D. A CNN-based classifier selected for the classification of PQ disturbances and a comparison of its results with SVM is presented in Section 6.

This network consists of the input layer, the convolution layer, network pooling layer, fully connected layer and outer layer. It is further connected with dropouts and regularization on the fully connected layer to refine the data. Backpropagation is introduced to tune the trainable parameters. The critical difference between 2-D and 1-D CNN is the utilization of a 1-D array instead of a 2-D matrix. Their 1-D convolution and counterparts replace 2-D convolution and rotation. The demonstration of a 1-D array is presented in Figure 3.

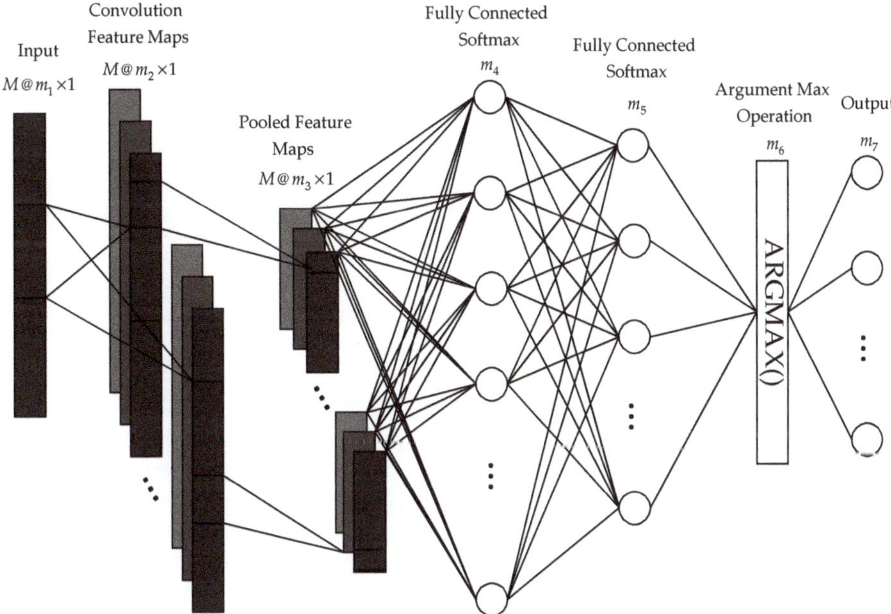

Figure 3. Detailed architecture of 1-D Convolutional neural network, where Input m_1 is 1-D feature vector followed by convolution feature map m_2, Max pooled feature map m_3 computes 20 feature maps (M) with Fully connected NN m_4 and Softmax layers m_5 and the output layer m_7.

The pooling layer can be thought of as a downsampling layer of the feature map. The pooling layer downsamples by a factor of two while keeping the features within their domain and passes the maximum value of the feature map onto the next layer. This process increased the number of trainable samples and reduced training time. The pooling feature map in this process is then transferred to the fully connected neural network, which consists of numerous hidden layers—this network regularized by 50% dropout and L2 regularization. The weights are connected to the pooled feature map and the fully connected neural network [61]. Moreover, output hidden layer is connected to the final Softmax output classifier layer. This detailed architecture is illustrated in Figure 3.

The training process consists of two steps: forward propagation and backpropagation. Forward propagation provides the actual classification information of input data, and backpropagation upgrades the trainable parameters to provide the desired classification results. In forward propagation, from the previous convolution layer, to the current input layer l, m_1 is an input unit in the input layer and m_7 is the output unit in the output layer and numerous hidden units in convolution layer, pooling layer, regularization layer and fully connected layers, which can be expressed as:

$$w_n^l = \sum_{i=1}^{M} w_i^{l-1} * j_{in}^l + b_n^l \tag{8}$$

$$w_n^l = \sum_{i=1}^{M} w_i^{l-1} * j_{in}^l + b_n^l \tag{9}$$

where matrix w_i^{l-1} is th feature map of the previous $(l-1)th$ layer, w_n^l is the lth feature map of current l layer and M is the number of input feature maps. b_n^l is the additive bias vector for the lth layer. $*$ is the convolution operation. The output can be expressed as:

$$z_n^l = f_n\left(w_n^l\right) \tag{10}$$

In the pooling layer, max(w) function is used. In this case, CNN is used as a multiclass classifier, so the fully connected layer is fed to the output layer via the Softmax function. The Softmax layer can be calculated as: where, z_n are the input from the fully connected layer and J is the number of Softmax layer units, i.e., number of classes

$$\delta(z)_n = \frac{e^{z_n}}{\sum_j^J e^{z_j}}, \; n = 1, \ldots, J \tag{11}$$

2.2.2. Backpropagation

The backpropagation is used to update the trainable parameters by the mean of the gradient descent method. The backpropagation is linked with the output of the pooling layer to detect the error. Let $l = 1$ and $l = L$ be the input and output layers respectively. l is the number of classes and q is the input vector and t_l^q is the corresponding target and $\left[z_1^l, \ldots, z_q^l\right]$ is the output vector. Mean square error in the output layer can be expressed as:

$$E_q = \sum_{l=1}^{N_c} \left(z_l^l - t_l^q\right)^2 \tag{12}$$

3. Feature Extraction by IPCA, 1-D-CNN and Statistical Analysis

The extraction of features from input signals is significant to decide the accuracy of classification [9]. The statistical parameters can be obtained from the literature [62]. The primary purpose of IPCA is to reduce the feature data and minimize the computational load. The use of all coefficients of input data into the classifier may enhance the computational load and decrease the classification accuracy.

Therefore, the selected coefficients from IPCA and 1-D-CNN can be used to reduce the data complexity, so feature extraction from IPCA, 1-D-CNN and statistical parameters are used to enhance the classification accuracy. The statistical parameters deployed such as mean (M), energy (En), root mean square (RMS), standard deviation (SD), skewness (SK), Shannon entropy (SE), range (R), kurtosis (K), crest factor (CF), form factor (FF), and log-energy entropy (LE). The PQ disturbances 1-D array is used as input data and performs a convolution operation on the input data with each kernel in the architecture (Figure 3). This filtering (first convolution layer) of input data with each kernel creates the features for the classification. In our implementation of 1-D-CNN, the number of initial kernels to be trained was set at 20. Features such as mean, energy, standard deviation, Shannon entropy, and, log-energy entropy are extracted. Principal components (PCs) coefficients are calculated from the equations in Table 1. In this paper, four PCs decomposition level is used. The total number of features obtained per phase is 24, and the statistical features such as Root mean square, skewness, range, kurtosis, crest factor, form factor are extracted. The total number of features obtained from three-phase waveforms are 132. Mean, standard deviation, RMS, entropy, and range are used to analyze the behavior of the signal. Energy and log-energy entropy describe the information about energy. The feature vector of these statistical parameters along with four PCs are formed in Equation (13), and four level decomposition is presented in Figure 4.

$$
\begin{aligned}
F_1 &= [RMS_{PC1} RMS_{PC2} \ldots RMS_{PC4}] \\
F_2 &= [Sk_{PC1} Sk_{PC2} \ldots Sk_{PC4}] \\
F_3 &= [R_{PC1} R_{PC2} \ldots R_{PC4}] \\
F_4 &= [K_{PC1} K_{PC2} \ldots K_{PC4}] \\
F_5 &= [CF_{PC1} CF_{PC2} \ldots CF_{PC4}] \\
F_6 &= [FF_{PC1} FF_{PC2} \ldots FF_{PC4}]
\end{aligned}
\tag{13}
$$

Table 1. Mathematical equations of statistical parameters.

Feature Extracted Methods			
Energy	$E_{ki} = \sum\limits_{j=1}^{N} \left(\left| X_{ij} \right|^2 \right)$	Log-energy Entropy	$LE_{ki} = - \sum\limits_{j=1}^{N} log\left(X_{ij}^2 \right)$
Entropy	$ET_{ki} = - \sum\limits_{j=1}^{N} X_{ij}^2 log\left(X_{ij}^2 \right)$	Mean	$M_{ki} = \frac{1}{N} \sum\limits_{j=1}^{N} X_{ij}$
Standard Deviation	$\sigma_{ki} = \left(\frac{1}{N} \sum\limits_{j=1}^{N} \left(X_{ij} - \mu_i \right)^2 \right)^{\frac{1}{2}}$	Root Mean Square Value	$RMS_{ki} = \frac{1}{N} \sum\limits_{j=1}^{N} \left(X_{ij} \right)^2$
Range	$R_{ki} = Max\left(X_{ij} \right) - Min\left(X_{ij} \right)$	Kurtosis	$KT_{ki} = \sqrt{\frac{N}{24}} \left(\frac{1}{N} \sum\limits_{j=1}^{N} \left(\frac{X_{ij} - \mu_i}{\sigma_i} \right)^4 - 3 \right)$
Crest Factor	$CF_{ki} = \frac{Xi_{max}}{RMS_{ki}}$	Skewness	$SN_{ki} = \sqrt{\frac{1}{6N}} \sum\limits_{j=1}^{N} \left(\frac{X_{ij} - \mu_i}{\sigma_i} \right)^3$
Form Factor	$FF_{ki} = \frac{\overline{X_i}}{RMS_{ki}}$	-	-

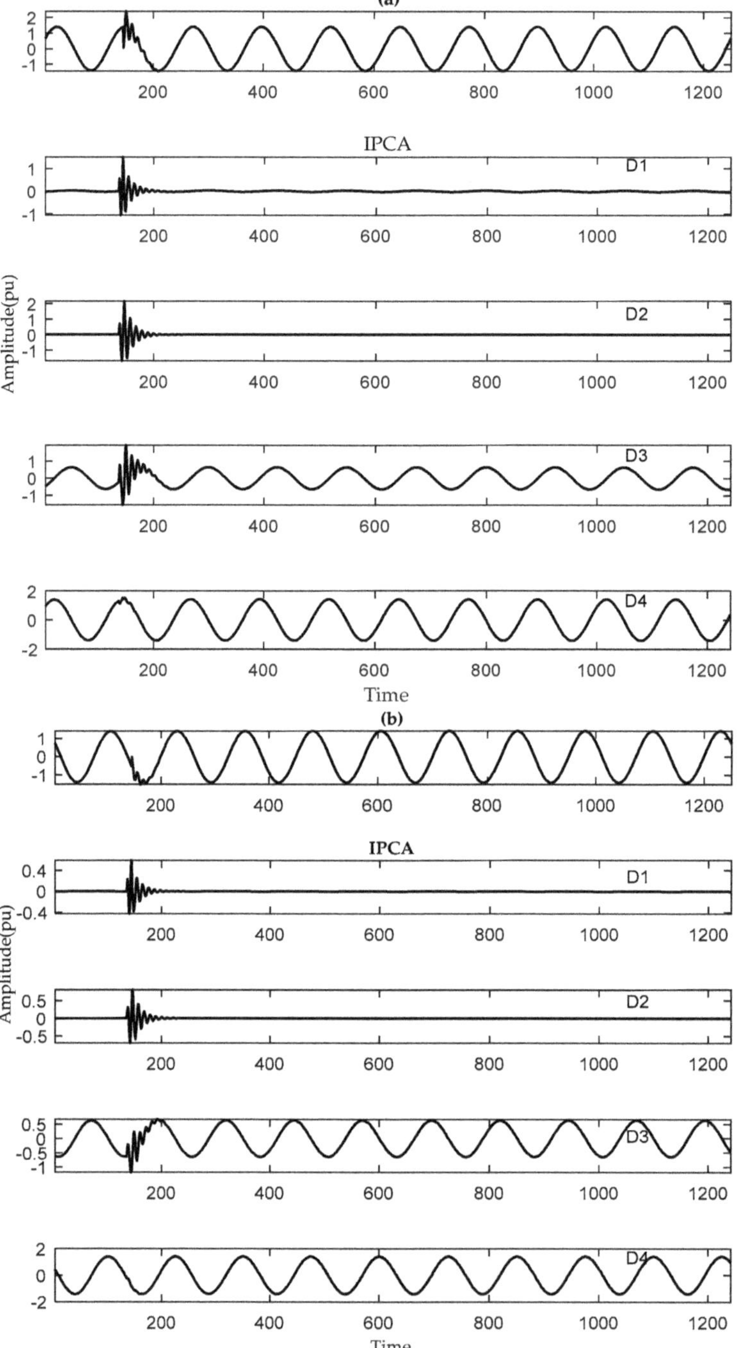

Figure 4. *Cont.*

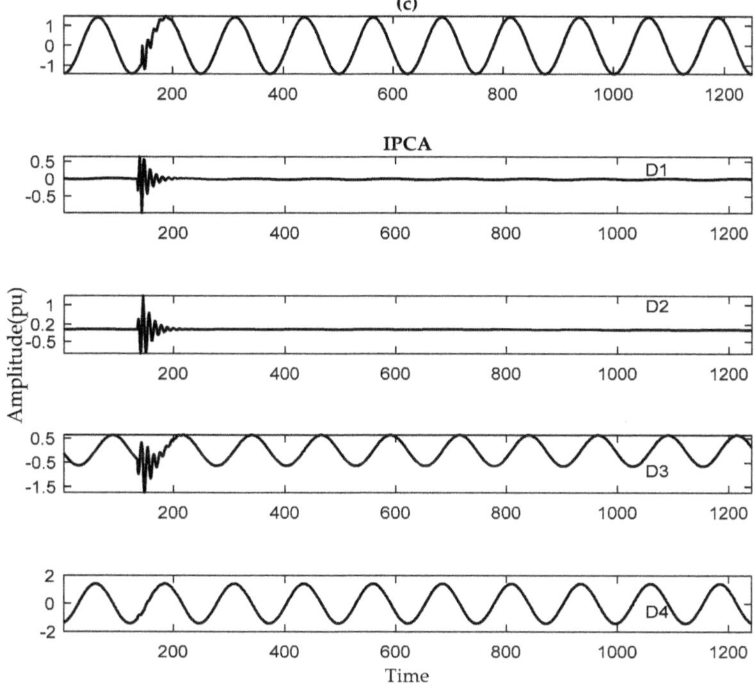

Figure 4. Four level signal decomposition by improved principal component analysis. (**a**) First phase oscillatory transients; (**b**) Second phase oscillatory transients; (**c**) Third phase oscillatory transient (modified IEEE bus system).

4. Proposed Algorithm

Generally, the proposed algorithm consists of two significant steps. Firstly, feature extraction of PQ events and secondly, classification of the extracted features. In this research, IPCA based on a 1-D-CCN intelligent system is employed for PQ classification. IPCA and 1-D-CNN are used for an optimal feature extraction that is further employed to 1-D-CCN based classifier for the classification. The flow hierarchy of proposed method is described in Figure 5. The detailed procedure is described as follows:

Aim: Classification of PQ events using improved principal component analysis and 1-D convolution neural network.

Input: Single and combined PQDs generated by synthetic and IEEE 13 bus based wind distribution system.

Step 1: The input signal is preprocessed and normalized by the following process,

Y_n is the 1-D training vector array of PQDs defined as $Y_n = [y_1, y_2, \ldots, y_n]^T \in \Re^N$ of length N. In the case of IPCA, the typical approach is described to normalize the raw data:

$$y'_{kl} = y_{kl} / \left(\frac{1}{n} \sum_{k=1}^{n} y_{kl} \right) \quad (k = 1, 2 \ldots, n \text{ and } l = 1, 2 \ldots, m)$$

where, "y'_{kl}" is the normalization value of "y_{kl}".

Step 2: The eigenvalue calculation is to decompose the covariance matrix "C". It can be diagonalized as:

$$C = \sum_{x=1}^{N} \lambda_x A_x A_x^T = A \Lambda A^T$$

Step 3: After normalization and set up of the covariance matrix "C" to calculate the eigenvalues and eigenvectors of matrix C. And the principal component "R_k" can be obtained as:

$$R_k = \begin{bmatrix} R_{k1} \\ R_{k2} \\ \vdots \\ R_{kn} \end{bmatrix} = \begin{bmatrix} \delta_{11} & \delta_{21} & \cdots & \delta_{q1} \\ \delta_{12} & \delta_{22} & & \delta_{q2} \\ \vdots & & \ddots & \vdots \\ \delta_{1n} & \delta_{2n} & \cdots & \delta_{qn} \end{bmatrix} y_k^T$$

Step 4: 1-D-CNN is also employed for feature extraction (convolution layer, max pooling layer) and the comparative results with other feature extraction techniques and detailed results are discussed in Section 6. Finally, optimal extracted features from these two methods are processed to fully connected layer and dropout, rectified linear units (ReLU) is employed for optimization. Each class is the label of the softmax output.

Output: Classification of PQ events is processed using IPCA and 1-D-CNN.

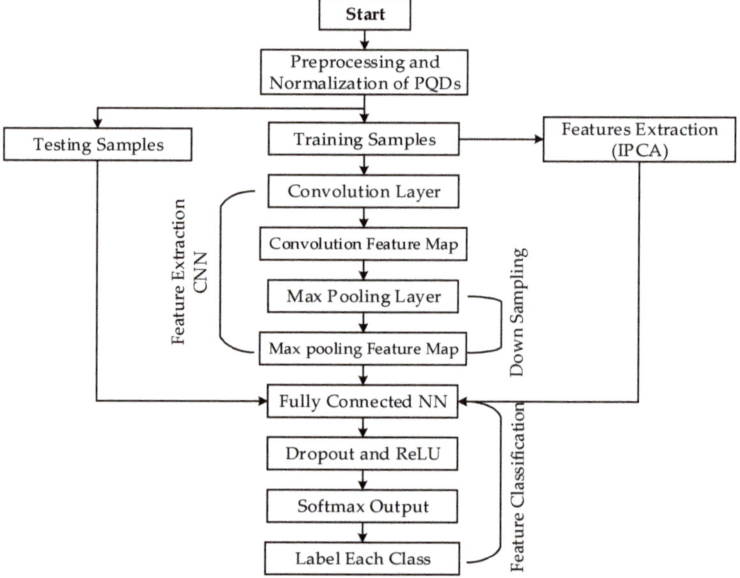

Figure 5. Flowchart hierarchy of proposed algorithm (IPCA-1-D-CNN).

5. Experiments

5.1. Generation of PQ Disturbances

5.1.1. Modified IEEE 13 Node Distribution Network

In this study, in order to test the PQ Events, a standard IEEE 13 node bus system [63] is considered with wind energy penetration in the balanced distribution system. The original system consists of 50 Hz, 5 MVA two voltage levels of 4.16 kV and 0.48 kV with balanced and unbalanced loads with no renewable energy (RE) sources. The original system is modified to extract the PQDs. In the modified system, two wind turbines of 1.5 MV each are integrated at bus 680 as shown in Figure 6 They connect through the transformer T-2 and an 8 km overhead transmission line. The transmission line has the following parameters. R_0 (zero sequences) resistance 0.413 Ohms/km, and R_1 (positive sequence) resistance 0.1153 Ohms/km, C_0 capacitance 5.09×10^{-9} F/km and C_1 capacitance 11.33×10^{-9} F/km and L_0 inductance 3.3×10^{-3} H/km and L_1 inductance 1.05×10^{-3} H/km, respectively. The loading, wind generators, capacitor banks, switching faults, and nonlinear loads location are mentioned in Table 2.

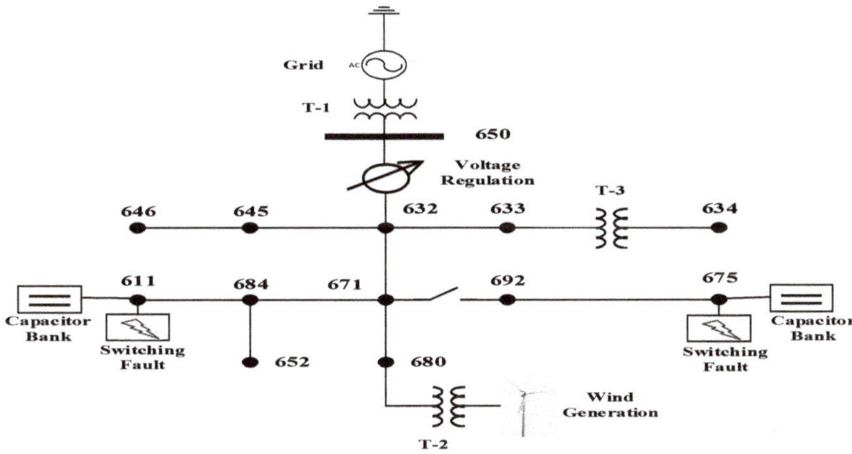

Figure 6. Modified IEEE 13-bus system with wind generation connected at node 680.

Table 2. Modified IEEE 13 bus system load and status data.

Bus Nodes	Load Model	Load		Capacitor Bank	Modified Data
-	-	kW	kVAr	kVAr	-
634					-
645	Y-PQ	400	290		-
646	Y-PQ	170	125	-	-
652	D-Z	230	132	-	-
671	Y-Z	128	86	-	-
675	D-PQ	1155	660	-	Switching Fault
692	Y-PQ	843	462	-	-
611	D-I	170	151	600	Switching Fault
632–671	Y-I	170	80	-	-
650	Y-PQ	200	116	100	Grid
680					WG/non-linear load

The three-phase overhead lines substitute single phase underground cables between the bus nodes 684 and 652 with configuration 601. The line parameters of these overhead lines are the same as the configuration of bus 601 in the original test system. In configuration 601, phase conductor type ACSR 556,500,26/7 and neutral conductor ACSR 4/0,6/1 is used with a spacing id 500. Three phase underground cables are between nodes 692 and 675. Node 650 is connected through the substation transformer T-1 via utility grid. Transformer T-3 is connected between bus nodes 633 and 634. According to the type of topology and conductor of the feeders, the series of impedance matrix of the test feeders is given in Equation (14). A capacitor bank and switching fault are connected at nodes 611 and 675. The nonlinear load is also connected to node 680. In the modified system voltage regulation between nodes 632 and 650 is not utilized. The transformer characteristics are presented in Table 3.

Table 3. Modified IEEE bus system transformer data.

Transformer	MVA	kV-High	kV-Low	HV Winding		LV Winding	
				$R(\Omega)$	$X(\Omega)$	$R(\Omega)$	$X(\Omega)$
Substation(T-1)	10	115	4.16	29.095	211.60	0.1142	0.8306
T-2	5	4.16	0.575	0.3807	2.7688	0.0510	0.0042
T-3	5	41.6	0.48	0.3807	2.7688	0.0510	0.0042

Standard grid voltage is free from any PQ disturbances. Frequency variation during the grid synchronization of a wind system is shown in Figure 7. The grid synchronization of a wind turbine of 1.5 MV drops suddenly to the frequency to 48.2 Hz. The standard frequency is restored within 0.2 s. High impact load and penetration degrades the frequency quality. Low magnitude oscillations are observed during this period. It is also observed that PQ disturbances occurred during a period of 0.2 s.

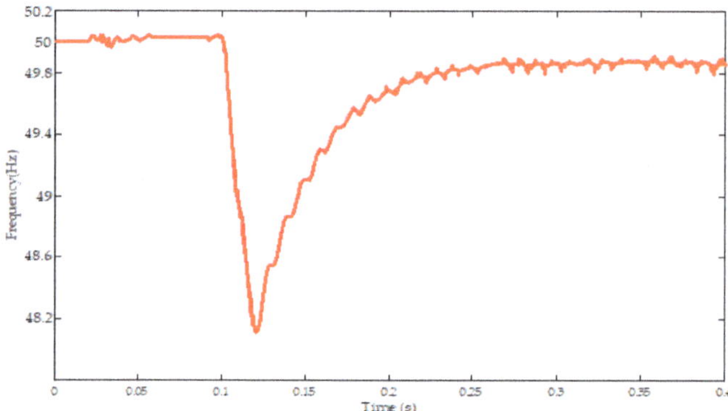

Figure 7. Frequency variations during grid synchronization of wind system.

The IEEE 13 bus system with a wind penetration model is simulated in MATLAB/Simulink, and generated waveforms are presented in Figure 8. Twelve types of three-phase PQDs are generated from this model. A standard waveform (sinusoidal) is generated at standard voltage amplitude and frequency. Three-phase voltage sag is generated using multistage and line to line faults at the generation end. Three phase voltage notch is generated by connecting three phases nonlinear load at the distribution end. Three phase voltage sag and swell is generated by line to line fault between two phases at generation end. Three phases oscillatory transient is generated by adding a three-phase capacitor bank at the distributed end. Three phase harmonics are generated by adding a nonlinear load at the distribution end. Arc furnace model is applied at the distribution end to generate three-phase voltage flickers. Three phase impulsive transients are generated by lightning at the distribution end. Three phase sag and harmonics are generated by multistage fault and line to line fault at generation end and nonlinear load. Three phase voltage sag and swell with harmonics are generated by line to line fault between two phases at generation end nonlinear switching load. Multistage fault and line to line fault generate three-phase voltage sag with oscillatory transients at generation end and capacitive load at distribution end. Three phase sag with oscillatory transients are generated by multistage fault and line to line fault at generation end and adding capacitive load distribution end.

$$Z_{601} = \begin{bmatrix} 0.3465 + j1.0179 & 0.1560 + j0.5017 & 0.1580 + j0.4236 \\ 0.1560 + j0.5017 & 0.3375 + j1.0478 & 0.1535 + j0.3849 \\ 0.1580 + j0.4236 & 0.1535 + j0.3849 & 0.3414 + j1.0348 \end{bmatrix} \tag{14}$$

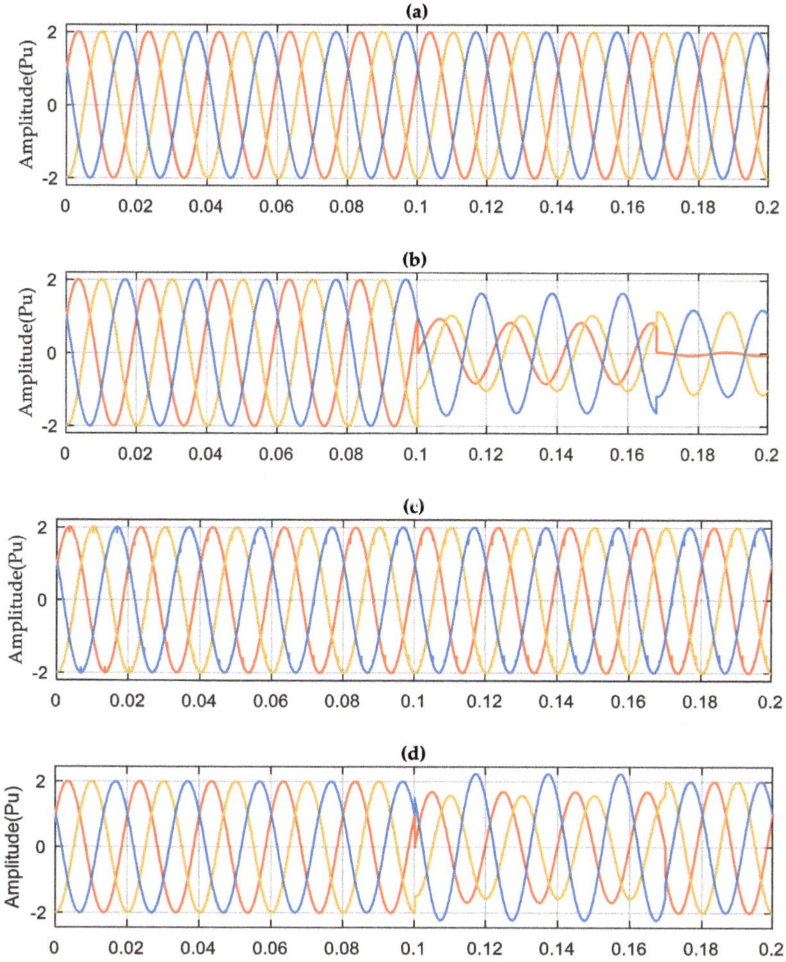

Figure 8. *Cont.*

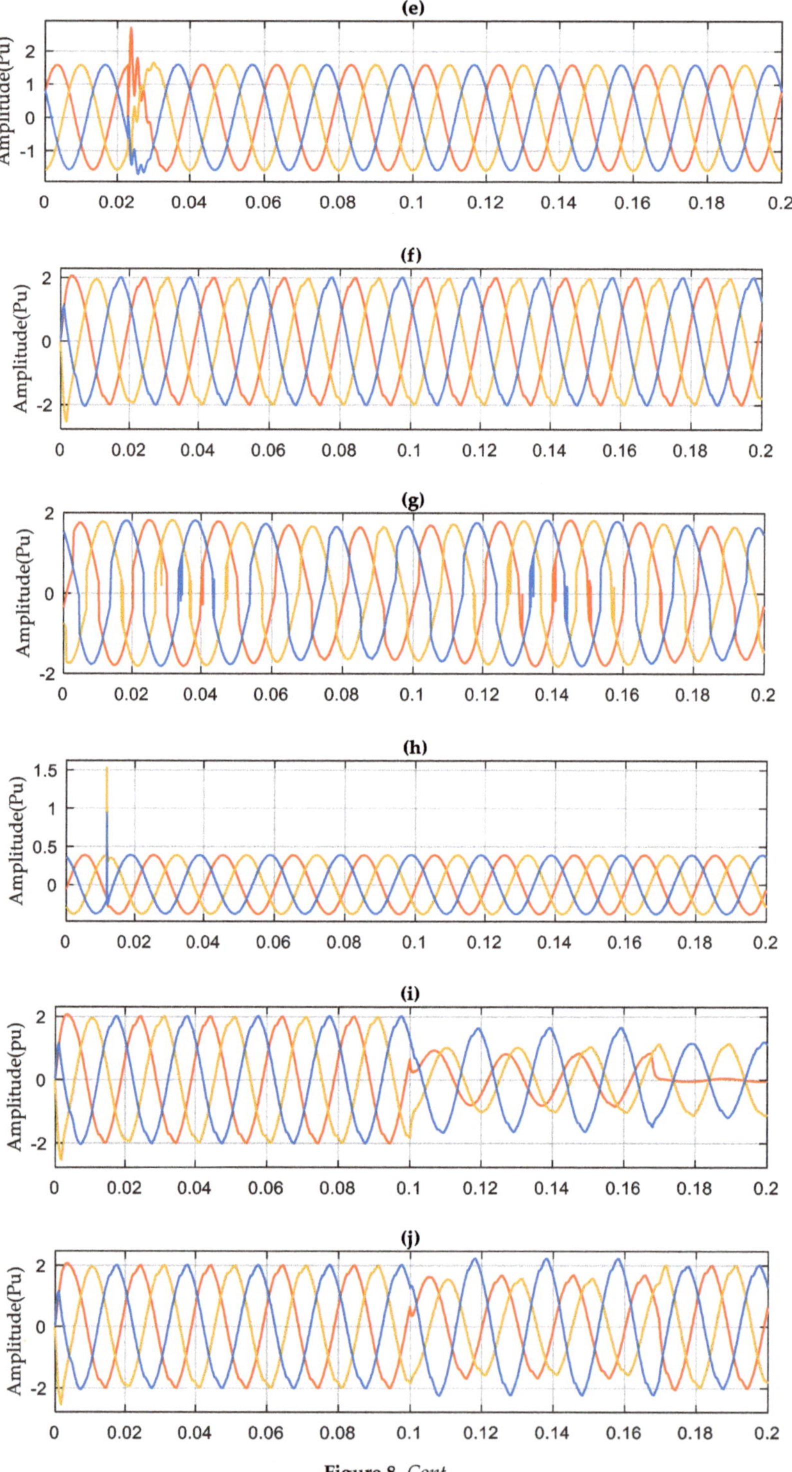

Figure 8. *Cont.*

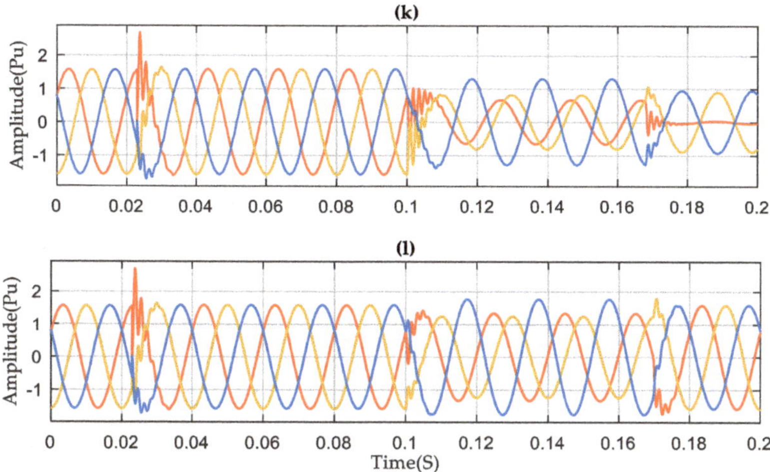

Figure 8. PQ disturbance generated from modified IEEE bus system. (**a**) Three phase normal; (**b**) Three phase voltage sag; (**c**) Three phase voltage notch; (**d**) Three phase voltage sag and swell; (**e**) Three phase oscillatory transient; (**f**) Three phase harmonics; (**g**) Three phase voltage flicker; (**h**) Three phase impulsive transients; (**i**) Three phase sag and harmonics; (**j**) Three phase sag, swell and harmonics; (**k**) Three phase sag and oscillatory transient; (**l**) Three phase sag, swell and oscillatory transient.

5.1.2. Synthetic PQ Disturbances

The synthetic PQ disturbances are generated based on a mathematical model shown in Table A1 using the MATLAB 2017b and PSCAD/EMTDC software. Synthetic PQ disturbance waveforms can be seen in Figure 9.

5.2. Dataset Generation

A two type dataset for single and multiple PQ disturbances has been obtained from synthetic parametric equations and a standard IEEE 13 bus system with wind penetration in the distribution system for the classification of PQ disturbances. The parametric equations shown in Table A1 were simulated in the MATLAB 2017b and PSCAD/EMTDC software to generate the first type of dataset. The waveforms with ten cycles at a sampling frequency of 10 kHz shown in Figure 9 are created for a total of 2000 samples points. The modified IEEE 13 node with wind penetration model is also simulated in MATLAB/Simulink to generate the second type of dataset. Considering the parameter variations of the different types of PQ disturbances, synthetic and simulated PQ disturbances datasets with 2400 and 5590 samples have been created for 12 types of PQ disturbances. In this case, the fundamental frequency was 50 Hz and the sampling frequency was 10 kHz.

Classification flow steps are presented in the form of a flowchart in Figure 5. Gaussian noise has been added at different SNR of 30, 40, and 50 dB to validate the classification performance. In this research, the optimal features are extracted by two methods. In the first method, PQ disturbance signals are preprocessed, normalized and converted it into a covariance matrix. Covariance matrix used for eigenvalues and eigenvector calculation.

Finally, principal components are calculated to four level decomposition. Figure 4 describes the four level decomposition using IPCA. Table 4 presents the detail about optimal feature extracted by IPCA and statistical parameters.

In the second method, after preprocessing of PQ disturbance, features are extracted by using 1-D-CNN. Detail of this process is presented in Sections 2 and 3. Finally, features extracted by these two methods are applied to the fully connected layer, and comparison accuracy of these different techniques are discussed in Section 6. This simulation was carried out more than 30 times to validate the proposed algorithm. The dataset is divided into training and testing samples with 70% and 30%, respectively.

Table 4. Features selected by statistical parameters.

Features Vectors	IPC Coefficients			
	D_1	D_2	D_3	D_4
RMS	F_1	F_7	F_{13}	F_{19}
Range	F_2	F_8	F_{14}	F_{20}
C-Factor	F_3	F_9	F_{15}	F_{21}
F-Factor	F_4	F_{10}	F_{16}	F_{22}
Kurtosis	F_5	F_{11}	F_{17}	F_{23}
Skewness	F_6	F_{12}	F_{18}	F_{24}

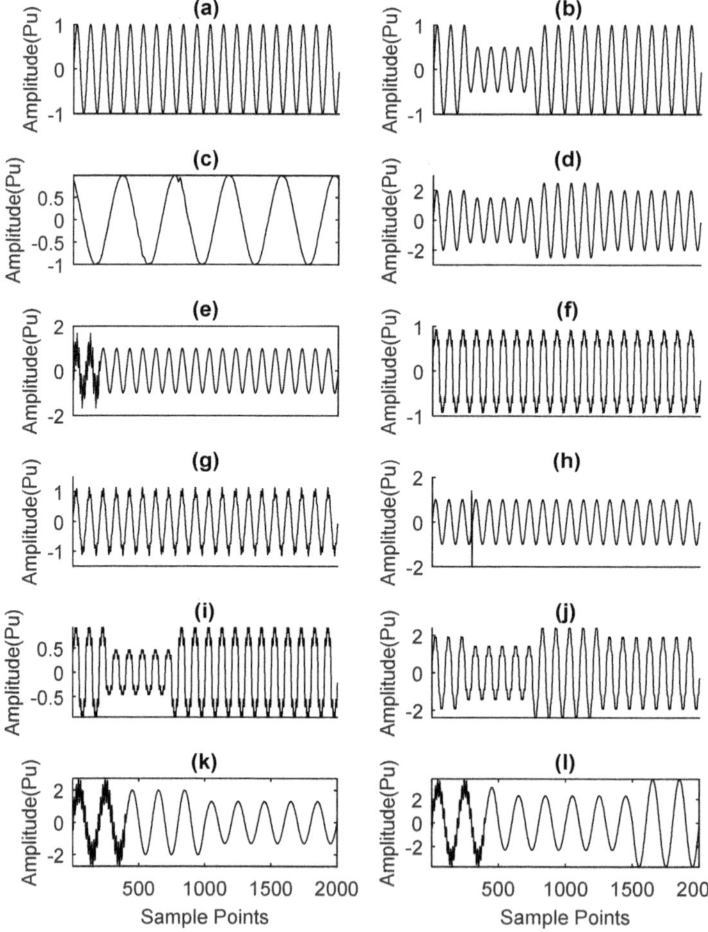

Figure 9. Synthetic PQ disturbances waveforms. (**a**) Normal waveform (50 Hz); (**b**) Voltage sag; (**c**) Voltage notch; (**d**) Voltage sag and swell; (**e**) Oscillatory transient; (**f**) Harmonics; (**g**) Voltage flicker; (**h**) Impulsive transient; (**i**) Sag and harmonics; (**j**) Sag, swell and harmonics; (**k**) Sag and oscillatory transients; (**l**) Sag, swell and oscillatory transient.

6. Results and Discussion

6.1. Classification Performance of Proposed Method

The results of the advanced IPCA-1-D CNN technique for classification of PQ disturbances are discussed in this section. Dataset 1 is based on synthetic PQ disturbances produced in MATLAB, while dataset 2 contains simulated PQ disturbances generated from the modified IEEE 13 bus distribution system. The average classification accuracy for single and multiple PQ Disturbances is 99.92% for the noiseless condition. The results in Table 5 show a high significance under noisy conditions. The proposed technique shows 100% results for normal, sag, notch and flickers. This confirms that the hybrid IPCA-1-D-CNN feature extraction method and 1-D CNN-based classifier is classified correctly as a single PQ disturbance. However, in the case of harmonics and transients, it becomes unpredictable due to the properties of high frequency and low magnitude. Sag and swell works fine in case of the synthetic dataset, but the accuracy somehow falls due to the handling of more samples in the simulated dataset. However, in case of complex signals, such as sag with harmonics, sag, swell with harmonics, sag with oscillatory transients and sag, swell with oscillatory transients the classifier cannot be able to differentiate among them correctly. Most of the sag amplitude is around 0.95 pu where the system has to work on the marginal area as a classifier, but it has considered the sag to happen around 0.94 to 0.96 pu. However, the system may confuse under 0.01 pu of the boundary area. Similarly, when the swell comes with different PQ disturbance, the classifier may have confused as most of the swell occurs at 1.1 pu. However, the classifier has classified the signal just above 1.0 pu. Therefore, the system has confused under 0.1 pu margin area. When sag and swell come with harmonics and oscillatory transients then the system has confused, it is the reason classifier accuracy precisely drops down. In the case of a simulated data set, accuracy decreases as compared to the synthetic data due to a large amount of data handling and model behavior.

Table 5. Classification accuracy results of PQ disturbances under noisy and noiseless conditions for proposed and IPCA-SVM algorithm.

| Power Quality Disturbances | | | Accuracy (%) Comparison between Proposed IPCA-1-D CNN Classifier and IPCA-SVM | | | | | | | |
| | | | IPCA-SVM | | | | IPCA-1-D CNN | | | |
	Class Labelled	Training/ Testing Sets	0 dB	20 dB	50 dB	Simulation Data	0 dB	20 dB	50 dB	Simulation Data
Normal	C_1	200	100	100	100	100	100	100	100	100
Sag	C_2	200	100	100	100	100	100	100	100	100
Notch	C_3	200	100	100	100	100	100	100	100	100
Flickers	C_4	200	100	100	100	100	100	100	100	100
Impulsive Transients	C_5	200	99.52	99	99.32	98.8	100	99.80	99.9	99.80
Oscillatory Transients	C_6	200	100	100	99.80	99.26	100	100	99.95	99.78
Harmonics	C_7	200	100	99.5	99.85	99.33	100	99.65	99.8	99.65
Sag with Swell	C_8	200	98.43	98	98.20	97.90	100	100	100	99.85
Sag with Harmonics	C_9	200	98.29	97.88	98	97.5	99.95	99.75	99.85	99.60
Sag, Swell with Harmonics	C_{10}	200	97.10	96.83	97	96.70	99.82	99.50	99.74	99.55
Sag with Oscillatory Transients	C_{11}	200	97.75	97.15	97.25	96.84	99.75	99.20	99.42	99.3
Sag, Swell with Oscillatory Transients	C_{12}	200	97.50	96.80	97	96.39	99.52	99.27	99.52	99.38
Accuracy (%)			99.05	98.76	98.87	98.55	99.92	99.76	99.85	99.75

The average percentage of correct classification of single and multiple PQ events are 99.76%, 99.85% and 99.75% with different noise level (20 dB, 50 dB) and simulated dataset respectively. The high classification accuracy under these conditions actively designates that it can be applied for the classification of single and multiple PQDs, especially in case of multiple PQ disturbances. In Figure 10, the feature recognition accuracy against the dimension of the feature database is shown. Feature extraction using PCA has the least accuracy among IPCA and CNN.

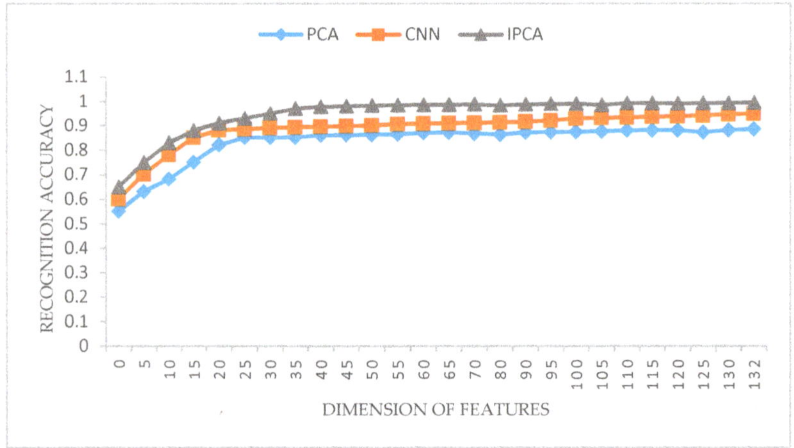

Figure 10. Feature recognition accuracy of PCA, CNN, and IPCA against the dimension of the features database.

6.2. Performance Comparison with SVM and Different Methods

In Figure 11, the combinations of different feature extraction techniques and classifiers have been analyzed and discussed. PCA-SVM has less than 95%, PCA-CNN has less than 96%, 1-D-CNN-SVM has less than 98%, feature extraction and classification using 1-D-CNN has less than 99%, IPCA-1DCNN with SVM classifier has about 99%, and the proposed method IPCA-1-D-CNN has 99.92% accuracies, respectively. The comparison of computational time between different feature extraction techniques for the proposed classifier and SVM is shown in Table 6. The computational time of 1D-CNN with CNN based classifier is relatively less among other methods, because of downsampling of the features in max pooling layer. Moreover, IPCA-1DCNN with CNN based classifier also actively responded, and the overall efficiency of the proposed method is much higher than the comparative methods.

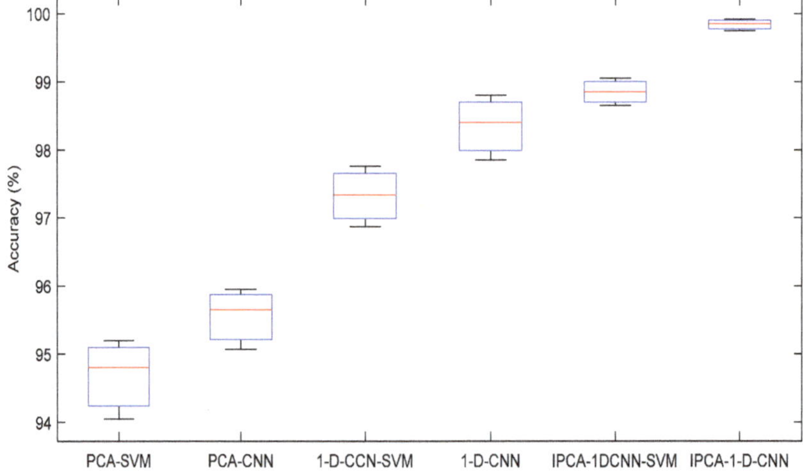

Figure 11. Classification accuracy comparison of proposed methods with different methods.

Table 6. Comparison of computational time (s) with difference feature extraction and classification methods.

Classifier	Feature Extraction Method			
	PCA	**IPCA**	**1D-CNN**	**IPCA-1DCNN**
SVM	1.565	0.859	0.725	0.792
1-D CNN	1.257	0.475	0.389	0.432

To validate the proposed classifier, the proposed hybrid feature extraction scheme with SVM classifier is also investigated. SVM is a supervised learning technique and is used for regression analysis and classification. The objective of SVM is to find the optimal hyperplane that separates the classes in the data. The optimal hyperplane found during the training phase allows only the smallest number of training error [42].

IPCA-1DCNN-SVM with 12 types of single and multiple PQ disturbances have been studied and 99.05%, 98.76%, 98.87% and 98.55% classification accuracy have been achieved for 0 dB, 20 dB, 50 dB and simulated data respectively. SVM classifier has the high precision of accuracy in case of single PQDs, but as for multiple and complex PQ disturbances, it does not compete for the accuracy of the proposed classifier. From the comparative results, it is clear that the proposed algorithm is best suited for multiple and complex PQ disturbances and as well as for the PQ disturbances generated from the wind distribution system.

6.3. Performance Comparison with Published Articles

A comparative study of published articles with the proposed method is presented in Table 7. WT and least squares SVM [64] considered only four types of three-phase PQ disturbances with simulated dataset having classification accuracy of 99.71%, but in the proposed method 12 types of three-phase PQ disturbance are considered with a significantly higher classification rate. Fast time-time transform (FTT) and small residual extreme learning machine (SR-ELM) were studied [4]. The classification rate for 12 types of three-phase simulated PQ disturbances was achieved as 99.59, and 107 features were selected, but no real data was evaluated in this paper. Optimal multi-resolution fast S-transform and cart algorithm were studied in [65]. Twelve types of single-phase simulated PQ disturbances were classified and a 98.92% classification rate was achieved. Therefore, the proposed method achieved higher classification for real and simulated datasets. Gabor transforms (GT), and a probabilistic neural network (PNN) method were used to improve the classification rate of PQ signals. Only seven types of single phase PQ disturbances were classified and 99.51% classification rate was obtained. However, this method evaluated less number of single phase PQ disturbances [66]. Hyperbolic S-transform (HST), Decision tree (DT) and SVM were proposed to improve the classification rate. Overall seven types of three-phase real and simulated PQ disturbances were classified, achieving an average classification accuracy of 99.5%. In this case, less number of PQ disturbances were considered [67]. Discrete WT, HST, and SVM were proposed in [68]. Nine types of three-phase PQ disturbances and 20 features were classified and a 99.44% classification rate was obtained. DWT, an artificial bee colony and PNN method were studied and sixteen types of single phase PQ disturbances classified. They achieved an average classification rate of 99.875%, but the proposed method has higher classification accuracy [69]. The comparative study shows that the proposed method based on IPCA and 1–D-CNN is best suited to classify multiple PQ disturbance real and simulated data and have a high precision of classification accuracy. This method is also well suited for practical implementation for classification of PQ disturbances.

Table 7. Performance comparison with the published article.

Feature Extraction and Classification Algorithms	No of PQ Disturbance	Data Type	PQDs Phase Type	Run Time	No of Features	Classification Accuracy (%)
WT + LSSVM [64]	4	Real	3-∅	–	21	99.71
FTT + SR − ELM [4]	12	Simulated	3-∅	0.029	107	99.59
OMFST + CA [65]	12	Simulated	Single ∅	–	67	98.92
GT + PNN [66]	9	Simulated and Real	Single ∅	–	6	99.51
HST + DT + SVM [67]	7	Simulated and Real	Single ∅	–	13	99.5
DWT + HST + S VM [68]	9	Simulated and Real	3-∅	0.0109	20	99.44
DWT + ABC + PNN [69]	16	Simulated and Real	Single ∅	1.2008	72	99.875
IPCA + 1-D-CNN (Proposed)	12	Simulated	3-∅	0.475	132	99.92

7. Conclusions

This study successfully presented an optimal method based on IPCA-1-D-CNN for accurate classification of PQ disturbances. An IEEE 13 node bus system was modified with wind-grid integration. Twelve types of three phases single and complex PQ disturbances were generated from this model. Synthetic PQ disturbances were generated from MATLAB R2017a. Three phase disturbance data is segmented into a single phase, and IPCA and 1-D CNN extracted the features. Finally, a 1-D-CNN-based classifier was used for the classification of PQ disturbances. 1-D-CNN performs convolution operations on input data with each kernel and creates the features for classification. IPCA normalized the results and could better interpret the data. The statistical parametric analysis was used to select the features vectors. Forty four different features were selected for each phase. 1-D-CNN was used to extract the statistical features such as mean, energy, standard deviation, Shannon entropy, and, log-energy entropy. IPCA was also employed to extract the statistical features such as root mean square, skewness, range, kurtosis, crest factor, and form factor. These optimal extracted features from hybrid IPCA and 1-D-CNN were then fed to the 1-D-CNN-based classifier to classify the PQ events at a high rate. The ReLU and dropout layers refined the data, which helps the classifier achieve better classification results. The proposed method was evaluated for different noise levels and a simulated data set. It was observed that the proposed method is prone to noise. The proposed method was also compared with different feature extraction methods and the SVM classifier to compare the classification accuracy. The comparative results with different feature extraction techniques, SVM classifier, and published methods depict that the proposed method is most suited for simulated data sets and noisy environments. This method can be applicable to other applications such as hyperspectral images, face recognition, fault detection, ECG signal detection, and classification.

Author Contributions: All the authors have contributed to this study in every step. Y.S. and H.L. contribute with funding resources and data analysis. M.A. contributes to the methodology, experiments and wrote the paper, and F.H. contribute with proofread the paper.

Funding: This work was supported by Jiangsu International Science and Technology Cooperation Project [BZ2017067], Jiangsu Provincial Key Research and Development Program [BE2018372], Jiangsu Natural Science Foundation [BK20181443], Zhenjiang City Key Research and Development Program [NY2018001], Qing Lan project of Jiangsu Province and the Priority Academic Program Development (PAPD) of the Jiangsu Higher Education Institutions, China.

Conflicts of Interest: The authors declare no conflict of interest.

Appendix A

Table A1. Mathematical model of PQ disturbances [9].

PQDs	Label	Equations	Parameter Constrains
Normal	C1	$x_1(t) = A\sin(\omega t)$	$A = 1(\mathrm{pu}), f = 50\,\mathrm{Hz}$
Sag	C2	$x_2(t) = [1 - \alpha_{sa}(u(t - t_1) - u(t - t_2))]\sin(wt)$	$0.1 \leq \alpha_{sa} \leq 0.9, T \leq t_2 - t_1 \leq 9T$
Notch	C3	$x_3(t) = \sin(wt) - sign(\sin(wt)) \sum\limits_{m=0}^{9} k[u(t - (t_1 + 0.02n)) - u(t - (t_2 + 0.02n))]$	$0.1 \leq k \leq 0.4, 0 \leq t_1, t_2 \leq 0.5T$ $0.01T \leq t_2 - t_1 \leq 0.05T,$
Sag with Swell	C4	$x_4(t) = [1 - \alpha_s(u(t - t_1) - u(t - t_2))] \times [1 + \alpha_s(u(t - t_1) - u(t - t_2))]\sin(wt)$	$0.1 \leq \alpha_s \leq 0.8, T \leq t_2 - t_1 \leq 9T$
Impulsive Transient	C5	$x_5(t) = [1 - \alpha_t\{u(t - t_1) - u(t - t_2)\}]\sin(\omega t)$	$0.1 \leq \alpha_t \leq 1; T \leq t_2 - t_1 \leq 9T$
Oscillatory Transient	C6	$x_6(t) = \alpha_t \sin(wt) + \alpha_t \exp\left(-\frac{t-t_1}{\tau}\right)(u(t - t_2) - u(t - t_1))\sin(2\pi f_t)$	$0.1 \leq \alpha_t 0.8, 300Hz \leq f_t \leq 900Hz,$ $8ms \leq \tau \leq 40ms, 0.5T \leq t_2 - t_1 \leq 3T$
Flicker	C7	$x_7(t) = \left(1 + \alpha_f \sin(3wt)\right)\sin(wt)$	$0.1 \leq \alpha_f \leq 2, 5 \leq \beta \leq 10$
Harmonic	C8	$x_8(t) = \alpha_1 \sin(wt) + \alpha_3 \sin(3wt) + \alpha_5 \sin(5wt) + \alpha_7 \sin(7wt)$	$0.05 \leq \alpha_3, \alpha_5, \alpha_7 \leq 0.15, \alpha_1 = 1$
Sag with harmonic	C9	$x_9(t) = [1 - \alpha_{sa}(u(t - t_1) - u(t - t_2))] \times \begin{bmatrix} \alpha_1 \sin(wt) + \alpha_3 \sin(3wt) + \\ \alpha_5 \sin(5wt) + \alpha_7 \sin(7wt) \end{bmatrix}$	$0.1 \leq \alpha_{sa} \leq 0.9, T \leq t_2 - t_1 \leq 9T,$ $0.05 \leq \alpha_3, \alpha_5, \alpha_7 \leq 0.15, \alpha_1 = 1$
Sag, Swell with harmonic	C10	$x_{10}(t) = [1 - \alpha_{sa}(u(t - t_1) - u(t - t_2))] \times$ $\left[1 + \alpha_{sa}\begin{pmatrix} u(t - t_1) - \\ u(t - t_2) \end{pmatrix}\right] \times \begin{bmatrix} \alpha_1 \sin(wt) + \alpha_3 \sin(3wt) + \\ \alpha_5 \sin(5wt) + \alpha_7 \sin(7wt) \end{bmatrix}$	$0.1 \leq \alpha_{sw} \leq 0.8, T \leq t_2 - t_1 \leq 9T,$ $0.05 \leq \alpha_3, \alpha_5, \alpha_7 \leq 0.15, \alpha_1 = 1$
Sag with Oscillatory Transient	C11	$x_{11}(t) = [1 - \alpha_{sa}(u(t - t_1) - u(t - t_2))] \times \alpha_t \sin(wt) +$ $\alpha_t \exp\left(-\frac{t-t_1}{\tau}\right)(u(t - t_2) - u(t - t_1))\sin(2\pi f_t)$	$0.1 \leq \alpha_{sa} \leq 0.9, T \leq t_2 - t_1 \leq 9T,$ $0.1 \leq \alpha_{sa} \leq 0.9, T \leq t_2 - t_1 \leq 9T$
Sag, Swell with Oscillatory Transient	C12	$x_{12}(t) = [1 - \alpha_s(u(t - t_1) - u(t - t_2))] \times [1 + \alpha_{sw}(u(t - t_1) - u(t - t_2))] \times$ $\alpha_t \sin(wt) + \alpha_t \exp\left(-\frac{t-t_1}{\tau}\right)(u(t - t_2) - u(t - t_1))\sin(2\pi f_t)$	$0.1 \leq \alpha_s \leq 0.8, T \leq t_2 - t_1 \leq 9T, 0.1 \leq \alpha_t \leq 0.8,$ $300Hz \leq f_t \leq 900Hz, 8ms \leq \tau \leq 40ms,$ $0.5T \leq t_2 - t_1 \leq 3T$

References

1. Jiang, J.N.; Tang, C.Y.; Ramakumar, R.G. *Control and Operation of Grid-Connected Wind Farms: Major Issues, Contemporary Solutions, and Open Challenges*; Springer: Berlin, Germany, 2016.
2. Mundaca, L.; Neij, L.; Markandya, A.; Hennicke, P.; Yan, J. *Towards a Green Energy Economy? Assessing Policy Choices, Strategies and Transitional Pathways*; Elsevier: Amsterdam, The Netherlands, 2016.
3. Lin, B.; Liu, Y.; Wang, Z.; Pei, Z.; Davies, M. Measured energy use and indoor environment quality in green office buildings in China. *Energy Build.* **2016**, *129*, 9–18. [CrossRef]
4. Saini, K.M.; Beniwal, R.K. Detection and classification of power quality disturbances in wind-grid integrated system using fast time-time transform and small residual-extreme learning machine. *Int. Trans. Electr. Energy Syst.* **2018**, *28*, e2519. [CrossRef]
5. Liu, H.; Hussain, F.; Shen, Y. Power quality disturbances classification using compressive sensing and maximum likelihood. *IETE Tech. Rev.* **2017**, *35*, 359–368. [CrossRef]
6. Shen, Y.; Hussain, F.; Shen, Y. Power quality disturbances classification based on curvelet transform. *Int. J. Comput. Appl.* **2017**, *40*, 192–201. [CrossRef]
7. Niitsoo, J.; Jarkovoi, M.; Taklaja, P.; Klüss, J.; Palu, I. Power quality issues concerning photovoltaic generation in distribution grids. *Smart Grid Renew. Energy* **2015**, *6*, 148. [CrossRef]
8. Bollen, H.M.; Gu, I.Y. *Signal Processing of Power Quality Disturbances*; John Wiley & Sons: Hoboken, NJ, USA, 2006; Volume 30.
9. Lee, C.-Y.; Shen, Y.-X. Optimal feature selection for power-quality disturbances classification. *IEEE Trans. Power Deliv.* **2011**, *26*, 2342–2351. [CrossRef]
10. Reid, W.E. Power quality issues-standards and guidelines. *IEEE Trans. Ind. Appl.* **1996**, *32*, 625–632. [CrossRef]
11. Heydt, G.; Fjeld, P.; Liu, C.; Pierce, D.; Tu, L.; Hensley, G. Applications of the windowed FFT to electric power quality assessment. *IEEE Trans. Power Deliv.* **1999**, *14*, 1411–1416. [CrossRef]
12. Liao, C.-C.; Yang, H.-T.; Chang, H.-H. Denoising techniques with a spatial noise-suppression method for wavelet-based power quality monitoring. *IEEE Trans. Instrum. Meas.* **2011**, *60*, 1986–1996. [CrossRef]
13. Poisson, O.; Rioual, P.; Meunier, M. Detection and measurement of power quality disturbances using wavelet transform. *IEEE Trans. Power Deliv.* **2000**, *15*, 1039–1044. [CrossRef]
14. Jurado, F.; Saenz, J.R. Comparison between discrete STFT and wavelets for the analysis of power quality events. *Electr. Power Syst. Res.* **2002**, *62*, 183–190. [CrossRef]
15. Santoso, S.; Powers, E.J.; Grady, W.M.; Hofmann, P. Power quality assessment via wavelet transform analysis. *IEEE Trans. Power Deliv.* **1996**, *11*, 924–930. [CrossRef]
16. Morsi, W.G.; El-Hawary, M. Novel power quality indices based on wavelet packet transform for non-stationary sinusoidal and non-sinusoidal disturbances. *Electr. Power Syst. Res.* **2010**, *80*, 753–759. [CrossRef]
17. Dash, P.; Panigrahi, G.P.B. Power quality analysis using S-transform. *IEEE Trans. Power Deliv.* **2003**, *18*, 406–411. [CrossRef]
18. Stockwell, R.G.; Mansinha, L.; Lowe, R. Localization of the complex spectrum: The S transform. *IEEE Trans. Signal Process.* **1996**, *44*, 998–1001. [CrossRef]
19. Dash, P.; Chilukuri, M. Hybrid S-transform and Kalman filtering approach for detection and measurement of short duration disturbances in power networks. *IEEE Trans. Instrum. Meas.* **2004**, *53*, 588–596. [CrossRef]
20. Reddy, M.J.B.; Raghupathy, R.K.; Venkatesh, K.; Mohanta, D. Power quality analysis using Discrete Orthogonal S-transform (DOST). *Digit. Signal Process.* **2013**, *23*, 616–626. [CrossRef]
21. Liu, H.; Hussain, F.; Shen, Y.; Arif, S.; Nazir, A.; Abubakar, M. Complex power quality disturbances classification via curvelet transform and deep learning. *Electr. Power Syst. Res.* **2018**, *163*, 1–9. [CrossRef]
22. Shukla, S.; Mishra, S.; Singh, B. Empirical-mode decomposition with Hilbert transform for power-quality assessment. *IEEE Trans. Power Deliv.* **2009**, *24*, 2159–2165. [CrossRef]
23. Li, T.-Y.; Zhao, Y.; Nan, L.; Fen, G.; Gao, H.-H. A new method for power quality detection based on HHT. *Zhongguo Dianji Gongcheng Xuebao Proc. Chin. Soc. Electr. Eng.* **2005**, *25*, 52–56.
24. Ozgonenel, O.; Yalcin, T.; Guney, I.; Kurt, U. A new classification for power quality events in distribution systems. *Electr. Power Syst. Res.* **2013**, *95*, 192–199. [CrossRef]

25. Cho, S.-H.; Jang, G.; Kwon, S.-H. Time-frequency analysis of power-quality disturbances via the Gabor–Wigner transform. *IEEE Trans. Power Deliv.* **2010**, *25*, 494–499.
26. Abdullah, A.R.; Sha'ameri, A.Z.; Saad, N.M. Asia-Pacific Conference on Power quality analysis using spectrogram and gabor transformation. In Proceedings of the 2007 Asia-Pacific Conference on Applied Electromagnetics, Melaka, Malaysia, 4–6 December 2007.
27. Manikandan, M.S.; Samantaray, S.; Kamwa, I. Detection and classification of power quality disturbances using sparse signal decomposition on hybrid dictionaries. *IEEE Trans. Instrum. Meas.* **2015**, *64*, 27–38. [CrossRef]
28. Lopez-Ramirez, M.; Ledesma-Carrillo, L.; Cabal-Yepez, E.; Rodriguez-Donate, C.; Miranda-Vidales, H.; Garcia-Perez, A. EMD-based feature extraction for power quality disturbance classification using moments. *Energies* **2016**, *9*, 565. [CrossRef]
29. Smith, L.I. *A Tutorial on Principal Components Analysis*; Technical Report OUCS: Dunedin, Otago, New Zealand, 26 February 2002.
30. Chawla, M.; Verma, H.; Kumar, V. ECG Modeling and QRS Detection Using Principal Component Analysis. In Proceedings of the IET 3rd International Conference MEDSIP 2006, Advances in Medical, Signal and Information Processing, Glasgow, UK, 17–19 July 2006.
31. Li, W.; Shi, T.; Liao, G.; Yang, S. Feature extraction and classification of gear faults using principal component analysis. *J. Qual. Maint. Eng.* **2003**, *9*, 132–143. [CrossRef]
32. Moon, H.; Phillips, P.J. Computational and performance aspects of PCA-based face-recognition algorithms. *Perception* **2001**, *30*, 303–321. [CrossRef] [PubMed]
33. Rodarmel, C.; Shan, J. Principal component analysis for hyperspectral image classification. *Surv. Land Inf. Sci.* **2002**, *62*, 115–122.
34. Ahila, R.; Sadasivam, V.; Manimala, K. Particle swarm optimization-based feature selection and parameter optimization for power system disturbances classification. *Appl. Artif. Intell.* **2012**, *26*, 832–861. [CrossRef]
35. Masoum, M.; Jamali, S.; Ghaffarzadeh, N. Detection and classification of power quality disturbances using discrete wavelet transform and wavelet networks. *IET Sci. Meas. Technol.* **2010**, *4*, 193–205. [CrossRef]
36. Biswal, B.; Mishra, S. Power signal disturbance identification and classification using a modified frequency slice wavelet transform. *IET Gener. Transm. Distrib.* **2014**, *8*, 353–362. [CrossRef]
37. Jamali, S.; Farsa, A.R.; Ghaffarzadeh, N. Identification of optimal features for fast and accurate classification of power quality disturbances. *Measurement* **2018**, *116*, 565–574. [CrossRef]
38. Kumar, R.; Singh, B.; Shahani, D.; Chandra, A.; Al-Haddad, K. Recognition of power-quality disturbances using S-transform-based ANN classifier and rule-based decision tree. *IEEE Trans. Ind. Appl.* **2015**, *51*, 1249–1258. [CrossRef]
39. Mishra, S.; Bhende, C.; Panigrahi, B. Detection and classification of power quality disturbances using S-transform and probabilistic neural network. *IEEE Trans. Power Deliv.* **2008**, *23*, 280–287. [CrossRef]
40. Gaing, Z.-L. Wavelet-based neural network for power disturbance recognition and classification. *IEEE Trans. Power Deliv.* **2004**, *19*, 1560–1568. [CrossRef]
41. Wang, H.; Wang, P.; Liu, T. Power quality disturbance classification using the S-transform and probabilistic neural network. *Energies* **2017**, *10*, 107. [CrossRef]
42. Weston, J.; Watkins, C. Multi-Class Support Vector Machines. Citeseer, 1998. Available online: http://citeseerx.ist.psu.edu/viewdoc/summary?doi=10.1.1.50.9594 (accessed on 1 April 2019).
43. Ucar, F.; Alcin, O.F.; Dandil, B.; Ata, F. Power quality event detection using a fast extreme learning machine. *Energies* **2018**, *11*, 145. [CrossRef]
44. Mehta, S.; Shen, X.; Gou, J.; Niu, D. A New Nearest Centroid Neighbor Classifier Based on K Local Means Using Harmonic Mean Distance. *Information* **2018**, *9*, 234. [CrossRef]
45. Hu, W.; Huang, Y.; Wei, L.; Zhang, F.; Li, H. Deep convolutional neural networks for hyperspectral image classification. *J. Sens.* **2015**, *2015*, 12. [CrossRef]
46. Hershey, S.; Chaudhuri, S.; Ellis, D.P.; Gemmeke, J.F.; Jansen, A.; Moore, R.C.; Plakal, M.; Platt, D.; Saurous, R.A.; Seybold, B. CNN architectures for large-scale audio classification. In Proceedings of the 2017 IEEE International Conference on CNN Architectures for Large-Scale Audio Classification, New Orleans, LA, USA, 5–9 March 2017.

47. Hu, G.; Yang, Y.; Yi, D.; Kittler, J.; Christmas, W.; Li, S.Z.; Hospedales, T. When face recognition meets with deep learning: An evaluation of convolutional neural networks for face recognition. In Proceedings of the IEEE International Conference on Computer Vision Workshops, Santiago, Chile, 7–13 December 2015.

48. Li, Q.; Cai, W.; Wang, X.; Zhou, Y.; Feng, D.D.; Chen, M. Medical image classification with convolutional neural network. In Proceedings of the 2014 13th International Conference on Medical Image Classification with Convolutional Neural Network, Singapore, 10–12 December 2014.

49. Fukushima, K. Nocognitron: A hierarchical neural network capable of visual pattern recognition. *Neural Netw.* **1988**, *1*, 119–130. [CrossRef]

50. LeCun, Y.; Bottou, L.; Bengio, Y.; Haffner, P. Gradient-based learning applied to document recognition. *Proc. IEEE* **1998**, *86*, 2278–2324. [CrossRef]

51. Ciresan, D.C.; Meier, U.; Masci, J.; Maria Gambardella, L.; Schmidhuber, J. *Proceedings-International Joint Conference on Artificial Intelligence Flexible, High Performance Convolutional Neural Networks for Image Classification*; IJCAI: Barcelona, Spain, 2011.

52. Sutskever, I.; Hinton, G.E. Deep, narrow sigmoid belief networks are universal approximators. *Neural Comput.* **2008**, *20*, 2629–2636. [CrossRef]

53. Rodriguez-Guerrero, M.A.; Jaen-Cuellar, A.Y.; Carranza-Lopez-Padilla, R.D.; Osornio-Rios, R.A.; Herrera-Ruiz, G.; Romero-Troncoso, R.D.J. Hybrid approach based on GA and PSO for parameter estimation of a full power quality disturbance parameterized model. *IEEE Trans. Ind. Inf.* **2018**, *14*, 1016–1028. [CrossRef]

54. Alorf, A.A. Performance evaluation of the PCA versus improved PCA (IPCA) in image compression, and in face detection and recognition. In Proceedings of the 2016 Future Technologies Conference (FTC), San Francisco, CA, USA, 6–7 December 2016.

55. Ince, T.; Kiranyaz, S.; Eren, L.; Askar, M.; Gabbouj, M. Real-time motor fault detection by 1-D convolutional neural networks. *IEEE Trans. Ind. Electron.* **2016**, *63*, 7067–7075. [CrossRef]

56. Sermanet, P.; LeCun, Y. The 2011 International Joint Conference on Traffic sign recognition with multi-scale convolutional networks. In *Neural Networks (IJCNN)*; IEEE: Piscataway, NJ, USA, 2011.

57. Krizhevsky, A.; Sutskever, I.; Hinton, G.E. Imagenet classification with deep convolutional neural networks. In *Advances in Neural Information Processing Systems*; ACM: New York, NY, USA, 2012.

58. Girshick, R.; Donahue, J.; Darrell, T.; Malik, J. Rich feature hierarchies for accurate object detection and semantic segmentation. In Proceedings of the IEEE Conference on Computer Vision and Pattern Recognition, Columbus, OH, USA, 24–27 June 2014.

59. Taigman, Y.; Yang, M.; Ranzato, M.A.; Wolf, L. Deepface: Closing the gap to human-level performance in face verification. In Proceedings of the IEEE Conference on Computer Vision and Pattern Recognition, Columbus, OH, USA, 24–27 June 2014.

60. Long, J.; Shelhamer, E.; Darrell, T. Fully convolutional networks for semantic segmentation. In Proceedings of the IEEE Conference on Computer Vision and Pattern Recognition, Santiago, Chile, 7–13 December 2015.

61. Srivastava, N.; Hinton, G.; Krizhevsky, A.; Sutskever, I.; Salakhutdinov, R. Dropout: A simple way to prevent neural networks from overfitting. *J. Mach. Learn. Res.* **2014**, *15*, 1929–1958.

62. Erişti, H.; Yıldırım, Ö.; Erişti, B.; Demir, Y. Optimal feature selection for classification of the power quality events using wavelet transform and least squares support vector machines. *Int. J. Electr. Power Energy Syst.* **2013**, *49*, 95–103. [CrossRef]

63. Kersting, W.H. Radial distribution test feeders. In Proceedings of the 2001 IEEE Power Engineering Society Winter Meeting, Conference Proceedings (Cat. No.01CH37194), Columbus, OH, USA, 28 January–1 February 2001; IEEE: Piscataway, NJ, USA, 2001.

64. Eristi, B.; Yildirim, O.; Eristi, H.; Demir, Y. A new embedded power quality event classification system based on the wavelet transform. *Int. Trans. Electr. Energy Syst.* **2018**, *28*, e2597. [CrossRef]

65. Huang, N.; Peng, H.; Cai, G.; Chen, J. Power quality disturbances feature selection and recognition using optimal multi-resolution fast S-transform and CART algorithm. *Energies* **2016**, *9*, 927. [CrossRef]

66. Moravej, Z.; Pazoki, M.; Niasati, M.; Abdoos, A.A. A hybrid intelligence approach for power quality disturbances detection and classification. *Int. Trans. Electr. Energy Syst.* **2013**, *23*, 914–929. [CrossRef]

67. Ray, P.K.; Mohanty, S.R.; Kishor, N.; Catalão, J.P. Optimal feature and decision tree-based classification of power quality disturbances in distributed generation systems. *IEEE Trans. Sustain. Energy* **2014**, *5*, 200–208. [CrossRef]

68. Hajian, M.; Foroud, A.A. A new hybrid pattern recognition scheme for automatic discrimination of power quality disturbances. *Measurement* **2014**, *51*, 265–280. [CrossRef]
69. Khokhar, S.; Zin, A.A.M.; Memon, A.P.; Mokhtar, A.S. A new optimal feature selection algorithm for classification of power quality disturbances using discrete wavelet transform and probabilistic neural network. *Measurement* **2017**, *95*, 246–259. [CrossRef]

 *energies*

Article

Power Quality Disturbances Assessment during Unintentional Islanding Scenarios. A Contribution to Voltage Sag Studies [†]

Alexandre Serrano-Fontova *, Pablo Casals Torrens and Ricard Bosch

Department of Electrical Engineering, Polytechnic University of Catalonia, ETSEIB - Av. Diagonal 647, 08028 Barcelona, Spain

* Correspondence: alexandre.serrano@upc.edu

† This paper is a technically extended version of a conference paper entitled "A novel voltage sag approach during unintentional islanding scenarios: A survey from real recorded events", presented at ICREPQ' 19, Tenerife, Spain, April 2019.

Received: 10 July 2019; Accepted: 14 August 2019; Published: 20 August 2019

Abstract: This paper presents a novel voltage sag topology that occurs during an unintentional islanding operation (IO) within a distribution network (DN) due to large induction motors (IMs). When a fault occurs, following the circuit breaker (CB) fault clearing, transiently, the IMs act as generators due to their remanent kinetic energy until the CB reclosing takes place. This paper primarily contributes to voltage sag characterization. Therefore, this novel topology is presented, analytically modelled and further validated. It is worth mentioning that this voltage sag has been identified in a real DN in which events have been recorded for two years. The model validation of the proposed voltage sag is done via digital simulations with a model of the real DN implemented in Matlab considering a wide range of scenarios. Both simulations and field measurements confirm the voltage sag analytical expression presented in this paper as well as exhibiting the high accuracy achieved in the three-phase model adopted.

Keywords: power quality; voltage sags; islanding operation; induction machines; modelling; distribution networks

---

1. Introduction

Over the last decade, the increasing interest in power quality (PQ) monitoring is becoming a crucial task for distribution operators (DSOs) and, presently, is among the issues of major concern. Since the DNs supply power to the costumers, monitoring the occurring events has become essential; therefore, presently, smart meters allocated in each customer node are playing a pivotal role [1,2]. Regarding the measurement and monitoring of PQ disturbances, several standards have been published within the EN-61000 series; these standards fall within the electromagnetic compatibility field. Particularly, EN61000-4-30 defines how to measure PQ disturbances and the measurement accuracy of the equipment and defines different classes of accuracy. A relevant summary and interpretation of these standards are given in References [3–5]. Furthermore, other research has been focused on how to identify the different type of disturbances from occurring events; for instance, in References [6–11]. Once PQ disturbances are detected, classifying them and compressing them into separate events is a challenging task and requires advanced tools; see References [12–15]. Some authors have considered fuzzy logic or neural networks to classify them [16–19], some others proceed based on clustering the data depending on its origin from offline and prescribed events [20,21], and lastly, certain studies have focused on real-time classification, which results in a successful classification; for instance, see References [22–24].

It is understood that PQ is one of the major concerns for DSOs, and probabilistic studies have been carried out to evaluate the grid reliability with statistical and stochastic assessments [25–29]. Such studies have concluded that protective devices and the weakness of the grid as well as the type of loads are crucial factors when carrying out PQ studies. Besides, with the growing interest in distributed generation (DG) and smart grids, their influence on the PQ field is currently under investigation [30–32]. It is also important to underline the importance of the PQ reliability indicators; for example, see the evaluations carried out in Reference [33–38].

All PQ disturbances are summarized in the standard EN-50160 (see Reference [39]). This standard defines all voltage disturbances as two types: (i) continuous phenomena (e.g., long-time disturbances in the range between ms and minutes), such as voltage drops, voltage swells or frequency oscillations; and (ii) short-duration disturbances such as voltage interruptions, voltage sags, voltage swells, and over-voltage transients (e.g., short-term disturbances in the range between µs to ms). An excellent review of PQ disturbances classification can be found in References [39,40].

Among all PQ disturbances, voltage sags have been the object of study for many years. At first, voltage sags were assumed to be rectangular. Afterwards, it was demonstrated that this was not strictly true, and the type of recovery, as well as the phase-angle-jump, played a significant role [41]. It is relevant to highlight the characterization done by Bollen in Reference [33], as well as in Reference [42]. Furthermore, immunity tests and studies of electrical devices' behavior towards voltage sags have been thoroughly analyzed in References [43–46], where the effects of the voltage sags are evaluated. In this direction, several standards have been published; for instance, EN 61000-3-2 (limits for harmonic current emissions, less than 16 A), EN-61000-4-34 (immunity test for high current equipment, more than 16 A), and voltage sags with statistical results are summarized in IEC-61000-2-8. Considering that electrical device sensitivity is an aspect of utmost importance, immunity tests for voltage sags and interruptions are still under study.

Even though standard indices for reliability such as standard average interruption frequency index (SAIFI), standard average interruption duration index (SAIDI) or the reliability curves published by the information technology industry council (ITIC) have been defined, each country has its own regulations regarding the number of interruptions or their duration [47,48]. To deal with voltage sags effects, well-known equipment such as dynamic voltage restorers or an uninterruptible power supply are used [49,50], or even the use of FACTS [51].

Mainly, the origin of voltage sags in DNs and transmission networks (TNs) are faults; given their random nature in time and location, their magnitude, type and recovery can change along the grid. A fault in the system can produce either a voltage sag or an interruption, depending on where the fault originates; that is, depending on the type of network and configuration which is feeding the fault. Thereby, if several lines are feeding the fault, and if we assume that each one has its own CB, the faulted line/section will experience an interruption, whereas the adjacent section will experience a voltage sag. Commonly, the DN configuration is radial (i.e., there is only one source); therefore, if a fault appears and no DG is considered, the only source is the main grid. Therefore, costumers located downstream of the CB will experience a voltage sag during the fault, and once it is cleared, a supply interruption is expected. It will be demonstrated that the last assumption in the presence of large IMs may not be strictly true because an unintentional IO can occur before the CB reclosing.

As a matter of fact, this unexpected IO is the guiding principle of the voltage sag studied in the present paper. The unintentional IO begins once a fault is cleared and ends when the CB recloses to restore the electrical supply; typically, this reclosing time in DNs is between 500 ms and 1 s. Once the fault is cleared, transiently, IMs act as generators due to their remnant kinetic energy until the CB recloses the circuit. This phenomenon is partially analyzed in bus transfer during industry operations; for example, see Reference [52]. To achieve seamless transitions in this fast bus transfer, a microprocessor-based scheme has been implemented in Reference [53]. Nevertheless, none of these studies has considered the produced IO in the DN due to the IMs, on which the current PQ assessment is focused. Therefore, this paper makes a contribution to voltage sag characterization studies.

The paper is structured as follows: Section 2 reviews the present voltage sag characterization. In Section 3, the main features of the proposed voltage sag are detailed as well as its analytical expression.

Section 4 shows the test system under investigation and the three-phase Matlab/Simulink model adopted. Section 5 displays the simulation results derived from the model. Section 6 discusses the model validation comparing the analytical form presented in Section 3, with both simulation results and field measurements. Lastly, Section 7 summarizes the main conclusions of the paper.

2. Voltage Sag Review

2.1. Introduction

Since voltage sags are a short-duration reduction of the root-mean-square (RMS) voltage, it can be said that their characterization depends on (i) their magnitude (i.e., voltage characteristic), (ii) their duration, (iii) their type, (iv) the type of recovery, (v) the phase-angle-jump, and (v) their origin. Nevertheless, the origin of the voltage sag will, in fact, dictate its type and its recovery. Moreover, voltage sags can change their type and become evolutionary voltage sags because the fault conditions can change (i.e., faults can evolve from one type of fault to another during the same event). Other parameters such as the point-on-wave in both the initiation and the recovery have been the object of study in Reference [54]. Besides this, an interesting analysis is done by Bastos et al. in Reference [55].

On the other hand, the data extracted from a measured sag are gained from the sag type characteristic voltage (i.e., the complex voltage) and by the positive–negative factor as published in References [56,57]. Particularly, according to Reference [56], there is a need to compress the three-phase voltage sag event into a single event, which is still under study by the IEEE project group P1159.2. Furthermore, a two-type contribution has been made in Reference [57], considering a three-phase–three-angle method. To obtain the main characteristics from a measured voltage sag, relevant methods are available in the literature, among which are the following:

1. Complex voltage: This method uses symmetrical components to evaluate the type of sag; thus, comparing the angle between the definite sequence and the negative sequence, the sag type is obtained. The voltage characteristic is computed with the complex sum and difference of the positive and negative sequence-voltage.
2. RMS voltage: These algorithms use the RMS voltage as input; thus, the computation of the six RMS values of the phase-to-phase and phase-to-neutral voltages is proposed. Therefore, the voltage characteristic, the voltage sag type and the positive-negative factor are extracted from these six voltage time-domain values.
3. Advanced features: Within this group, a considerable number of articles have been published, and the main aim is to extract features from the voltage waveform considering a wide range of measurements and classify them with an algorithm that helps to compress the data. From all studies, the most appropriate methods are summarized in Reference [58]. Furthermore, Cai et al. in Reference [59] developed a mapping strategy for multiple disturbances.

In Reference [40], the authors focus on voltage sag characterization; thus, a general distinction is made considering balanced and unbalanced voltage sags originating from faults (i.e., short-circuits). Nonetheless, in this study, other sources of voltage sags, such as induction or synchronous motors during their starting process, are considered. Moreover, a further classification is included in Reference [60], where the voltage sag produced during transformer energization is evaluated.

Previously, in faults originated by short-circuits, between the fault inception and the fault recovery, the voltage characteristic has been considered constant; see Figure 1. However, this voltage characteristic may not be constant during the voltage sag; for instance, see Figure 3 of Reference [61]. One reason for this voltage characteristic increasing or decreasing during the fault could be that the impedance seen from the source changes during the disturbance. Another cause for a decrease in the residual voltage during the voltage sag can occur in the presence of IMs. In such a way, immediately following a fault, the IMs tend to contribute towards the fault (e.g., roughly one or two cycles) so that the drawn current decreases in time and the voltage value also decreases; this phenomenon is thoroughly detailed in Reference [62].

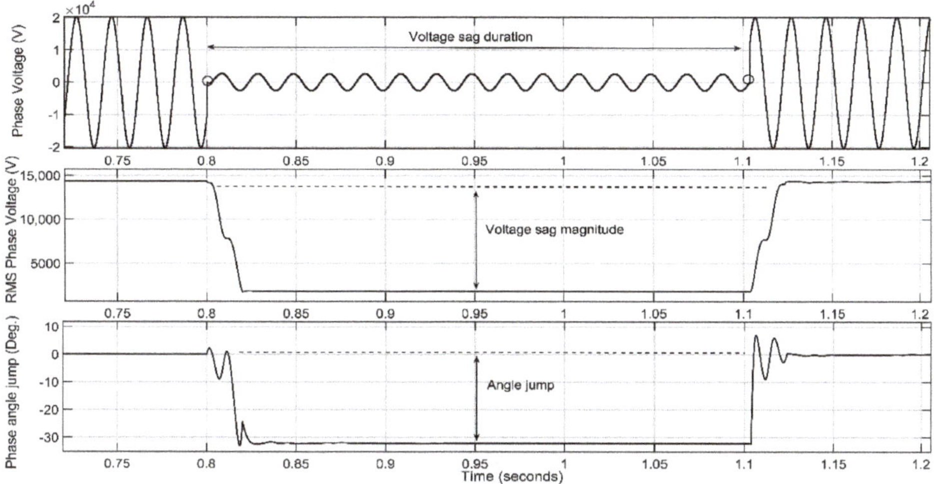

Figure 1. Example of a constant voltage sag produced by a single-line-to-ground (SLG) fault and its characteristics.

To summarize, the studied sources of voltage sags are listed as follows:

1. Short-circuits (constant retained voltage);
2. Motors start—either synchronous or induction motors (non-constant retained voltage);
3. Transformer energization (non constant retained voltage);
4. Short-circuits in the presence of IMs (non-constant retained voltage);
5. Sudden load changes—e.g., composite load connections (non-constant retained voltage).

2.2. Voltage Sag Main Features

This subsection will show the main characteristics of voltage sags graphically in order to compare these parameters with the proposed voltage sag of the subsequent Section 3. Although the voltage sag representation of Figure 1 is not innovative, this figure and its analysis are necessary to highlight the improvement made with the proposed voltage sag as the object of study.

In Figure 1, the three principal magnitudes are shown in different plots. The first plot depicts the phase voltage waveform of the faulted phase. In this waveform, the beginning and the end of the fault can be observed as well as the point-on-wave in both the initiation and clearing (these two points are within two circles in the voltage waveform). The retained voltage—that is, the magnitude of the voltage sag—is shown in the second plot of Figure 1, and lastly, the phase-angle-jump is observable in the third plot of the same figure. For this particular event, the duration is 300 ms, and it originated due to a single-line-to-ground fault (SLG), which belongs to voltage sag type A in the characterization done by Bollen [63], the retained voltage is 0.12 pu, and the phase-angle-jump is −30 degrees, which means that the grid is predominantly inductive.

Figure 2 show the voltage sag produced by IMs starting; in this case, a group of 4 IMs of 160 kW have been simulated (all the IM data are detailed in Appendix A). It can be seen that the duration of the event is 500 ms, the voltage depth is 0.96 pu, and the voltage sag is balanced. In these cases, the strength of the grid plays a pivotal role. The first plot depicts the RMS phase voltage, and the second depicts the current.

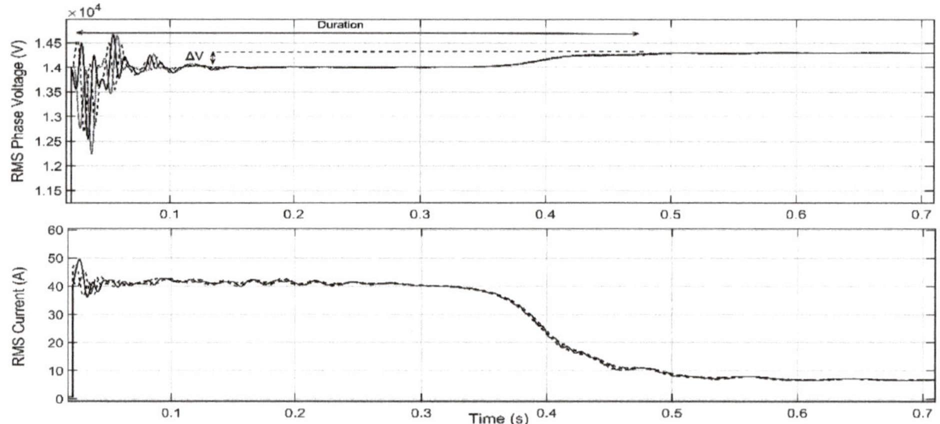

Figure 2. Example of a non-constant voltage sag produced by an induction motor (IM) start.

In Figure 3a, an example of the voltage sag produced when an IM contributes to a three-phase fault is shown. In the figure, the three-phase fault (Type A) is plotted. Nevertheless, to properly observe the voltage variation, in Figure 3b, the voltage sag is enlarged. In this figure, the decrease in the voltage characteristic during the event can be observed.

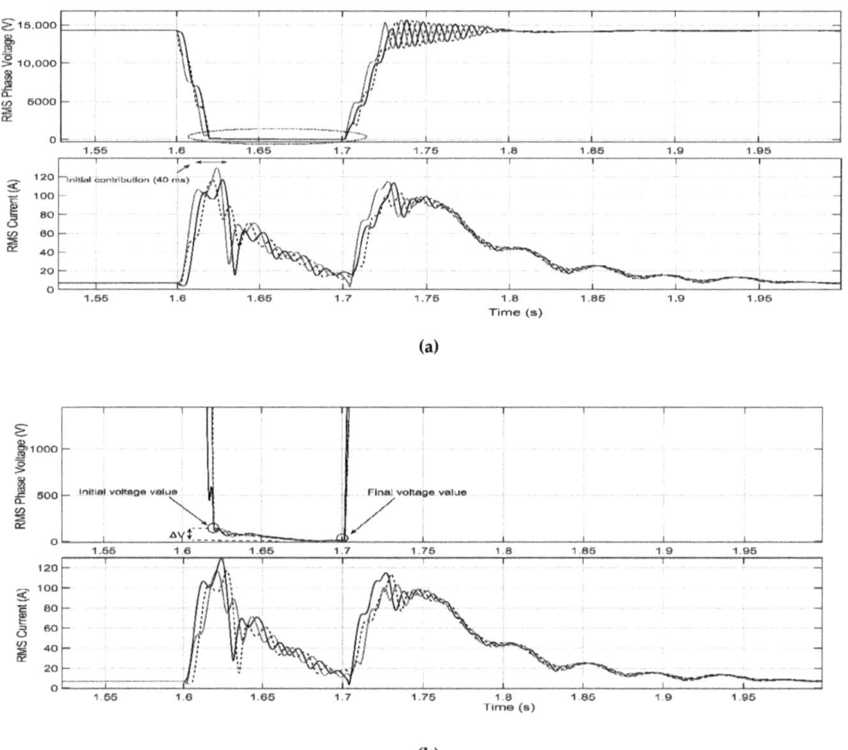

Figure 3. Example of a non-constant voltage sag produced for a three-phase fault with IMs. (**a**) RMS voltage and current during the event; (**b**) enlarged section of the voltage sag characteristic.

This voltage reduction during the voltage sag is due to the brief IM contribution towards the fault, which starts at t = 1.62 s, and in 2 cycles, the current starts to decrease. In these cases, the voltage variation due to this effect is ΔV = 125 V.

Taking into consideration that the loss of speed during the fault is low, the voltage sag due to the post-fault reacceleration is negligible (see the voltage characteristics between t = 1.7 s and t = 1.9 s in Figure 3a).

Figure 4 depicts the voltage sag produced when a transformer is energized; due to the current drawn by the transformer, transiently, the voltage drops to 0.94 pu. As can be seen, the voltage characteristic is not constant; it reaches a minimum value at t = 0.82 and starts to increase until t = 1s when the voltage sag is absolutely recovered. This voltage sag is unbalanced due to the current unbalance produced by the high content of current harmonics. Observing the third plot of Figure 4, the high content of harmonics of the current waveform can be seen; these harmonics are a direct consequence of the transformer core saturation [64].

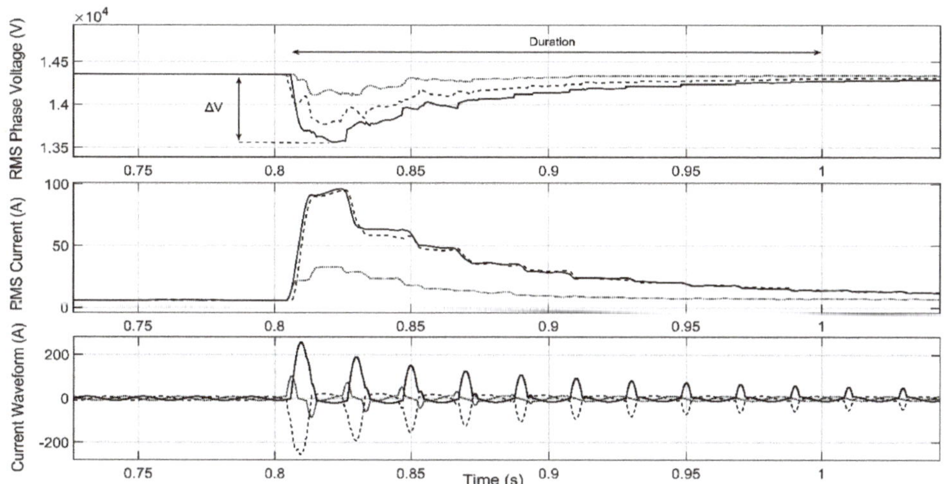

Figure 4. Example of a non-constant voltage sag produced by transformer energization.

To summarize, in this section, voltage sags with constant and non-constant retained voltage have been evaluated, which helps to clarify the voltage sag topology presented in this paper.

3. Voltage Sag Modelling

3.1. General Overview

The current section is divided into four parts. The first introduces a general overview of the proposed voltage sag, the second analyses the analytical expression that follows the voltage sag magnitude, the third part examines its duration, and finally, the fourth part focuses on the recovery process once the CB recloses the circuit. Before the voltage sag modelling guidelines are analyzed in the further subsections, to observe the importance of this phenomenon clearly, Figure 5 shows the difference between one scenario with IMs and another one with the same rated apparent power when removing the IMs.

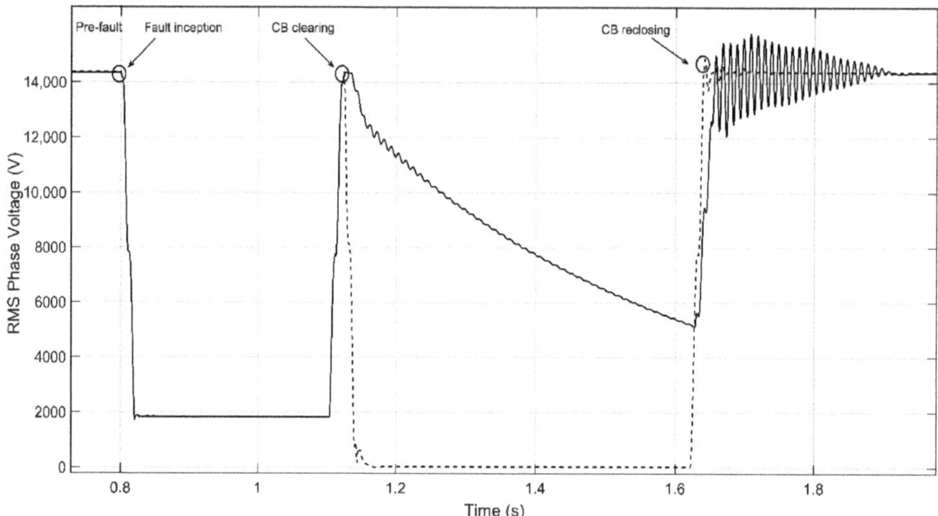

Figure 5. Comparison between scenarios, distribution network (DN) with IMs (solid line) and without IMs (dashed line).

As can be seen in Figure 5, during the process between the clearing time (t = 1.1 s) and the CB reclosing (t = 1.6 s), in the case in which high IMs are present, a voltage sag appears. Meanwhile, when the IMs are removed, as expected, an interruption occurs.

3.2. Voltage Sag Magnitude: Analytical Expression

The analytical expression that follows the voltage sag during the island is, in fact, the IM stator voltage, which in turn depends on the stator current and its mechanical speed. Therefore, to obtain this expression, the IM equations have to be recalled. In a steady state, the IM acts as a motor and is being supplied by the grid and running at a specific point of operation. This point of operation is defined by the slip, the mechanical load torque coupled with the rotor shaft and the mechanical speed. These parameters dictate the initial kinetic energy of the machine before the fault appears. When a fault occurs, depending on the characteristics of the originated voltage sag (i.e., voltage sag magnitude and duration), the IM will lose speed according to this severity. Once the CB clears the fault, IMs, due to their stored kinetic energy, transiently act as generators. The pre-islanding conditions will dictate the voltage sag magnitude during the IO. Hence, the IMs' initial kinetic energy is expressed as Equation (1)

$$E_{Kin\_initial} = \frac{1}{2} J_T \, \omega_{pre\_islanding}^2 \tag{1}$$

where $E_{Kin\_initial}$ is the initial kinetic energy prior to the CB opening, J_T is the total moment of inertia (considering the rotor inertia and mechanical load) and $\omega_{pre\_islanding}$ is the mechanical speed before the island is formed. The IM set of electrical equations are considered in stator reference and expressed in Park's transformation to avoid the dependence of the coupling inductance with the mechanical angle θ_m Equations (2)–(5):

$$v_{sd} = (R_s + L_s \frac{d}{dt}) i_{sd} + M \frac{d\,i_{rd}}{dt} \tag{2}$$

$$v_{sq} = (R_s + L_s \frac{d}{dt}) i_{sq} + M \frac{d\,i_{rq}}{dt} \tag{3}$$

$$0 = (R_r + L_r \frac{d}{dt}) i_{rd} + M \frac{d\,i_{sd}}{dt} + L_r \, (\, \omega_m \wp) \, i_{rq} + M \, (\, \omega_m \wp) \, i_{sq} \tag{4}$$

$$0 = (R_r + L_r \frac{d}{dt}) i_{rq} + M \frac{d\,i_{sq}}{dt} - L_r\,(\,\omega_m \wp)\,i_{rd} - M\,(\,\omega_m \wp)\,i_{sd} \tag{5}$$

where V and i represent the voltage and current, respectively, M is the coupling inductance, L represents the inductance of the windings, ω_m is the mechanical speed, ψ is the reference angle (stator reference in this case), $\wp$ is the IM pole pairs, θ_m is the rotor angle, the sub-indices s and r denote stator and rotor, respectively and dq refers to the direct and quadrature components of the Park's transformation. The relation between abc and dq can be expressed as in Equation (6)

$$\begin{aligned} v_S &= \sqrt{v_{sd}{}^2 + v_{sq}{}^2} \\ i_S &= \sqrt{i_{sd}{}^2 + i_{sq}{}^2} \end{aligned} \tag{6}$$

Following the CB opening, due to the lack of supply, the equilibrium between torques is broken, its speed derivative becomes negative, and therefore only the dynamic torque drives the machine. Moreover, the feeder loads which remain within the island represent a load torque for the IMs when acting as generators. Thereby, the electrical loads act as a torque decelerating the machines. The mechanical expressions acting as generators once the IMs are removed from the grid are computed as Equations (7) and (8):

$$\frac{d\omega_m}{dt} = \frac{(\Gamma_{res} + \Gamma_b + \Gamma_{eload})}{J_T} \tag{7}$$

$$\omega_m = \frac{d\theta_m}{dt} \tag{8}$$

where Γ_{em} is the electromagnetic torque, Γ_b is the friction torque, Γ_{res} is the mechanical load torque, θ_m is the angular position, J_T is the total moment of inertia (considering the rotor inertia and mechanical load), and Γ_{eload} represents the equivalent torque developed by the feeder-load electrical power. If we consider load voltage dependence, constant impedance and constant current load models have to be considered; thereby, electrical loads are divided into P_1, which is assumed to be the sum of all constant impedance loads, and P_2, the sum of all constant current loads within the island. It follows that

$$P_{eload} = \frac{P_1}{|\Delta V|^2} + \frac{P_2}{|\Delta V|} \tag{9}$$

where the sum of all constant impedance and constant currents considering n low voltage (LV) nodes is computed as Equation (10):

$$\begin{aligned} P_1 &= \sum_{i=0}^{n} P_i \\ P_2 &= \sum_{j=0}^{n} P_j \end{aligned} \tag{10}$$

By integrating the IM speed derivative and mechanical angle with respect to time (7) and (8) during the island, the temporal evolution of $\omega_m\,(t)$ and $\theta_m\,(t)$, are given by Equations (11) and (12):

$$\int_{t_i}^{t_f} \frac{d\omega_m}{dt}\,dt = \int_{t_i}^{t_f} \frac{(\Gamma_{res} + \Gamma_b + \Gamma_{eload})}{J_T}\,dt = \omega_m(t) = \int_{t_i}^{t_f} \frac{(\Gamma_{res} + \Gamma_b + \Gamma_{eload})}{J_T}\,dt \tag{11}$$

$$\int_{t_i}^{t_f} \frac{d\theta_m}{dt}\,dt = \int_{t_i}^{t_f} \frac{d\omega_m}{dt}\,dt = \theta_m(t) = \int_{t_i}^{t_f} \omega_m\,dt \tag{12}$$

where t_i is the instant when the CB opens and t_f when the CB recloses. Similarly, by integrating the current derivatives with respect to time in Equations (2)–(5), the temporal evolution of $isd(t)$, $isq(t)$, $ird(t)$, and $irq(t)$ is given by Equations (13)–(16):

$$\int_{t_i}^{t_f} (v_{sd} - i_{sd}R_s)\, dt = \int_{t_i}^{t_f} \left(L_s\frac{di_{sd}}{dt}q + M\frac{d\,i_{rd}}{dt}\right) dt = L_s\cdot i_{sd}(t) + M\cdot i_{rd}(t) = \int_{t_i}^{t_f} (v_{sd} - i_{sd}R_s)\, dt \tag{13}$$

$$\int_{t_i}^{t_f} (v_{sq} - i_{sq}R_s)\, dt = \int_{t_i}^{t_f} \left(L_s\frac{di_{sq}}{dt}q + M\frac{d\,i_{rq}}{dt}\right) dt = L_s\cdot i_{sq}(t) + M\cdot i_{rq}(t) = \int_{t_i}^{t_f} (v_{sq} - i_{sq}R_s)\, dt \tag{14}$$

$$\int_{t_i}^{t_f} -(i_{rd}\cdot R_r) - (L_r\cdot\omega_m\wp\cdot i_{rq}) - (M\omega_m\wp i_{sq})dt = \int_{t_i}^{t_f} \left(L_r\frac{di_{rd}}{dt} + M\frac{d\,i_{sd}}{dt}\right) dt$$

$$L_r\cdot i_{rd}(t) + M\cdot i_{sd}(t) = \int_{t_i}^{t_f} -(i_{rd}\cdot R_r) - (L_r\cdot\omega_m\wp\cdot i_{rq}) - (M\omega_m\wp i_{sq})dt \tag{15}$$

$$\int_{t_i}^{t_f} -(i_{rq}\cdot R_r) + (L_r\cdot\omega_m\wp\cdot i_{rd}) + (M\omega_m\wp i_{sd})dt = \int_{t_i}^{t_f} \left(L_r\frac{di_{rq}}{dt} + M\frac{d\,i_{sq}}{dt}\right) dt$$

$$L_r\cdot i_{sq}(t) + M\cdot i_{sq}(t) = \int_{t_i}^{t_f} -(i_{rq}\cdot R_r) + (L_r\cdot\omega_m\wp\cdot i_{rd}) + (M\omega_m\wp i_{sd})dt \tag{16}$$

By solving the described set of Equations (11) and (12) and Equations (13)–(16) between t_i and t_f, the function that follows the voltage sag with respect to time is given by the IM stator voltages V_{sd} and V_{sq}. Furthermore, by solving Equations (11) and (12), one can observe that the IM mechanical speed decays exponentially due to the lack of supply; as a consequence, this decay is also observed in the stator voltages. Since the voltage sag magnitude is given by the IM stator voltage, its function has to be a negative exponential. In such a way, the voltage sag analytical expression with respect to time follows:

$$V_{sag}(t) = V_{pre\_fault}\cdot t^b; \ \forall t \in [t_i, t_f] \tag{17}$$

where $V_{sag}(t)$ is the RMS phase voltage value, V_{pre_fault} is the RMS pre-fault phase voltage value, and b is a non-dimensional coefficient. Given the natural tendency of the IM to decelerate due to the lack of grid supply, this coefficient has to be below zero. Note that b values close to zero correspond to low loaded feeders, whereas large negative b values correspond to large loaded feeders. Figure 6 plots the function that follows the analytical expression for a given b value. It is crucial to highlight the fact that, as mentioned earlier, the analytical expression of the voltage sag magnitude is given by the IM stator voltages, which result from solving the differential Equations (11)–(16). Therefore, the b factor of Equation (17) can be obtained after this set of equations have been solved. Hence, Equation (17) provides a general expression to model this voltage sag which is performed after solving the differential equations of the system. Note, however, that depending on the terms in Equation (11) (i.e., the feeder-load power, the load torque and the friction torque), several b values can be obtained. Particularly, in Figure 6, $t_i = 1.04$ s, $t_f = 1.54$ s and $b = -1.52$. It is relevant to highlight the fact that, as discussed above, this curve expression is only valid between t_i and t_f, where in fact the voltage sag characteristic is of interest.

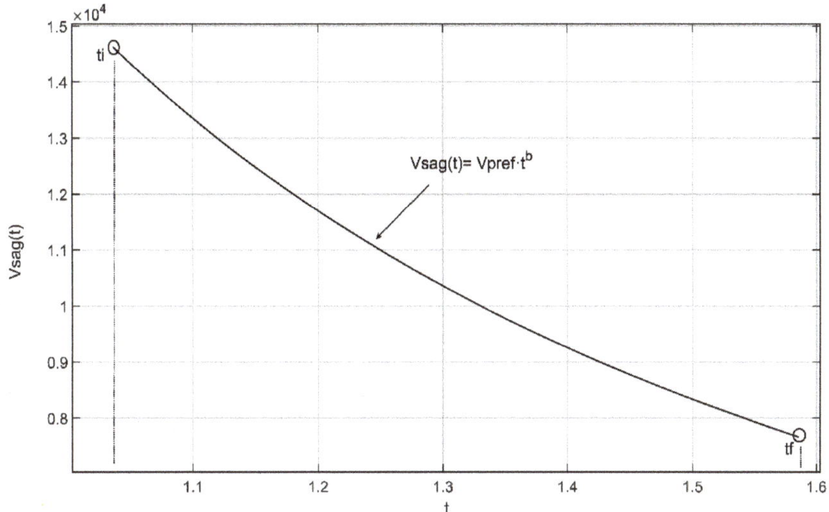

Figure 6. Voltage sag analytical function for a given *b* parameter.

3.3. Voltage Sag Duration

Voltage sag duration is defined as the period between the fault inception and the fault clearing; see the first plot of Figure 1. The voltage sag duration depends directly on the protective device's settings. Particularly for the voltage sag described here, the voltage sag duration depends on the preset reclosing that takes place after a fault is cleared. This preset reclosing operation is widely implemented in DN relays, principally to avoid the need to operate the CB for non-permanent faults manually (e.g., self-extinguished faults). Commonly, DSOs set the first reclosing between 0.5 and 1 s.

3.4. Voltage Sag Recovery

Voltage recovery takes place at the instant when the CB clears a fault, as previously mentioned; depending on the type of voltage sag, it can be either abrupt (one stage recovery) or discrete (multi-stage recovery). Nonetheless, in the present voltage sag topology, the recovery is dictated by the CB reclosing following a preset time after the CB clearing. It is worth pointing out that, in this case, unlike the recovery due to short-circuits, the recovery can be defined in two steps.

The first step is due to the CB zero-crossing, which has to occur at the same instant for the three-phases, and the second is the recovery until the voltage reaches the pre-fault value. In fact, this is due to the high current drawn by the IM as a result of the out-of-phase reclosing and the subsequent IM reacceleration. The transient due to the out-of-phase reclosing is dictated by the difference in voltage amplitude between the grid and islanding as well as by the difference in frequency between the grid and islanding [52].

4. System Under Investigation

The system by which measurements have been recorded is a real DN which operates radially with an infeeder from a transmission system at 120 kV. The single-line diagram of the system is depicted in Figure 7a. As shown in the figure, the substation wye/delta transformer supplies three medium voltage (MV) feeders, Feeder 2 and Sub-Feeders A and B, forming the system analyzed in this work. It is relevant to point out that Feeders 1 and 3 are not detailed, because they are not relevant to the task due to their radial configuration and also because Feeder 1 is the faulted one in which the unintentional IO object of study occurs. The end-users are supplied at an LV of 0.4 kV through their MV/LV Dyn 11 transformers.

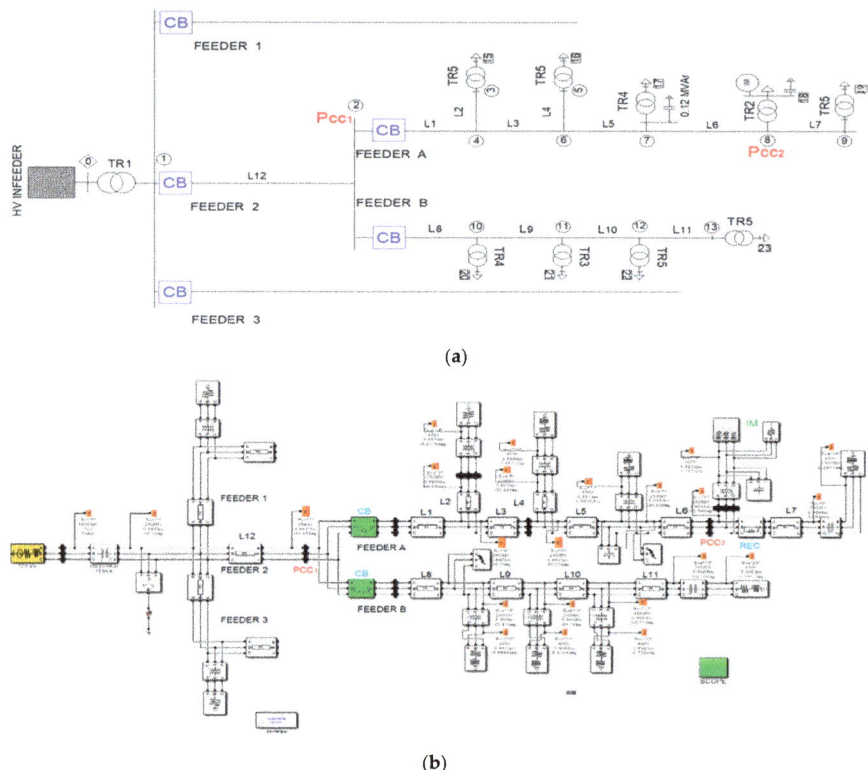

Figure 7. Test system under study. (**a**) One-line diagram of the test system.; (**b**) Matlab/Simulink implementation of the test system.

The MV system is grounded via a zig-zag transformer with a resistance of 23 Ω that limits the fault current to 630 A. The short circuit capacity at the 120 kV busbar is 1000 MVA. The main system parameters and the IMs data are detailed in Appendix A.

The equipment used to record the events comprise the protective relays installed at both PCC_1 and PCC_2. The fault recorder registers voltage and current signals with 32 samples per cycle. The overcurrent relays are equipped with an oscillography function; in PCC_1, the relay registration is set to 400 ms, while in PCC_2, this value is set to 600 ms. More information about the relays and their recording characteristics can be obtained from the manufacturer data available in Reference [65]. These relays record the voltage, currents, powers and frequency during the event. For each event, a pre-trigger of 200 ms is used in order to obtain the pre-fault active and reactive powers; thus, it is possible to compute the feeder-load power before the fault.

5. Simulation Results

This section will summarize the results of the simulations carried out in the three-phase Matlab/Simulink model adopted based on the single-line diagram of the real DN; see Figure 7b. The result discussion will be done by analysing the main features of the proposed voltage sag separately. Therefore, the further three subsections are as follows: (Section 5.1) gives the voltage sag magnitude, (Section 5.2) gives the voltage sag duration, and lastly (Section 5.3) gives the voltage recovery. The main parameters of the simulated scenarios are detailed in each subsection; nevertheless, the relay settings, fault characteristics and IM point of operation, which are the same for all simulations, are detailed below:

- An SLG fault with a fault resistance of 5 Ω is simulated at node 7; the implemented settings of the CB located at PCC_1 (header of Feeder A, see Figure 7b) are listed in Table 1, where I_{of} represents an admissible overload factor; $I_{threshold}$ is the steady-state current threshold; and n, k, t are parameters of the curve. Particularly, the equation where the parameters mentioned above are used is defined in Equation (18). Besides, the most frequent curves for overcurrent protection are detailed in Reference [66]. The operation time T of the phase overcurrent relay is defined by Equation (18).

$$T(I) = \frac{t \cdot k}{\left(\frac{I}{I_{threshold} \cdot I_{of}}\right)^n - 1} \tag{18}$$

where factor k is adjustable and typically ranges between 0.05 and 1.6, and n and t factors are used to choose the slope of the relay curves.

- IM is running under no-load torque at the time the fault occurs.

Table 1. Overcurrent relay parameters.

ANSI Code	k	n	t	$I_{threshold}$ (A)	I_{of}	ANSI Curve
50	0.05	0.02	0.14	100	1.25	Standard inverse

5.1. Voltage Sag Magnitude

This part evaluates the simulation results with different values of feeder loads in order to analyze the different originated voltage sag characteristics and compare them with the previously described analytical expression. Figure 8 shows, as an example, one particular scenario and the analytical function that follows the voltage sag characteristic. In this simulation, the load feeders have been set to 53 kW. In the figure, the solid black line represents the simulation results, the solid blue line represents the analytical function of the voltage sag magnitude, and solid blue double arrows indicate the magnitude of the sag, which increases between t_i ($t = 1.05$ s) and t_f ($t = 1.55$ s).

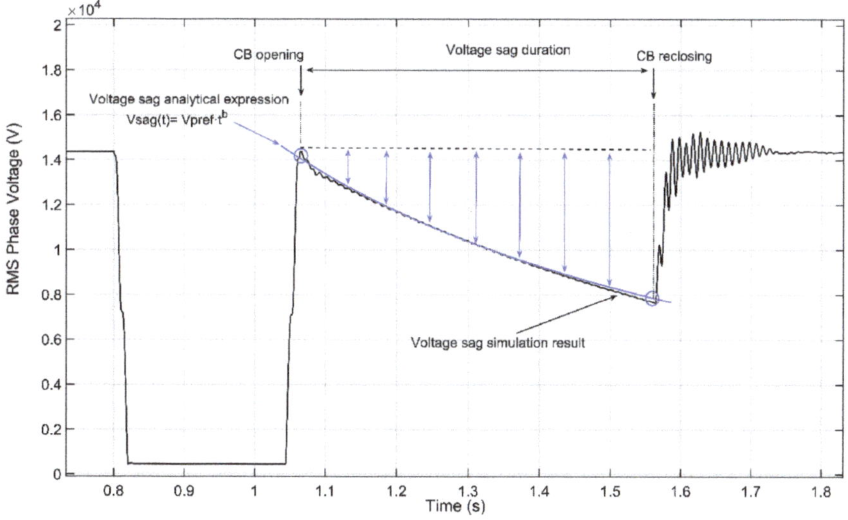

Figure 8. Voltage sag magnitude comparison between simulation results and the proposed analytical curve.

Moreover, four simulations have been performed with different feeder-load values in each one in order to observe different curve slopes. In Figure 9, four curves are displayed: the solid curves are the voltage magnitude of four voltage sags which result from simulations, and the dashed lines represent the function which fits the voltage sag magnitude according to each particular b value. Table 2 summarizes the main values of the simulated scenarios.

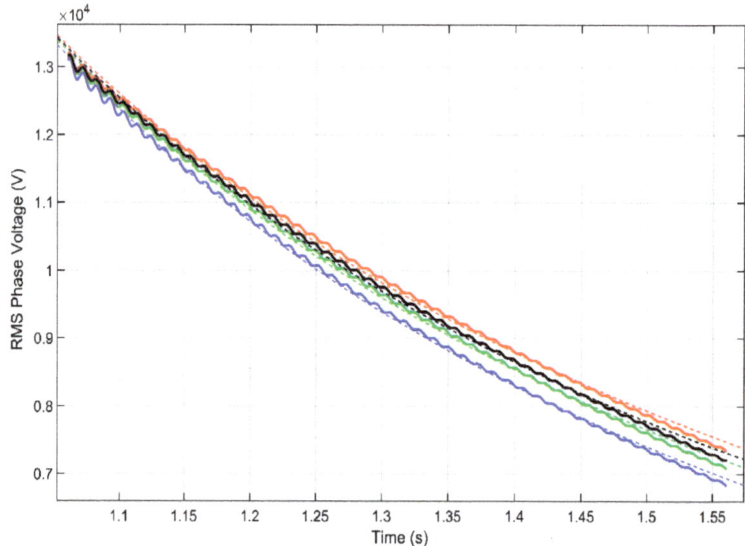

Figure 9. Voltage sag magnitude of the four simulated scenarios with different feeder-load values.

Table 2. Simulated scenarios and resulting parameters of Section 5.1.

Scenarios	Color	Feeder Loads (kW)	b	ΔV *(pu)*	*Duration (ms)*
1	Red	64	−1.492	0.49	500
2	Black	80	−1.54	0.5	500
3	Green	100	−1.576	0.51	500
4	Blue	128	−1.655	0.53	500

5.2. Voltage Sag Duration

In this section, four CB reclosing times have been tested to observe the duration of the voltage sag with a fixed value of feeder loads of 140 kW. From Figure 10, it can be seen that, depending on the reclosing time, as expected, the duration is higher, and consequently the voltage magnitude decreases with time. For all simulations, the initial time corresponds to 0.4 s when the CB clears the fault, and the sag recovery takes place depending on the reclosing time for each scenario.

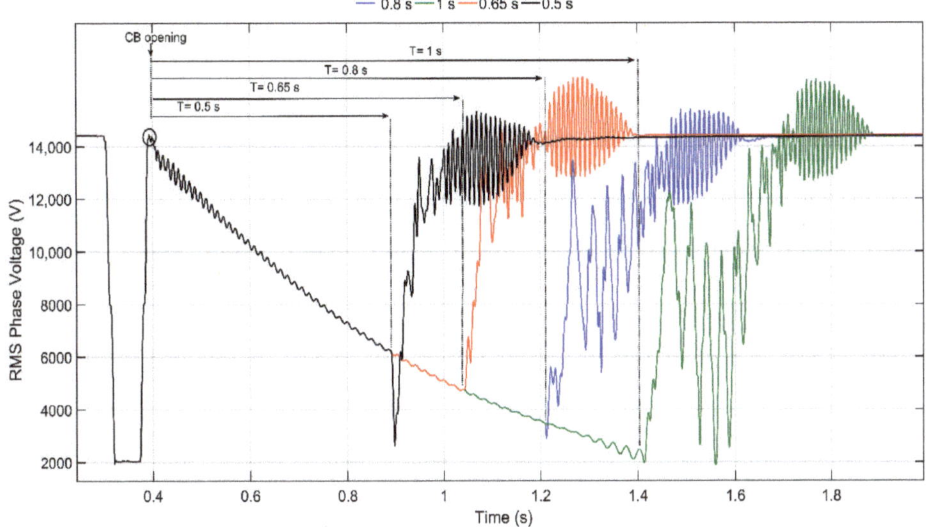

Figure 10. Voltage sag duration for the four simulated reclosing times.

5.3. Voltage Sag Recovery

By observing Figure 11, as expected, the recovery takes place for all phases at the same instant, t = 1.63, when the CB recloses the circuit; however, the voltage amplitude is not recovered until 80 ms after this. Thereby, it has been demonstrated that defining the recovery process of this voltage sag accurately has to be done with two steps: the first is when the CB recloses the circuit, and the second is when the voltage reaches the pre-fault value.

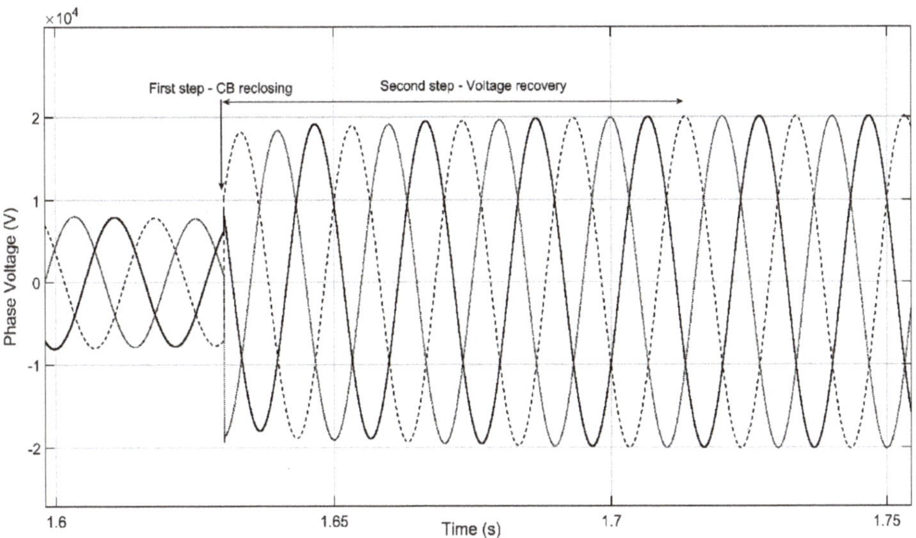

Figure 11. Two-step voltage sag recovery.

6. Voltage Sag Model Validation

This section is focused on validating the proposed voltage sag, which has been analytically modelled in Section 3 by comparing simulation results with field measurements. For reasons of brevity, among all available real events, only two have been selected. Each event will be discussed separately detailing the voltage sag characteristics, and Table 3 summarizes the main features of these two events. Note that, since the voltage magnitude is non-constant, the voltage value of the fifth column of Table 3 is referred to the last RMS value before the reclosing takes place. Monitors have been installed at nodes PCC_1 and PCC_2 (See Figure 7a), and field measurements have been recorded at the highest voltage side of the MV/LV transformer in PCC_2 where the IMs are allocated. In both Figures 12 and 13, the three-phase voltages are compared, and simulation results are depicted in solid lines while field measurements are plotted in dashed lines.

Table 3. Voltage sags characteristic summary for Events *I* and *II*.

Event	Beginning (h)	Ending (h)	b	*Sag Magnitude (pu)*	*Sag Duration (ms)*
I	17:01:09.776	17:01:10.636	−1.05	0.61	800
II	11:49:04.702	11:49:05.456	−1.7	0.41	800

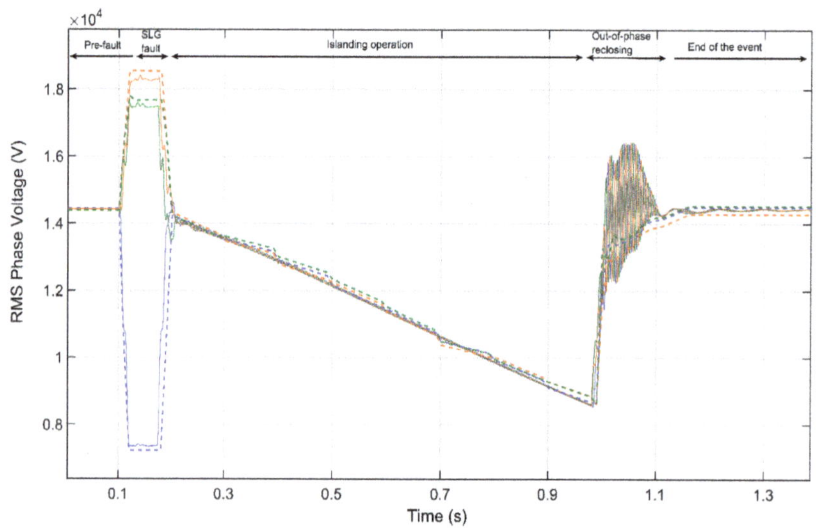

Figure 12. *Event I* comparison. Solid (simulation)/dashed (measurements): orange (Phase A), blue (Phase B) and green (Phase C).

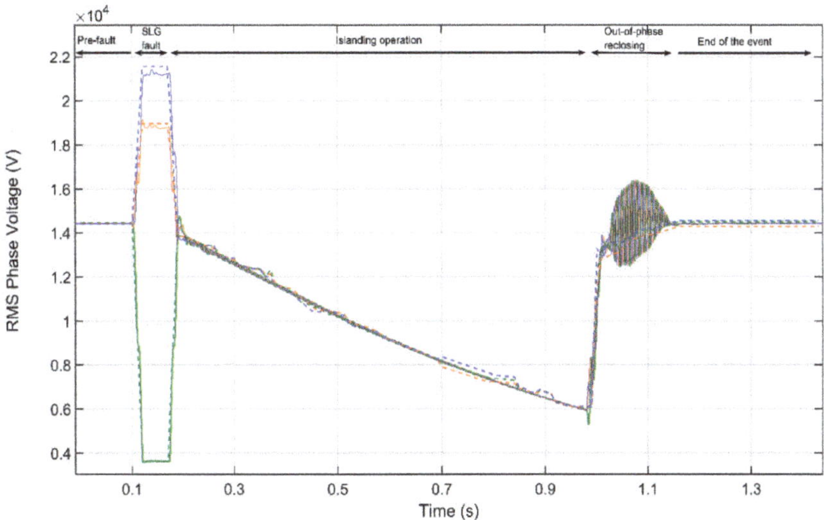

Figure 13. *Event II* comparison. Solid (simulation)/dashed (measurements): orange (Phase A), blue (Phase B) and green (Phase C).

Event I: In Figure 12, the event starts in pre-fault conditions under a steady state until the SLG fault occurs at phase b in node 8 (See Figure 7a) at 0.1 s, and it is cleared within 100 ms. At that moment, the IMs were running under no-load torque, and the active-power value of the feeder loads was 810 kW. The voltage sag duration was 800 ms, coinciding with the CB reclosing time; at that time, the phase voltage was 8.78 kV for all phases in both simulation and field measurements. In this event, the voltage sag characteristic is defined by $b = -1.05$.

Event II: In Figure 13, the event starts in pre-fault conditions under a steady state until the SLG fault occurs at Phase C between Nodes 7 and 8 (Line 6) at 0.1 s, and it is cleared within 100 ms. At that moment, the IMs were running under no-load torque, and the active-power value of the feeder loads was 900 kW. The voltage sag duration was 800 ms, coinciding with the CB reclosing time; at that time, the phase voltage was 6 kV for all phases in both simulation and field measurements. In this event, the voltage sag characteristic is defined by $b = -1.7$. As in *Event I*, the recovery takes place in two stages: firstly, at the same instant for the three-phases due to the CB reclosing ($t = 1$ s), and the second step takes place at $t = 1.12$ s when the voltage is fully recovered.

It is essential to point out that after the first step of the recovery, between field measurements and simulations, a slight difference in the RMS phase voltage can be seen; this difference is due to the frequency dependence of the phase locked loop (PLL) measurement unit used in the simulation model and is also a result of the frequency oscillation during the recovery. Therefore, if the voltage waveform is taken into consideration instead of its computed RMS value, the simulation fits the recorded measurements equally for both events, and no oscillation appears. By observing the voltage of the waveform of Figure 11 in Section 5, one can observe that no oscillation occurs.

7. Conclusions

This paper makes a contribution to voltage sags characterization studies by presenting a novel voltage sag typology which occurs during unintentional IOs. Hence, the voltage sag has been analytically modelled, simulated and subsequently validated with field measurements recorded in a real DN.

As discussed in this paper, this voltage sag is defined as balanced, a non-constant retained voltage characterizes its magnitude, its duration is dictated by the CB reclosing, and its recovery takes place in a two-step process. One of the main goals of the present contribution is to describe the voltage sag non-constant voltage magnitude, which expands the present voltage sag typology. An analytical expression to model this voltage sag has been provided. It is seen that the voltage sag magnitude follows a negative exponential curve, and its slope is defined by b; this parameter is linked with the IM's mechanical loss of speed. It has also been found that the higher the feeder loads value, the higher b is, and consequently, the higher the slope of the voltage sag is.

In Section 6, the voltage sag model validation has been carried out with two real events; the comparison between the simulations and these events undoubtedly underscore the appropriateness of the presented analytical expression, as well as evidencing the accuracy of the three-phase Matlab/Simulink model adopted.

Further work will be focused on the equipment sensitivity analysis for this particular voltage sag via hardware-in-the-loop testing.

Author Contributions: All the authors have contributed equally to this study.

Funding: This research was funded by the Catalonian College of engineers, grant number EICPhD-2018.

Acknowledgments: The authors want to acknowledge the financial support and the willingness of sharing field measurements by the Spanish DSO Electrica Serosense Distribuidora. The authors would also like to thank the Catalonian College of Engineers for the grant received to carry out this research.

Conflicts of Interest: The authors declare no conflicts of interest.

Appendix A

Table A1. Test system parameters.

Transformers					
		V_p/V_s (kV)	S (MVA)	ε (%)	L_m/R_m (H/kΩ)
Substation transformer	TR1	120/25 kV	10	10.4	78; 313
	TR2	25/0.4 kV	1	6	180; 370
Distribution transformers	TR3	25/0.4 kV	0.63	5.1	350; 480
	TR4	25/0.4 kV	0.4	4.3	404; 694
	TR5	25/0.4 kV	0.25	4.1	811; 1666

Distribution Lines				
Lines	Nodes	Length (km)	Parameters: Z_1/Z_0 (Ω/km); C_0 (μF/km)	Loadability (A)
L1	2–4	5	0.306+j0.405/0.38+j1.62	300
L2	4–3	3.3	1.07+j0.441/1.3+j1.76,	120
L3	4–6	4.3	0.306+j0.405/0.38+j1.62	300
L4	6–5	4	1.07+j0.441/1.3+j1.76	120
L5	6–7	2.5	0.306+j0.405/0.38+j1.62	300
L6	7–8	2	0.306+j0.405/0.38+j1.62	300
L7	8–9	7	0.687+j0.416/0.8+j1.66	200
L8	2–10	3.8	0.127+j0.114/0.17+j0.45; 0.229	389
L9	10–11	4.43	0.208+j0.123/0.278+j0.492; 0.192	320
L10	11–12	4	0.687+j0.416/0.8+j1.66	200
L11	12–13	3.3	0.687+j0.416/0.8+j1.66	200
L12	1–2	0.28	0.306+j0.405/0.38+j1.62	300

Table A2. Induction motor (IM) data.

Internal Parameters	Values	Rated Parameters	Values
L_s	0.0078 H	P	160 kW
L_r	0.0078 H	$\wp$	1
R_r	0.0077 Ω	f	50 Hz
R_s	0.0137 Ω	V	400 V
L_m	0.0076 H	J	2.9 kgm^2

References

1. Moreno-Munõoz, A.; González de la Rosa, J.J. Integrating power quality to automated meter reading. *IEEE Ind. Electron. Mag.* **2008**, *2*, 10–18. [CrossRef]
2. Ali, S.; Wu, K.; Weston, K.; Marinakis, D. A machine learning approach to meter placement for Power Quality Estimation in Smart Grid. *IEEE Trans. Smart Grid* **2016**, *7*, 1552–1561. [CrossRef]
3. Broshi, A. Monitoring Power Quality Beyond EN 50160 and IEC 61000-4-30. In Proceedings of the 2007 9th International Conference on Electrical Power Quality and Utilisation, EPQU, Barcelona, Spain, 9–11 October 2007.
4. Gosbell, V.J.; Baitch, A.; Bollen, M.H. The Reporting of Distribution Power Quality Surveys. In Proceedings of the CIGRE/IEEE PES International Symposium Quality and Security of Electric Power Delivery Systems, CIGRE/PES 2003, Montreal, QC, Canada, 8–10 October 2003; pp. 48–52.
5. Florencias-Oliveros, O.; Agüera-Pérez, A.; González-De-La-Rosa, J.-J.; Palomares-Salas, J.-C.; Sierra-Fernández-Álvaro, J.-M.; Montero, J. Cluster Analysis for Power Quality Monitoring. Qualitative Analysis Based in Higher-Order Statistics for the Smart Grid. In Proceedings of the 2017 11th IEEE International Conference on Compatibility, Power Electronics and Power Engineering (CPE-POWERENG), Cadiz, Spain, 4–6 April 2017.
6. Mei, F.; Ren, Y.; Wu, Q.; Zhang, C.; Pan, Y.; Sha, H.; Zheng, J. Online Recognition Method for Voltage sags based on a deep belief network. *Energies* **2018**, *12*, 43. [CrossRef]
7. Guerrero-Rodríguez, J.M.; Cobos-Sánchez, C.; González-de-la-Rosa, J.J.; Sales-Lérida, D. An Embedded Sensor Node for the Surveillance of. *Energies* **2019**, *12*, 1561. [CrossRef]
8. de la Rosa, J.J.; Muñoz, A.M.; de Castro, A.G.; López, V.P.; Sánchez Castillejo, J.A. A web-based distributed measurement system for electrical power quality assessment. *Meas. J. Int. Meas. Confed.* **2010**, *43*, 771–780. [CrossRef]
9. Cisneros-Maga, R.A.; Medina, A.; Anaya-Lara, O. Time-domain voltage sag state estimation based on the unscented kalman filter for power systems with nonlinear components. *Energies* **2018**, *11*, 1411. [CrossRef]
10. Otcenasova, A.; Bodnar, R.; Regula, M.; Hoger, M.; Repak, M. Methodology for determination of the number of equipment malfunctions due to voltage sags. *Energies* **2017**, *10*, 401. [CrossRef]
11. Bollen, M.H.; Styvaktakis, E.; Gu, I.Y. Categorization and analysis of power system transients. *IEEE Trans. Power Deliv.* **2005**, *20*, 2298–2306. [CrossRef]
12. de la Rosa, J.; Sierra-Fernández, J.; Palomares-Salas, J.; Agüera-Pérez, A.; Montero, Á. An application of spectral kurtosis to separate hybrid power quality events. *Energies* **2015**, *8*, 9777–9793. [CrossRef]
13. Florencias-Oliveros, O.; González-de-la-Rosa, J.J.; Agüera-Pérez, A.; Palomares-Salas, J.C. Reliability monitoring based on higher-order statistics: A scalable proposal for the smart grid. *Energies* **2019**, *12*, 55. [CrossRef]
14. Shen, Y.; Abubakar, M.; Liu, H.; Hussain, F. Power Quality Disturbance Monitoring and Classification Based on Improved PCA and Convolution Neural Network for Wind-Grid Distribution Systems. *Energies* **2019**, *12*, 1280. [CrossRef]
15. Saini, M.K.; Kapoor, R. Classification of power quality events—A review. *Int. J. Electr. Power Energy Syst.* **2012**, *43*, 11–19. [CrossRef]
16. Reaz, M.B.; Choong, F.; Sulaiman, M.S.; Mohd-Yasin, F.; Kamada, M. Expert system for power quality disturbance classifier. *IEEE Trans. Power Deliv.* **2007**, *22*, 1979–1988. [CrossRef]
17. Youssef, A.M.; Abdel-Galil, T.K.; El-Saadany, E.F.; Salama, M.M.A. Disturbance Classification Utilizing Dynamic Time Warping Classifier. *IEEE Trans. Power Deliv.* **2004**, *19*, 272–278. [CrossRef]

18. Valtierra-Rodriguez, M.; De Jesus Romero-Troncoso, R.; Osornio-Rios, R.A.; Garcia-Perez, A. Detection and classification of single and combined power quality disturbances using neural networks. *IEEE Trans. Ind. Electron.* **2014**, *61*, 2473–2482. [CrossRef]
19. Huang, C.H.; Lin, C.H.; Kuo, C.L. Chaos synchronization-based detector for power-quality disturbances classification in a power system. *IEEE Trans. Power Deliv.* **2011**, *26*, 944–953. [CrossRef]
20. Manikandan, M.S.; Samantaray, S.R.; Kamwa, I. Detection and classification of power quality disturbances using sparse signal decomposition on hybrid dictionaries. *IEEE Trans. Instrum. Meas.* **2015**, *64*, 27–38. [CrossRef]
21. Massignan, J.A.; London, J.B.; Bessani, M.; Maciel, C.D.; Delbem, A.C.; Camillo, M.H.; de Lima Soares, T.W. In-field Validation of a Real Time Monitoring tool for distribution feeders. *IEEE Trans. Power Deliv.* **2017**, *33*, 1798–1808. [CrossRef]
22. Cai, L.; Thornhill, N.F.; Kuenzel, S.; Pal, B.C. Real-Time Detection of Power System Disturbances Based on k-Nearest Neighbor Analysis. *IEEE Access* **2017**, *5*, 5631–5639. [CrossRef]
23. Perez, E.; Barros, J. A proposal for on-line detection and classification of voltage events in power systems. *IEEE Trans. Power Deliv.* **2008**, *23*, 2132–2138. [CrossRef]
24. Borghetti, A.; Bosetti, M.; Nucci, C.A.; Paolone, M.; Abur, A. Integrated use of time-frequency wavelet decompositions for fault location in distribution networks: Theory and experimental validation. *IEEE Trans. Power Deliv.* **2010**, *25*, 3139–3146. [CrossRef]
25. Faried, S.O.; Member, S.; Aboreshaid, S.; Member, S. Stochastic Evaluation of Voltage Sags in Series Capacitor Compensated Radial Distribution Systems. *IEEE Trans. Power Deliv.* **2003**, *18*, 744–750. [CrossRef]
26. Quality, E.P. Stochastic Estimation of Voltage Sag Due to Faults in the Power System by Using PSCAD/EMTDC Software as a Tool for Simulation. *Power Qual.* **2007**, *13*, 59–63.
27. Martínez-Velasco, J.A. Computer-based voltage Dip Assessment in transmission and distribution networks. *Electr. Power Qual. Util. J.* **2008**, *14*, 31–38.
28. Martinez-Velasco, J.A.; Martin-Arnedo, J. Voltage sag studies in distribution networks—Part I: System modeling. *IEEE Trans. Power Deliv.* **2006**, *21*, 1670–1678. [CrossRef]
29. Martinez-Velasco, J.A.; Martin-arnedo, J. Voltage Sag Studies in Distribution Networks—Part II: Voltage Sag Assessment. *IEEE Trans. Power Deliv.* **2006**, *21*, 1679–1688. [CrossRef]
30. Martinez-Velasco, J.A.; Martin-arnedo, J. EMTP model for analysis of distributed generation impact on voltage sags. *IET Gener. Transm. Distrib.* **2007**, *1*, 112–119.
31. Bollen, M.H.J.; Bahramirad, S.; Khodaei, A. Is There a Place for Power Quality in the Smart Grid? In Proceedings of the International Conference on Harmonics and Quality of Power, ICHQP, Bucharest, Romania, 25–28 May 2014.
32. Bollen, M.H.J.; Olguin, G.; Martins, M. Voltage dips at the terminals of wind power installations. *Wind Energy* **2005**, *8*, 307–318. [CrossRef]
33. Sabin, D.D.; Bollen, M.H. Overview of IEEE Std 1564-2014 Guide for Voltage Sag Indices. In Proceedings of the International Conference on Harmonics and Quality of Power, ICHQP, Bucharest, Romania, 25–28 May 2014.
34. Dash, P.K.; Padhee, M.; Barik, S.K. Estimation of power quality indices in distributed generation systems during power islanding conditions. *Int. J. Electr. Power Energy Syst.* **2012**, *36*, 18–30. [CrossRef]
35. Martinez-Velasco, J.A.; Martin-Arnedo, J. Voltage Sag Studies in Distribution Networks—Part III: Voltage Sag Index Calculation. *IEEE Trans. Power Deliv.* **2006**, *21*, 1689–1697. [CrossRef]
36. Otcenasova, A.; Bolf, A.; Altus, J.; Regula, M. The Influence of Power Quality Indices on Active Power Losses in a Local Distribution Grid. *Energies* **2019**, *12*, 1389. [CrossRef]
37. Golovanov, N.; Lazaroiu, G.C.; Roscia, M.; Zaninelli, D. Power quality assessment in small scale renewable energy sources supplying distribution systems. *Energies* **2013**, *6*, 634–645. [CrossRef]
38. Markiewicz, H.; Klajn, A. EN 50160—Voltage Characteristics in Public Distribution Systems. Cenelec, 1 July 2001. Available online: https://standards.globalspec.com/std/9943573/EN2050160 (accessed on 9 July 2019).
39. Bollen, M.H. *Understanding Power Quality Problems*; IEEE press: Piscataway, NJ, USA, 2000.
40. Bollen, M.H.; Gu, I.Y. *Signal Processing of Power Quality Disturbances*; John Wiley & Sons: Hoboken, NJ, USA, 2005.
41. Rönnberg, S.; Bollen, M. Power quality issues in the electric power system of the future. *Electr. J.* **2016**, *29*, 49–61. [CrossRef]

42. Yalçinkaya, G. Characterization of voltage sags in industrial distribution systems. *IEEE Trans. Ind. Appl.* **1998**, *34*, 682–688. [CrossRef]
43. CIGRE/CIRED/UIE Joint Working Group C4.110. Voltage Dip Immunity of Equipment and Installations. CIGRE. Publ., 2010; pp. 1–18. Available online: http://www.uie.org/sites/default/files/CIGRE%20TB412%20voltage%20dip%20immunity%20of%20equipment%20and%20installations.pdf (accessed on 9 July 2019).
44. Bollen, M.H.; Cundeva, S.; Gordon, J.M.; Djokic, S.Z.; Stockman, K.; Milanovic, J.V.; Neumann, R.; Ethier, G. Voltage dip immunity aspects of power-electronics equipment—Recommendations from CIGRE/CIRED/UIE JWG C4.110. Proceeding of the International Conference on Electrical Power Quality and Utilisation, EPQU, Ohrid, Macedonia, 6–8 September 2011.
45. Pedra, J.; Bogarra, S.; Córcoles, F.; Monjo, L.; Rolán, A. Testing of three-phase equipment under voltage sags. *IET Electr. Power Appl.* **2015**, *9*, 287–296.
46. Bollen, M.H. Characterisation of voltage sags experienced by three-phase adjustable-speed drives. *IEEE Trans. Power Deliv.* **1997**, *12*, 1666–1671. [CrossRef]
47. Milanović, J.V.; Gupta, C.P. Probabilistic assessment of financial losses due to interruptions and voltage sags—Part II: Practical implementation. *IEEE Trans. Power Deliv.* **2006**, *21*, 925–932. [CrossRef]
48. Dolara, A.; Leva, S. Power quality and harmonic analysis of end user devices. *Energies* **2012**, *5*, 5453–5466. [CrossRef]
49. Yang, Y.; Xiao, X.; Guo, S.; Gao, Y.; Yuan, C.; Yang, W. Energy storage characteristic analysis of voltage sags compensation for UPQC based on MMC for medium voltage distribution system. *Energies* **2018**, *11*, 923. [CrossRef]
50. Tien, D.; Gono, R.; Leonowicz, Z. A multifunctional dynamic voltage restorer for power quality improvement. *Energies* **2018**, *11*, 1351. [CrossRef]
51. Roncero-Sánchez, P.; Acha, E. Design of a control scheme for distribution static synchronous compensators with power-quality improvement capability. *Energies* **2014**, *7*, 2476–2497. [CrossRef]
52. Beckwith, T.R.; Hartmann, W.G. Motor bus transfer: Considerations and methods. *IEEE Trans. Ind. Appl.* **2006**, *42*, 602–611. [CrossRef]
53. Muralimanohar, P.K.; Haas, D.; McClanahan, J.R.; Jagaduri, R.T.; Singletary, S. Implementation of a Microprocessor-Based Motor Bus Transfer Scheme. *IEEE Trans. Ind. Appl.* **2018**, *54*, 4001–4008. [CrossRef]
54. Varadarajan, M.; Swarup, K.S. Advanced Voltage Sag Characteristics: Point on Wave. *Gener. Transm. Distrib. IET* **2007**, *1*, 324.
55. Bastos, A.F.; Lao, K.W.; Todeschini, G.; Santoso, S. Accurate Identification of Point-on-Wave Inception and Recovery Instants of Voltage Sags and Swells. *IEEE Trans. Power Deliv.* **2019**, *34*, 551–560. [CrossRef]
56. Wang, Y.; Bagheri, A.; Bollen, M.H.; Xiao, X.Y. Single-Event Characteristics for Voltage Dips in Three-Phase Systems. *IEEE Trans. Power Deliv.* **2017**, *32*, 832–840. [CrossRef]
57. Madrigal, M.; Rocha, B.H. A contribution for characterizing measured three-phase unbalanced voltage sags algorithm. *IEEE Trans. Power Deliv.* **2007**, *22*, 1885–1890. [CrossRef]
58. Bollen, M.H. Algorithms for characterizing measured three-phase unbalanced voltage dips. *IEEE Trans. Power Deliv.* **2003**, *18*, 937–944. [CrossRef]
59. Cai, D.; Li, K.; He, S.; Li, Y.; Luo, Y. On the application of joint-domain dictionary mapping for multiple power disturbance assessment. *Energies* **2018**, *11*, 347. [CrossRef]
60. Zhang, L.; Bollen, M.H. Method for Characterisation of Three-Phase Unbalanced Dips (Sags) from Recorded Voltage Waveshapes. In Proceedings of the 21st International Telecommunications Energy Conference, INTELEC'99 (Cat. No. 99CH37007), Copenhagen, Denmark, 9 June 1999; p. 188.
61. Bollen, M.H.; Styvaktakis, E. Characterization of Three-Phase Unbalanced Dips (as Easy as One-Two-Three?). In Proceedings of the International Conference on Harmonics and Quality of Power, ICHQP, Orlando, FL, USA, 1–4 October 2000; Volume 1, pp. 81–86.
62. Bollen, M.H. Influence of motor reacceleration on voltage sags. *IEEE Trans. Ind. Appl.* **1995**, *31*, 667–674. [CrossRef]
63. IEEE Power Engineering Society; IEEE Power Electronics Society; IEEE Industry Applications Society; Bollen, M.H. *Understanding Power Quality Problems: Voltage Sags and Interruptions*; IEEE Press: Piscataway, NJ, USA, 2000.
64. Martinez, J.A.; Walling, R.; Mork, B.A.; Martin-Arnedo, J.; Durbak, D. Parameter determination for modeling system transients-Part III: Transformers. *IEEE Trans. Power Deliv.* **2005**, *20*, 2051–2062. [CrossRef]

65. Ormazabal Protection and Automation. EkorRPS-Multifunctional Protection. Available online: https:
     //www.ormazabal.com/es/tu-negocio/productos/ekorrps (accessed on 9 July 2019).
66. Benmouyal, G.; Meisinger, M.; Burnworth, J.; Elmore, W.A.; Freirich, K.; Kotos, P.A.; Leblanc, P.R.; Lerley, P.J.;
     McConnell, J.E.; Mizener, J.; et al. IEEE standard inverse-time characteristic equations for overcurrent relays.
     IEEE Trans. Power Deliv. **1999**, *14*, 868–871. [CrossRef]

Article

An Embedded Sensor Node for the Surveillance of Power Quality

José-María Guerrero-Rodríguez [1],*, Clemente Cobos-Sánchez [1], Juan-José González-de-la-Rosa [2] and Diego Sales-Lérida [1]

[1]   Area of Electronics, Engineering School, University of Cádiz, Av. de la Universidad, S/N, 11510 Puerto Real, Spain; clemente.cobos@uca.es (C.C.-S.); diego.lerida@uca.es (D.S.-L.)
[2]   Area of Electronics, Research Unit PAIDI-TIC-168, Polytechnic School of Algeciras, University of Cádiz, Av. Ramón Puyol, S/N, E-11202 Algeciras, Spain; juanjose.delarosa@uca.es
*   Correspondence: josem.guerrero@uca.es; Tel.: +34-956-483303

Received: 6 March 2019; Accepted: 19 April 2019; Published: 24 April 2019

Abstract: The energy supply of office buildings and smart homes is a key issue in the global energy system. The growing use of microelectronics-based technology achieves new devices for a more comfortable life and wider use of electronic office equipment. On the one hand, these applications incorporate more and more sensitive electronic devices which are potentially affected by any external electrical transient. On the other hand, the existing electrical loads, which generally use electronic power systems (such as different types of battery chargers, ballasts, inverters, switching power supplies, etc.), generate different kinds of transients in their own electrical internal network. Moreover, improvements in the information of the state of the mains alternating current (AC) power line allows risk evaluation of any disturbance caused to permanently connected electronic equipment, such as computers, appliances, home security systems, phones, TVs, etc. For this reason, it is nowadays more important to introduce monitoring solutions into the electrical network to measure the level of power quality so that it can protect itself when necessary. This article describes a small and compact detector using a low-cost microcontroller and a very simple direct acquiring circuit. In addition; it analyzes different methods to implement various power quality (PQ) surveillance algorithms that can be implemented in this proposed minimum hardware platform. Hence; it is possible to achieve cheap and low-power monitoring devices that can become nodes of a wireless sensor network (WSN). The work shows that using a small computational effort; reasonable execution speed; and acceptable reliability; this solution can be used to detect a variety of large disturbance phenomena and spread the respective failure report through a 433 MHz or 2.4 GHz radio transmitter. Therefore, this work can easily be extended to the Internet of Things (IoT) paradigm. Simultaneously, a software application (*PulsAC*) has been developed to monitor the microcontroller's real-time progress and detection capability. Moreover, this high-level code (C++ language), allows us to test and debug the different utilized algorithms that will be later run by the microcontroller unit. These tests have been performed with real signals introduced by a function generator and superimposed on the true AC sine wave

Keywords: power quality (PQ); embedded microcontroller; low cost monitor; sensor node; wireless sensor network; IoT

---

1. Introduction

It is well known that the increasing introduction of electronic loads (e.g., ballasts, electric vehicle chargers, switching power supplies, etc.) in a worldwide scenario where the distributed energy resources are gaining greater presence is provoking a major concern in power quality (PQ) degradation.

However, the energy supply is not only affected by atmospheric phenomena or external power switching affecting power lines, but also by internal transients or nearest electronic equipment perturbations.

From the final consumer's point of view, it is doubtless that the monitoring and processing of the instantaneous electrical parameters of the supply power chain are fundamental for the improvement of the electrical service, the producer–consumer relation, and the quality of the energy. The transmission and manipulation in real time of this information are one of the greatest immediate challenges in the present decade. It is also worth mentioning the technological effort and diverse challenges within the European program 'Horizon 2020' (H2020), or the proliferation of scientific contributions into the new electrical engineering fields as the smart grid paradigm.

As a matter of fact, during the last few years, there has been increasing concern about the smart measurement systems to report the actual state of the power lines or to anticipate an imminent electrical power supply failure. Most of this work has been supported using high-level processing on personal computers and in many cases, running powerful software solutions requiring considerable hardware resources. Moreover, if these monitors are finally implemented, the definitive solutions can be very expensive or the required equipment is too complex to be handled by the end customer, who lacks sufficient technical knowledge for the correct interpretation of the results.

Along with the development of measurement instruments, the PQ readings procedures and recommendations have been standardized and contained in the IEEE 1159-1995 (Institute of Electrical and Electronics Engineers) and subsequently revised, resulting in the standard IEEE 1159-2009 [1] as well as the international IEC 61000-4-30 introduced in its first edition in 2003 [2]. These standards set test conditions, tolerances, and uncertainties of measures that provide insight into the quality of electricity supply. In addition, the IEC 61000-4-30 class A, establishes test methods and procedures to be met by various manufacturers of instrumentation for harmonizing measurements and reaches the same results regardless of the test equipment used [3]. The IEC 61557-12, respect to the performance measuring and monitoring devices (PMD), also establish test methods for low-voltage distribution and equipment restrictions [4].

There are many commercial smart meters and units of monitoring equipment provided by reputable manufacturers. In general, these devices meet the aforementioned test standards. However, there is also a large field of research about techniques and algorithms to make PQ measurements more approachable or to obtain a new dimension of the results. Many of these techniques are computer-based, use commercial software tools (e.g., MATLAB®, LabVIEW®) or are developed on specific acquiring equipment. On the other hand, other researchers have focused on developing specific meter microcontrollers or those that are field programmable gates array (FPGA)-based. In this direction, previous articles have proposed many analysis techniques derived from specific mathematical algorithms or implementing particular digital signal-processing methods, along with hybrid processing techniques between other alternatives. Some of these include the use of fast Fourier transform (FFT), spectral kurtosis [5], fourth-order spectrogram [6], higher-order statistics (HOS) classification [7], neural network [8] and trees decision [9], least squares as Prony's method [10], wavelets [11] or short-time Fourier transform (STFT) [12] and Kalman filters [13]. Since the target platform will be an embedded system, these algorithms and their respective codes are discarded due to the high computational burden required, impossible to implement in the case of a simple low-performance microcontroller.

Granados et al. [14] have reviewed a multitude of algorithms designed for the detection of non-stationary transient and power disturbances. In this analysis, they have attached a complete list of references regarding these detection methods and highlighted its advantages and disadvantages. Barros et al. [15] evaluated also three groups of fundamental techniques for transient detection: the comparative method, the sliding windows method and wavelet transform (WT).

An important step is the measurement of the PQ within the consumer's power lines (home, office, small industries, etc.) for monitoring the efficiency of alternating current (AC) supply and detecting real-time incidents due to characteristic transients (swelled amplitude, voltage sags, micro-interruptions, impulsive transient, oscillations, fluctuations, etc.). Owing to the increased interest in low-cost solutions (much cheaper than dedicated equipment) to monitor and control critical variables, other scientists and engineers have proposed specific electronics designs to detect some PQ perturbations, using different methods and algorithms. In this direction, it is important to point out the works conducted for obtaining specifically designs for the measurement of specific parameters of interest.

Previous articles contributed showing medium and low-cost hardware projects as described by Artoli et al. [16] (DSP-based solution) but they do not provide enough information about the hardware system or communication methods. In [17], Gallo et al. have presented a microcontroller-based design (SMT32F107VCT6 from ST Microelectronics™) but mainly oriented as a power meter; Granados et al. [18] have implemented a Hilbert-based smart-sensor on a commercial FPGA platform (Spartan3E XC3S1600), and tested with a synthetic signal. Yingkayuk et al. [19] have proposed a measuring device based on one processor, but supported on two commercial development cards (LPC2368 and ADUC7024 microcontroller boards), with direct connection to TCP-IP and to PC software. Quiros et al. [20] approach the problem from the spectral kurtosis physically implemented in an FPGA development board (Artix-7 FPGA from Xilinx™) and not on a specific microcontroller.

None of the cases proposed in the commented works address the design and simplification to obtain adequate size and low-power measurement devices and do not consider the formation of the sensor network.

Kiranmai and Laxmi [21] have presented a microcontroller-based design which performs an AC to DC voltage conversion of each phase to determine morphological changes in every input voltage. However, using the circuit design they propose, it is impossible to measure high-frequency transients due to the filtering effect introduced by the full-wave rectification and regulation of the AC input voltage. Khairul et al. [22] have developed another measurement platform using a microcontroller (PIC series from Microchip™), but only consider the frequency evaluation for PQ analysis. Malekpour et al. [23] have used a commercial STM32F4 DISCOVERY board to implement a higher-frequency transient detector applying the Kalman filter algorithm.

At the same time, new technological paradigms are emerging around PQ surveillance, such as wireless sensor networks (WSN). They involve significant advantages regarding capture and reporting data, and in a broader sense, the increasing Internet of Things (IoT) technology.

Since the appearance of the first sensor nodes in Berkeley about 1999 (Mote WEC) the catalog has been continuously increasing, adopting popular names like Mote, Telos, Mica, Mica2Dot, iMote, Waspmote or WiSense among others [24].

Although efforts are focused on obtaining more powerful platforms, by integrating ARM architecture processors or FPGA (field programmable gates array) devices, some more simple designs would achieve more efficient nodes, in terms of price and low-power energy [25]. This paradigm is augmented towards a new concept when a WSN has access to the web [26].

In this way, considering the WSN or the IoT networks, it is essential to develop simple-in-conception measurement electronic modules to achieve a high degree of penetration. In this sense, low-cost microcontroller-based systems (or embedded systems) offer acceptable solutions, maintaining at the same time enough computation power to ensure autonomous in-node calculations, and consequently avoiding high data volume from the measurement nodes to the main station.

A measurement network where the base station (BS) is the only decision-making element collapses the transmission channel if many nodes report a large amount of data to be processed by the BS. Besides, limiting radio-emitting time, the energy consumed by the node is significantly reduced and also increases in the transmission efficiency of all sensors and the BS. Conversely, when measurements are processed inside every sensor node, it is necessary to report a minimum data-set (conclusions) to another node or a BS.

Alonso et al. [27] have proposed a microcontroller-based sensor node (IoT), but this platform uses a dedicated integrated circuit (ADE7758). This high-accuracy, 3-phase electrical energy measurement component detects and evaluates directly the different disturbances. The design presented in [28] may be a complete sensor-node solution for IoT, where Viciana presents a circuit to read voltage and current, using an STM32F373 microcontroller to real-time processing on a Linux-based ARM board (core A7, Nanopi Air). They use Hall-effect current sensors (TLI4970) and isolated amplifiers to voltage measurement (ACPL-C79). The device also includes Wi-Fi communication.

Although an IoT sensor network allows a global data spread, the access delay to the Internet due to TCP-IP protocol could be slow in risk situations by electrical transients. This could be a disadvantage with respect to the wireless sensor network, where every node implements a radio-transmitter. A sensor-to-sensor connection via radio achieves more efficient coordination through the network.

Our main goal is to produce a simple estimator device which can report any basic information about the line state or alert when a dangerous disturbance occurred. The proposed module is oriented to detect the fundamental parameters related to the external electrical disturbances, as the transients which are generated in the consumer's own line. In this sense, the surveillance and protection of the state of the power line may be accomplished by a cooperative interaction among different nodes, which can be protection elements (e.g., a special circuit breaker node remotely operated) or simply measurement elements. Another possible application is to incorporate the sensor module inside the electrical appliances themselves, for self-protection and protection of adjacent equipment connected to the electrical network.

Considering the conclusions of the previous works our interest is in the design of devices that are as simple as possible. For this case, reduced use of electronics components is supposed, but it requires taking full advantage of the microcontroller resources. Considering the limited internal resources of this microcontroller, the algorithms must be implemented through coding in assembly language. In this first version, the proposed circuit measures only the voltage. Our purpose is to measure voltage transients in the electrical installation. Optionally, a current Hall sensor could be incorporated, but that was not the initial objective. However, the internal resources of this microcontroller cannot execute real-time routines, sampling voltage, current, and computing its respective results.

Therefore, the circuit proposed is a microcontroller-based voltage meter, directly connected to the LV power line, and running dedicated algorithms (low-level programming implemented).

Following this line, this article introduces, first, the sensor-node architecture and second, analyzes several algorithms to detect a lot of typical electrical transients. This proposal is the base of development of a low-cost node solution, conceived to implement an IoT node or a wireless sensor network. It is structured as follows. Section 2 shows a brief review of the detector circuit, the electrical parameters measurement, several processing techniques and the description of the electronic circuit to accomplish these actions. In Section 3, the physical system implementation is described. In Section 4, the results are detailed. Finally, the conclusions are presented in Section 5.

2. Sensor Node Architecture

The solution was conceived to facilitate the network installation and it consists of a sensor module connected to AC power sockets (Figure 1a). Likewise, it could be installed inside the appliances, actuating as a self-protection mechanism, disconnecting when a dangerous transient is detected. The sensor node block diagram is depicted in Figure 1b. Each node is used as a real-time failure data logger. In this case, it is compulsory to include an independent real-time clock and to provide enough memory to implement an efficient data-logger for failure reporting. Moreover, wireless data communication is required for a couple of reasons. Firstly, the complexity and the installation cost are strongly reduced. Secondly, users and processing equipment must be electrically isolated from the sensor node.

Due to the widespread use of open hardware prototyping platforms such as Beagle-Bone, Raspberry-Pi or Arduino, we adopted the ATmega328 microcontroller [29] that is also the core of the various development boards in the market. Thus, the results can run on an Arduino or an equivalent platform (assembly language directly) or on other similar processors of the same Atmel manufacturer series. It is also possible the immediate transfer to other analogous micro-controllers from other manufacturers. In addition, many sensor node platforms have also integrated ATmega devices (Waspmote, Arduino-BT, WiSmote, Tnode, NanoQplus, PanStamp, Mica, Mica2Dot, BTnode, AVRaven, Cricket, Iris, etc.) [25].

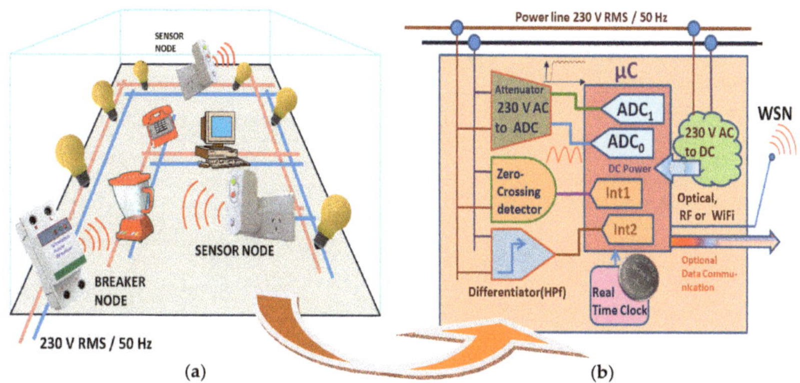

Figure 1. (**a**) Wireless sensor network for transient detection and self-protection: a proposal; (**b**) diagram of the presented node platform.

2.1. Acquiring and Conditioning of Electrical Parameters

The sensor node involves the microcontroller and its internal resources (analog-digital converter (ADC), memory, timers, interruption mechanisms and comparators). Each node is situated at one concrete point of the installation and monitors local voltage transients.

2.1.1. Alternating Current (AC) Power Line Adapter

As depicted in Figure 2, the analog-to-digital conversion (ADC) is realized from an attenuator and rectifier adaptor (adding overvoltage protection, as Transil™ or TransZorb™ devices). Thus, the voltage at ADC input is a scaled value, smaller with respect to the no-rectified voltage present on the AC power line. The AC voltage divider introduces an attenuation factor of 57.5. In this way, a 230 V AC is represented in the ADC by 4 V DC. Regarding the IEC 61557-12 standard, referring to root mean square (RMS) and swell and dip measurement, the '120% Un' limit is equivalent to 276 V RMS. The maximum value measurable with this attenuator is 287 V RMS (ADC full scale). However, concerning IEEE STD 1159-2009, the maximum RMS voltage (swell cases) is higher than 400 V. To increase the range, a solution could be to use an additional attenuator, with a factor divider higher than 57.5 and connected to other ADC input available (e.g., a factor of around 100).

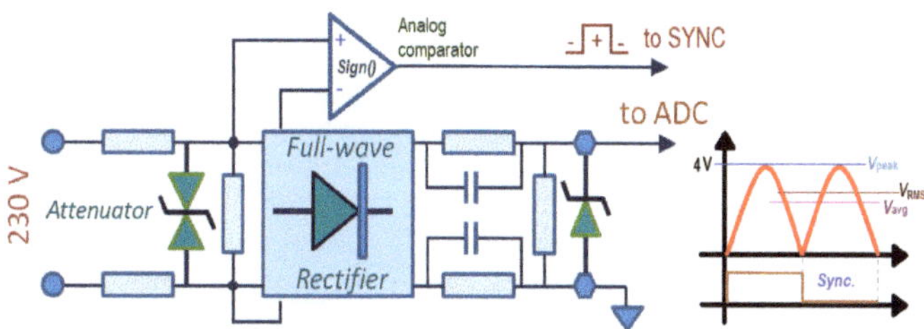

Figure 2. Alternating current (AC) power line attenuator and high-frequency equalization proposed for interfacing. A full-wave rectifying of a sine-wave permits an easy V_{peak}, V_{RMS} and V_{avg} interrelation.

Processing from rectified voltage is proposed because the dynamic range of the ADC converter is doubled. In this sense, all the programming codes of the microcontroller will take into account the absolute value of the sinusoidal voltage, abs[$V(t)$], but it is necessary to know the original sign of every rectified half-wave. This function is achieved by utilizing an analog-comparator.

2.1.2. Average Voltage Measurement

Considering a sine-wave input, if it is applied an averaging algorithm, we can obtain the RMS value using the V_{RMS} and the average voltage, V_{avg}, ratio given by $V_{RMS}/V_{avg} = \pi/\left(2\sqrt{2}\right) = 1.111$ (Figure 3a).

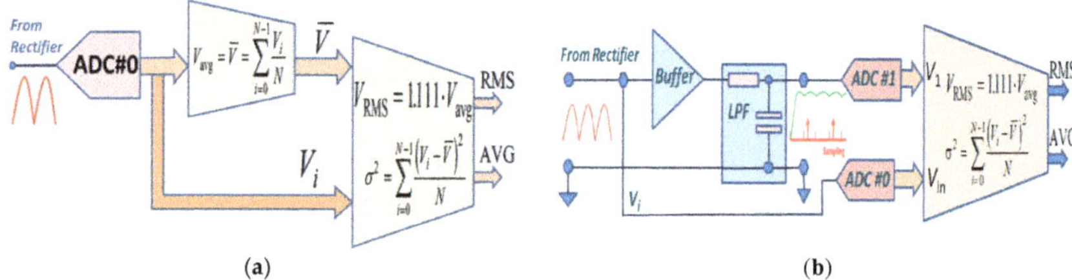

Figure 3. (**a**) Average voltage and root mean square (RMS) value applying digital processing. (**b**) A fast calculation method to get V_{RMS} and V_{avg} values from the peak value acquired (applying V_{peak}, V_{RMS} and V_{avg} relationships).

To get the arithmetic average value only a simple summation processing applied during the preceeding complete cycle is required, which for 50 Hz or 60 Hz is needed to just analyze one frame of 20 ms (50 Hz) or 16.7 ms (60 Hz). To simplify, it is not necessary to divide by N, the samples number: the final sum is just proportional with regard to N to the average amplitude. Any deformation of the signal shape will affect equally this final result and could be a failure indicator. However, to speed up the computation, it is possible to get the V_{RMS} and V_{avg} taken as the result of the peak value measurement. This value can be acquired sequentially from a second ADC (Figure 3b). We assume that the input voltage is sinusoidal, without anomalies most of the time, except during transient events.

Table 1 shows a list of classified phenomena according to IEEE-1159 recommendations. In the IEC 61557-12 are established test methods to assess them [4]. In general, a lot of these disturbances can be detected using time-domain methods instead of the frequency domain. On the other hand, high-frequency disturbances or oscillations require implementation of the Fourier transform or complex digital filtering, resulting in a high computational cost when we are using reduced feature processors. The evaluation of the peak and average voltage allows quick detection of some common electrical supply failures such as sag, swell and interruptions.

Table 1. Different categories of interruptions (**a**) and disturbances (**b**), according to the duration of the phenomenon and their amplitude (IEEE STD 1159-2009, regarding to the 50 Hz and 230 V AC case).

a) Interruptions (Categories)	Amplitude (230 V RMS/ 50 Hz)	Duration	b) Transients /Disturbances	Typical Spectral	Typical Duration	Relative Duration (D) (230 V RMS/50 Hz)
Instantaneous:			*Impulsive:*			
-Sag	207 V to 23 V	0.5 to 30 cycles or	-Short	5 ns rise	< 50 ns	$T = 50\text{ ns} \rightarrow D = 25 \times 10^{-5}\%$
-Swell	253 V to 414 V	10 ms to 0.5 s	-Medium	1 µs rise	0.05 to 1 ms	$T = 0.5\text{ ms} \rightarrow D = 2.5\%$
Momentary:			−10% cycle	0.1 ms	< 1 ms	$T = 1\text{ ms} \rightarrow D = 5.0\%$
-Interruption	<23 V	0.5 s to 3.0 s or 25 to	*Oscillatory:*			
-Sag	207 V to 23 V	150 cycles				$T = 0.3\text{ ms} \rightarrow D = 1.5\%$
-Swell	253 V to 414 V		-Low freq.	<5 kHz	0.3 to 50 ms	$T = 50\text{ ms} \rightarrow D = 250\%$
Temporary:						0 to 920 V
-Interruption	<23 V	3.0 s to 1.0 min or 150	-Med. freq.	5 to 500 kHz	20 µs	$T = 20\text{ µs} \rightarrow D = 0.1\%$
-Sag	207 V to 23 V	to 3000 cycles				0 to 1840 V
-Swell	253 V to 414 V		-High freq.	0.5 to 5 MHz	5 µs	$T = 5\text{µs} \rightarrow D = 25 \times 10^{-3}\%$
Long duration:						0 to 920 V
-Sust. Interr.	≈0 V	>1.0 min				
-Under-volt.	207 V to 184 V	or				
-Over-voltage	253 V to 414 V	3000 cycles				

2.2. Signal Processing Techniques

Our goal is to achieve a compact sensor node employing minimal hard ware equipment. For this reason, the use of C++ language or machine code would allow accelerating the algorithms to real-time detection. We will focus on analyzing only easily implementable algorithms on a microcontroller-based platform. Mainly, the statistical techniques are easy to implement. Unquestionably, it is very difficult to obtain real-time FFT computed in a microcontroller with limited resources. For this reason, the selection of the algorithms must be the main purpose.

2.2.1. Statistical Processing Techniques

Frequently, statistical variables from first-order moments (such as variance and the mean) are used. This does not exclude the possibility to apply higher-order statistics (HOS) and Kurtosis procedures. However, the HOS evaluation is more oriented to harmonic analysis (spectral kurtosis) [5–7]. Basically, the arithmetic average value per cycle and the previous average calculated during many cycles provide information about the voltage waveform integrity and the supply continuity (interruptions). Therefore, it is possible to apply this detection method to some of the phenomena listed in Table 1a.

The application of statistical analysis requires then calculating moments over a whole period. As N (samples number) is a constant value and the sampling is synchronized every period, the variance of the rectified signal is given by:

$$\sigma^2 \propto \sum_{i=0}^{N-1}\left(V_i - \overline{V}\right)^2 \quad , \text{ where } \quad V_{\text{avg}} = \overline{V} = \sum_{i=0}^{N-1} \frac{|V_i|}{N} \tag{1}$$

The knowledge of the arithmetic average value reveals indirectly the effective total surface value of the waveform previous cycle and, thus, large voltage variations as well as micro-interruptions. It would not be applied to detect sudden transients that momentarily affect to the sinusoidal waveform. Although the algorithm is very simple, the HOS computational effort is really notable.

2.2.2. Detection Algorithm Using Kalman Filter

A Kalman filter [30] behaves as a predictive element to develop the effective comparison of the regular signal, with a momentary perturbed sine wave. This algorithm attempts to minimize the covariance error between every captured input sample and the corresponding expected output value with respect to a previous wave form.

For the prediction of the sine wave voltage given $V(t) = U\sin(\omega t)$ (representing an AC voltage with $U = \sqrt{2} \cdot V_{\text{RMS}}$ the peak value, without disturbances), can be established as an equation in the state space considering the amplitude $U(t)$ as the main parameter to be estimated and, secondly, the frequency variations of this waveform. In order to form a system of at least two matrix equations, we compose the orthogonal state equations according to the signal itself (sine) and 90 degrees out of phase (cosine) obtaining both equations:

$$\left.\begin{array}{l} x_1(k) = U\sin(2\pi \cdot f \cdot t_k + \theta) \\ x_2(k) = U\cos(2\pi \cdot f \cdot t_k + \theta) \end{array}\right\} \quad \begin{array}{l} \text{to } t_{k+1} = t_k + \Delta t \\ \text{and } \theta = 0 \end{array} \quad \Rightarrow \quad \left.\begin{array}{l} x_1(k+1) = U\sin(\omega \cdot t_k + \omega \cdot \Delta t) \\ x_2(k+1) = U\cos(\omega \cdot t_k + \omega \cdot \Delta t) \end{array}\right\} \tag{2}$$

where $\omega = 2\pi f$ is the sine wave frequency, U is the estimated voltage value and $x_1(k)$ and $x_2(k)$, state functions at time k. In order to simplify, it can be considered an initial null delay ($\theta = 0$). The next state $(k + 1)$ can be established using the differential Δt, after of t_k time. By making use of the incremental time $t_{k+1} = t_k + \Delta t$, $x_1(k + 1)$ and $x_2(k + 1)$ will be the next states. Employing trigonometric relationships for sums of sine and cosine, the Equation (2) can be simplified as follows:

$$\left.\begin{array}{l} x_1(k+1) = U\cos(\omega \cdot t_k)\sin(\omega \cdot \Delta t) + U\sin(\omega \cdot t_k)\cos(\omega \cdot \Delta t) \\ x_2(k+1) = U\cos(\omega \cdot t_k)\cos(\omega \cdot \Delta t) - U\sin(\omega \cdot t_k)\sin(\omega \cdot \Delta t) \end{array}\right\} \Rightarrow \left.\begin{array}{l} x_1(k+1) = x_2(k)\sin(\omega \cdot \Delta t) + x_1(k)\cos(\omega \cdot \Delta t) \\ x_2(k+1) = x_2(k)\cos(\omega \cdot \Delta t) - x_1(k)\sin(\omega \cdot \Delta t) \end{array}\right\} \tag{3}$$

and finally, we have:

$$\begin{pmatrix} x_1(k+1) \\ x_2(k+1) \end{pmatrix} = \begin{pmatrix} \cos(\omega \cdot \Delta t) & \sin(\omega \cdot \Delta t) \\ -\sin(\omega \cdot \Delta t) & \cos(\omega \cdot \Delta t) \end{pmatrix} \begin{pmatrix} x_1(k) \\ x_2(k) \end{pmatrix} \tag{4}$$

Since the output variable of interest is precisely the state $x_1(k)$, the equation indicating the measured value output, $z(k)$, is exactly the state $x_1(k)$ itself and the superimposed noise $v(k)$:

$$z(k) = \begin{pmatrix} 1 & 0 \end{pmatrix} \begin{pmatrix} x_1(k) \\ x_2(k) \end{pmatrix} + v(k) \tag{5}$$

where $v(k)$ is white noise with null mean value. When a transient is superimposed, then this disturbance $v(k)$ is now not Gaussian, which causes discrepancy with respect to the previously estimated voltage. The Kalman filter allows a sufficient prediction of sinusoidal voltage from the previously sampled data. By comparing a recently sampled period with abnormalities (disturbances) with a previously predicted sine wave without apparent irregularities, it will generate a differential value when detecting any mismatch between the two periods (with and without transient pulses).

2.2.3. Detection Algorithm Using the Helmholtz Equation

Given that function $V(t) = U\sin(\omega t)$ representative of the AC power line is an n-times differentiable function, it is possible to set it into an one-variable ordinary differential equation, as the one-dimensional Helmholtz mathematical expression, and this can by written as:

$$F = f''(t) + K \cdot f(t) = 0 \tag{6}$$

Annulment of F in Equation (6) is a method to find variations of $f(t)$ when this function is a periodic signal such as the analyzed sine wave. The proof is obvious in the linear version case of F: given $f(t) = U\sin(\omega t)$, we only need to derive the sinusoidal function consecutively to reach the conclusion that $K = \omega$; under normal conditions, the contrasting of the sine function with its second derivative yields a null value. Once a value of K has been found for the undistorted sinusoidal function, any shape deviation with respect to the original (due to some disturbance) will affect to the function $f(t)$ which will not be now a perfect sine wave, and consequently, Equation (6) is not verified.

When a discrimination threshold appropriate is exceeded, an alarm signal is notified announcing the disturbances presence. For the discrete case, utilizing the discrete-time variable ($t \rightarrow \Delta t$) and where is now applied the voltage $V(n \cdot \Delta t) = U\sin(\omega \cdot n \cdot \Delta t)$, with peak value $U = \sqrt{2} \cdot V_{RMS}$, without disturbances, for every instant $n = 0,1,2,\ldots$. the derivative function is given by:

$$fn(t) = \frac{U[\sin(\omega \cdot (t + \Delta t)) - \sin(\omega \cdot t)]}{\Delta t} = \frac{U}{\Delta t}[\sin(\omega \cdot t + \omega \cdot \Delta t) - \sin(\omega \cdot t)] \tag{7}$$

and the second derivative function:

$$fnn(t) = \frac{U}{\Delta t}\left[\frac{(\sin(\omega \cdot t + \omega \cdot \Delta t + \omega \cdot \Delta t) - \sin(\omega \cdot \Delta t + \omega \cdot \Delta t)) - (\sin(\omega \cdot t + \omega \cdot \Delta t) - \sin(\omega \cdot t))}{\Delta t} \right] \tag{8}$$

which can be simplified in the following expression:

$$f''(t) = \frac{U}{(\Delta t)^2}[\sin(\omega \cdot t + 2\omega \cdot \Delta t) - 2\sin(\omega \cdot t + \omega \cdot \Delta t) + \sin(\omega \cdot t)] \tag{9}$$

Applying the equality expressed in Equation (6), we have:

$$f''(t) - Kf(t) = \frac{U}{(\Delta t)^2}[\sin(\omega \cdot t + 2\omega \cdot \Delta t) - 2\sin(\omega \cdot t + \omega \cdot \Delta t) + \sin(\omega \cdot t)] + KU\sin(\omega \cdot t) = 0 \tag{10}$$

If we consider $K = -1$, the function is canceled when the two terms are equal, or expressed otherwise, the function $g(t)$ given by:

$$g(t) = \frac{1}{(\Delta t)^2} [\sin(\omega \cdot t + 2\omega \cdot \Delta t) - 2\sin(\omega \cdot t + \omega \cdot \Delta t) + \sin(\omega \cdot t)] \equiv \sin(\omega \cdot t) \tag{11}$$

that is, equal to $\sin(\omega \cdot t)$, which is equivalent to the initially sine wave function. Therefore, when the original $f(t)$ slightly differs from an ideal $\sin(\omega \cdot t)$ function, the homologous, $g(t)$, will not verify the sinusoidal relationship. Since $\sin(w \cdot t)$ represents a previous cycle sampled from the next $\sin(\omega \cdot (t + \Delta t))$, we can express the function $g(t)$ as an equation in differences, $G(n)$, with a sampling interval $(n \cdot \Delta t)$ of the acquired power sequence $V(n \cdot \Delta t)$ that can be rewritten as:

$$V(n) = U\sin[\omega \cdot (n \cdot \Delta t)] \quad \Rightarrow \quad G(n) = V(n+2) - 2 \cdot V(n+1) + V(n) \equiv (\Delta t^2) \cdot V(n) \approx 0 \tag{12}$$

where $\Delta t = 1 / F_S$ is a very small number. Using the Z^{-2} operator, the definitive equation to be algorithmically implemented is:

$$G_1(n) = V(n) - 2 \cdot V(n-1) + V(n-2) \approx 0 \tag{13}$$

2.2.4. Detection Algorithm by Contrasting with Successive Previous Integrations

If during a long time interval, the acquired sinusoidal input is integrated (each sample with its respective previous one of similar phase), the $V_{acc}[n]$ array of the resulting waveform should contain a filtered AC power sine wave without any transient.

$$V_{acc}(n)\big|_{current} = \frac{1}{2}\left[V_i(n) + V_{acc}(n)\big|_{previous}\right] \qquad n = 0, 1, 2, \ldots, N \tag{14}$$

Now, if the power line is unexpectedly affected (transients, disturbances, etc.) different instantaneous input samples, $V_i(t)$, will differ with respect to the previous arithmetic averages, $V_{acc}(t)$, for those same array indexes. These differences can detect any anomaly when implementing this comparison algorithm:

$$If \quad \left|V_i(n)\right| - \left|V_{acc}(n)\right|\big|_{n=0,1,2,\ldots,N} \quad > \quad threshold \quad \Rightarrow \quad Transient/disturbance \tag{15}$$

3. Hardware Platform

Our contribution is a low-cost and compact sensor circuit. Therefore, it is required to use the smallest number of components. The prototype has a small size (two cards, staked, 80×60 mm) that can be inserted into a plastic box and directly connected to the electrical socket. This purpose suggests designing an optimized hardware platform, capable of detecting disturbances in real time and avoiding the inclusion of specific integrated circuits. This sensor node is mainly oriented to measure the voltage of the power line directly using a balanced AC voltage divider. The circuit is auto-powered by the same power line where it performs the monitoring work (Figure 4b) and optionally, it can integrate supercapacitors to provide autonomy in long interruption failures.

The processor ATmega328 from Atmel (up to 20 millions of instructions per second (MIPS) is adequately fast for that price range) has been selected. This integrated circuit is the core component of the Arduino prototyping platforms. This choice allows us to translate the validated conclusions from our prototype toward others similar open hardware platforms. A real time clock (Maxim DS1307 powered by a back-up battery) drives the timing for the failure data-logger. The prototype also includes a buzzer and two light-emitting diodes (LEDs) to show the current state of the node and to alert of any recent failure.

A wireless communication emitter (AM modulation at 433 MHz) is added to transmit the results. Optionally, a BlueTooth or WiFi module can be connected. This communication method allows the formation of the network and the galvanic isolation since the modules work directly with the AC line voltage. Additionally, an infra-red communication receiver has been included only for purposes of wireless configuration from a computer. A second circuit board contains the attenuators, edge detector (synchronism), rectifier and equalizer for the direct measurement of the main power voltage. Both cards are connected in a sandwich mode (stacked) to reduce the overall size of the equipment and achieve greater galvanic isolation from the AC power card.

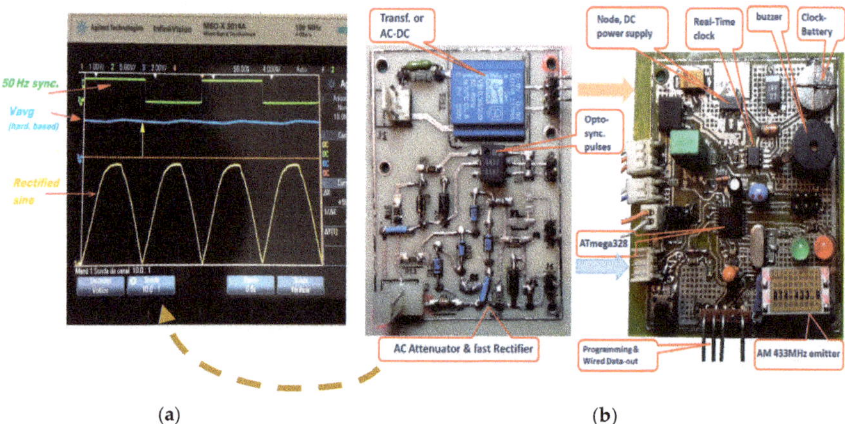

Figure 4. (**a**) Oscilloscope image showing the sync pulses (green) and the average value (blue) of the full-wave rectified voltage (yellow), from the power board (left); (**b**) a prototype photograph: the microcontroller unit (at right) and the high voltage board (rectifier + attenuator, at center).

In order to avoid electric risk during the preliminary debug and tests, a transformer has been inserted to isolate the main power voltage and it allows direct connection from the prototype to the computer, through the universal serial bus (USB) port.

3.1. Measurement Setup

The sensor node must be connected directly to main power line. The attenuator adapts the AC power voltage to the microcontroller ADC converter. Figure 4a shows an oscilloscope image of the rectified sine wave (yellow) at the ADC input. This ADC voltage is scaled over 4 V peak when a 230 V RMS nominal voltage is present in the power line. It can also be seen the sync pulses (green, every zero crossing) and the peak value, V_{peak}, using hardware method (rectifier and RC-filter).

3.1.1. Dynamic Range

Dynamic range is optimized for the ADC full-range (10 bits and 5 V DC input max) to represent the rectified full-wave. It is planned that this maximum ADC range, 5 V, represents a 25% higher input than the service voltage (230 V AC previously scaled by the input attenuator, see Figure 2) and 287 V RMS (ADC full scale).

With this factor, we have 4 V DC (equivalent 230 V AC) into the ADC converter. Regarding the IEC 61557-12 standard, the '120% Un' limit referred to is equivalent to 276 V RMS. For specified cases in IEEE-1159 (Table 1) as swell and over-voltage, this recommendation considers a maximum amplitude of 1.8 p.u., that regarding 230 V power line, can estimate shall rise around 414 V RMS (or 585 V peak). This will develop a peak value into the AD converter of 7.2 V (the maximum permissible ADC input level is 5 V). Therefore, a Zener protection circuit must limit it to a maximum voltage of 5.1 V. Voltage peaks above the 1.25 p.u. will not be evaluated quantitatively by our detector (only qualitatively). In Section 2.1.1, an optional measurement range increase is proposed.

3.1.2. Synchronous Conversion and Sampling Rate

The conversion synchronized every zero crossing of the input voltage avoids the requirement of windowing techniques and leads to simple comparisons of 256 samples per block, sequentially captured every period. Thus, similar samples successively occupy the same phase locations in a conversion data array (a solution already proposed by Artioli et al. [16]). The crossing detection is very easy using an analog-comparator or an opto-coupler. This provides constantly an array size of $N = 2^8 = 256$ numbers in all applied algorithms. Therefore, the conversion speed is set to 256 samples per period of the input voltage (50 or 60 Hz). This quantity is a compromise between resolution and memory size, and affects both data arrays of samples and of intermediate results. For a power line of 50 Hz, this supposes a sampling rate $F_S = 256 \times 50\ \text{Hz} = 12.8\ \text{kHz}$. This frequency allows an accurate representation of the full-wave rectified sine wave of 50 Hz, but limits the bandwidth of transients at about 6.4 kHz according to Nyquist limit.

3.1.3. The Absence of the Anti-Aliasing Filter

For simplicity, the essential anti-aliasing filter was not included in the data acquisition sequence. Since the monitor mission is to detect disturbances qualitatively, we expect to capture easily high-frequency transients outside the Nyquist range. We take advantage of the overlapping phenomenon (a consequence of Shannon's theorem itself) considering the mirrored copies of the higher spectral components, to the low side, with respect to the sampling frequency point, F_S.

Thus, any high-frequency transient that exceeds the requirements of band-pass framed below Nyquist frequency, F_S (e.g., some perturbation frequencies above of $F_S/2 = 12800\ \text{Hz}/2 = 6400\ \text{Hz}$) should generate errors in the conversion process and shall appear as spectral components into the sampling frequency band due to overlap effect (lower sideband in a heterodyning phenomenon, Figure 5). Consequently, this mirrored component, below than F_S frequency, can also reveal the disturbances.

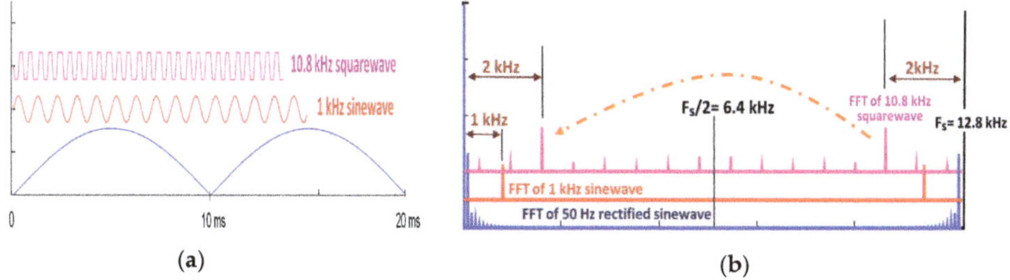

(a) (b)

Figure 5. (a) Input voltage (a full-wave rectified 50 Hz sine wave) and two oscillatory signals (1 kHz sine wave and 10.8 kHz square wave) sampled to 12.8 kHz. (b) FFT of each waveform: when the disturbance spectrum is greater than Fs, the spectrum is reflected and overlapped; transients may be qualitatively identified into the below zone from the Nyquist frequency. (Simulated by Octave™ (GNU) to $N = 256$ points/period, $F_S = 12.8$ kHz and 50 Hz equivalent FFT spectral resolution).

By Shannon's theorem (graphically expressed in Figure 5), generally if aliasing exists, the transient component located above $F_S/2$ is now reflected in the Nyquist sampling area, within the spectrum band conversion between 0 Hz and F_S (overlap). Under these circumstances, some disturbance components will be qualitatively processed and the algorithms can indicate the presence of anomalies. However, although this simplification allows detection of transient pulses above the sampling frequency, cannot correctly determine the event frequency: it is only possible to determinate that this disturbance had occurred.

3.1.4. Application Software for Essays of Algorithms and Test-Bench

An application called '*puls.AC*' was developed to verify the algorithm implementations and the prototype electronic board (Figure 6). This software tool has been programmed in C++ language employing the LabWindows/CVI environment from National Instrument^{TM}. If the procedures are written using exclusively an integer-numbers based programming code, the detection algorithms can be directly translated and tested in assembler language for ATmega328 processor (Atmel™ company also provides a C free compiler for the developer community using its own microcontrollers).

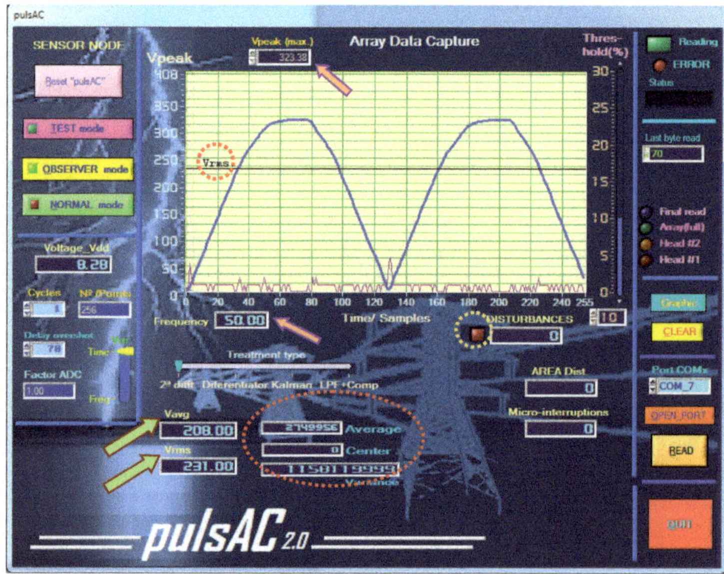

Figure 6. LabWINDOWS™ /CVI-C ++ test bench application that shows real-time acquired data from the hardware prototype.

3.1.5. Data Acquisition State Machine

To maximize the processing speed, a fast state machine was implemented using the incorporated features into the microcontroller, mainly the internal interrupting architecture. The acquiring process starts every detected zero crossing. An internal clock counter is configured to generate software interruptions during an equivalent time to the sampling period. This interruption routine forces the system to acquire an input voltage sample and save it. A counter variable allows for the conversion of a frame of up to 256 sampled data. The available free time between adjacent-acquisitions permits the processor to run the required detection algorithms (Figure 7). When the synchronism is different to 20 ms (50 Hz), the internal counter generates a count overflow. This event marks an anomaly situation.

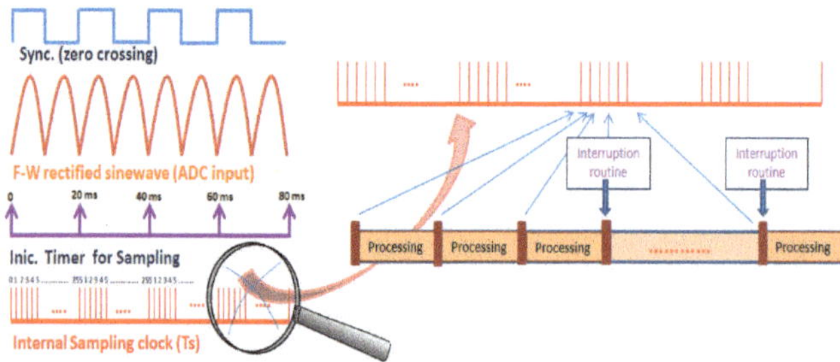

Figure 7. A timing diagram of the acquisition state machine. Each zero-crossing (positive slope) starts the acquisition of 256 samples per block.

3.2. Algorithms' Implementation Codes

Three computational strategies have been considered:

3.2.1. Detection Algorithm Based on the Helmholtz Equation

In this case, it was implemented the comparison algorithm of the second derivative of the input sine wave sampled. For a given sampling rate of $\Delta t = 78.12$ μs, is $\Delta t^2 \approx 0$. The method was previously exposed and here is implemented (pseudocode) according to the conclusion given by Equation (13):

```
X2, X1;                                         // old samples previous period, variables
For i = 0 to 255:                               // X[] current samples and 256 samples
 a = abs( X[i] - 2 * X1 + X2);                   //algorithm equation (13) implementation(is a=0??)
 if (a > threshold) {goto Pulse_Detection_Proc();}    // error report(if a >adjustable threshold)
 X2 = X1;                                         // save sample (n-2) for next iteration
 X1 = X[i];                                       // save sample (n-1) for next iteration
```

where the $X[i]$ array contains $N = 256$ ADC samples captured during an interval of $T = 20$ ms (50 Hz period); thereby it is $\Delta t = 78.12$ μs.

3.2.2. Kalman Filter Implementation

As discussed in Section 2.2.1., the minimized Kalman filter to estimate simply amplitude A of the sinusoid $A\sin(\omega \cdot t)$, leads to the implementation of Equations (4) and (5), implementing the following pseudocode sequence:

```
Cx, Sx                              //previously defined SIN,COS Const(interg  scale factor)
X_Old[i]; a = 63;                   // Input array(old samples sine previous period )
For i = 0 to 255:                   // X[] current samples
 a = a + 1; if a > 255 then {a = 63}    // 90º Phase-shift to compute X1(k) and X2(k) 2D array
 K[i] = (int)(Cx * K[i] + Sx * K[a] );  // K, Kalman out array(intergers, with a scale factor)
 D[i] = abs( X_old[i] - K[i] );         // matching: estimated – acquired signal mismatch
 if ( D[i] > threshold ) { goto Pulse_Detection_Proc.(); }//error report(adjustable)
 K[i] = X_Old[i];                       // to compute next state for 2D Kalman array
 X_Old[i] = X[i];                       // (optinonal) save sample TWO period delayed
```

where, $X[i]$ is the array of the input voltage sampled, and Cx, Sx, the coefficients of the resulting Kalman-2D array for sinusoidal prediction case (3). The $K[i]$ array are estimates of the ideal sine wave (without interference) and are compared with the $X[i]$ array, resulting in the differences array, $D[i]$. Three arrays are required: one to contain the instantaneous samples input voltage (256 words) and two more to the consecutive previous periods (512 words); two fixed-point multiplications are required too.

3.2.3. Detection Algorithm by Contrasting with Successive Previous Integrations

This method was previously shown and requires the successive integration of the input sine-wave captured. In fact, an averaging algorithm successively and repeatedly applied to a sinusoidal voltage with transients leads towards a pure waveform without disturbances. Every voltage sample newly acquired is compared continuously with the accumulative previous average value, with an identical phase, by making use of the expression:

$$D[i] = X[i] - \overline{X}[i - 256] \tag{16}$$

where $X[i]$ is the current samples array and $\overline{X}[i - 256]$ contains the average samples of previous periods. The general equation for any integration algorithms is $\overline{X}[i] = (X[i] + X_{old}[i])/\beta$. This algorithm acts as a low pass filter and can to remove any previous transient. For this case, we have used the averaging formula $(X[i] + X_{old}[i]) / 2$, since it is easier compute a base-2 binary division that can be solved with a simple shift to the left, following the addition. The basic pseudocode is:

```
X_Old[i];                                      // old samples sine previous period, array
For i = 0 to  255:                             // X[] current acquired samples
D[i] = (X[i]-X_Old[i]);                         //"signal - average(previous)" D[] Difference
if ( D[i] > threshold ) {goto Pulse_Detection_Proc(); } // error report(Adjust.thresh.)
X_Old[x] = (int)(( X[i] + X_Old[i] )/2);        //average computed using shift register.
```

4. Results

To verify the sensor-node prototype performance and to check the different detection algorithms, three routines only were implemented. Therefore, a basic set of different transient pulses and oscillations has been applied according to some conditions included in the IEEE-1159 recommendation (Table 1). Specifically, the Helmholtz equation method, the Kalman predictor, and the successive integration (LPF) and comparison method have been compared. These three techniques have shown that they are easily implementable and give an efficient response to the failure detection.

To generate real transients, we used a standard analog function generator Tektronix CFG250 and an ad-hoc circuit board with manual switches (Figure 8) that allow us to enter the following scaled disturbances:

- V_{RMS} decreased by approximately half of the nominal 230 V AC (−54.7%) (0.6 p.u. momentary sag).
- Half-cycle missing or a sag in an amount bigger than 10% (instantaneous sag, 0.5 cycles > 0.1 p.u.).
- Swelled peak voltage at 13.35% over the 311 V nominal peak (1.13 p.u. momentary swell).
- Permanent or intermittent disturbance signals, as sine or square waves, with different amplitude levels and frequencies, from the function generator.
- To emulate impulsive transients of different duration, square pulses of low, medium and high frequency (10 Hz, 1 kHz, 10 kHz and 100 kHz) and amplitude up to 10% of V_{peak} (corresponding to 0.1 p.u., with respect to the rectified voltage and using a voltage divider) were injected. To emulate harmonic distortion a 200 Hz sine low-amplitude component can also be superimposed. Different sinusoidal functions (10 Hz to 10 kHz), approximately 10% (0.1 p.u.) amplitude with respect to the rectified voltage peak value were used for oscillations simulation.

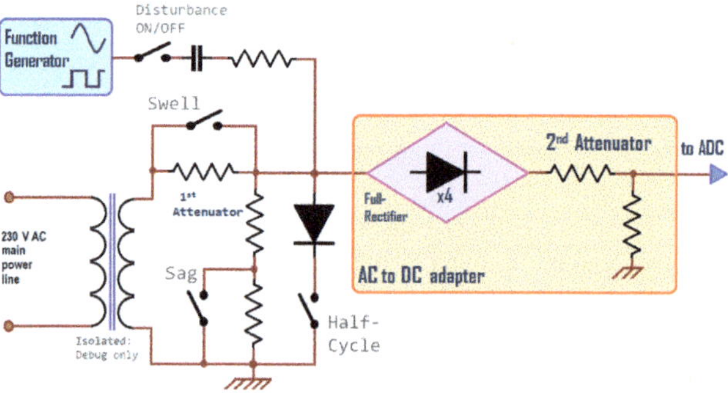

Figure 8. Transients pattern generator: auxiliary circuit arrangement diagram modified from the acquiring circuit shown in Figure 2.

The adjustable 'threshold' level permits us to modify the failure detection sensitivity. To achieve the maximum, sensitivity should be taken to the lowest level possible, but avoiding the occurrence of false alarms.

4.1. Helmholtz-Based Method

1. Voltage fluctuations testing: the slow voltage variation generates small differences values in the derivative algorithm given by Equation (13). This method is insensitive for phenomena such as sag or swell voltage.
2. The process is not sensitive to brief changes such as micro-interruption or sudden half-cycle missing failures (Figure 9a) because the algorithm requires a full period to compute completely the array of the sampled input voltage.

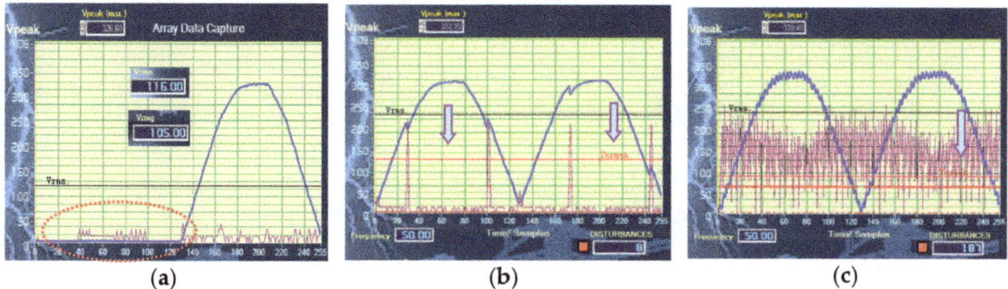

Figure 9. Helmholtz equation method detection test (Blue line: sampled voltage; Magenta line: magnified value resulting from (13); red line: threshold level. The V_{RMS} value is computed from V_{peak} measurement): (**a**) half-cycle missing, V_{avg} = 105 V (with respect to 326 V peak). No detection in this case. (**b**) Superimposed square wave of A = 10% of V_{peak}, F = 100 Hz, with 8 detections/cycle as resulting. (**c**) Oscillation, sine wave, 10% of V_{peak} (0.1p.u.), adjust. th. = 5%, F = 10 kHz, with 187 detections/cycle.

3. When a square voltage (1 kHz or 100 Hz) is superimposed (Figure 9b) the algorithm presents a good detection response (relative threshold = 5% adjustable from 0 up to 30%). It also works correctly for high frequencies (10 kHz and 100 kHz, even greater than F_S (Section 4.4)

To detect up to 1 kHz order sinusoidal oscillation injected, it is necessary to decrease the detection threshold up to 2%. This causes any false alarms even when the AC voltage is normal (undistorted). In this situation, it is impossible to detect the 4th harmonic distortion using this method. However, it works properly for higher frequencies (as 10 kHz and 100 kHz > F_S) (Figure 9c).

4.2. Kalman Filter Detection

1. Voltage fluctuations testing: Figure 10a shows a half-cycle sudden failure (blue). Thus, the filter predicts the half-cycle missing (magenta) and consequently detects the failure (the green line show the error value; the red one is the computed comparator result). Likewise, it also detects slow disturbances with a presence for a longer time than the period of the signal.
2. Since detection is a strategy based on prediction, the code correctly detects the square transients (Figure 10b). When a square wave voltage is superimposed (frequencies of 10 Hz to 1 kHz), the algorithm presents a good differentiation (up to a relative threshold level of 6%, adjustable).
3. Injecting a sine wave disturbance higher than 10 Hz, the system shows a good response, for different threshold values (Figure 10). In this case, it is possible to detect the harmonic distortion (4th-harmonic of the 50 Hz sine wave).

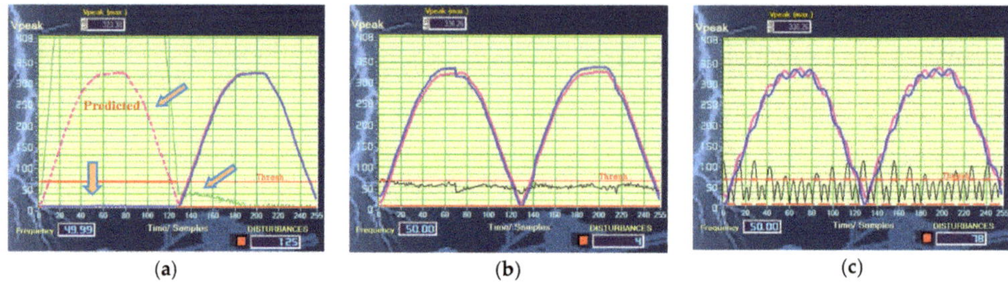

(a)   (b)   (c)

Figure 10. Kalman filter (Blue line: sampled input voltage; Magenta line: Kalman prediction; Green line: error) (**a**) Half-cycle missing, V_{avg} = 104 V, $V_{effective}$ = 115 V RMS and 125 disturbance detections (aprox. 256/2). (**b**) Superimposed square wave, 10% of V_{peak}, F = 100 Hz, relative threshold adjusted to 6%, with 4 failure detections/cycle. (**c**) Oscillation, sine wave, 10% of V_{peak}, F = 1 kHz and relative threshold 6%, with 78 detections/cycle.

4.3. Method of Successive Integrations (LPF) and Comparison

1. Voltage fluctuations testing: since the algorithm obtains an arithmetic average of each similar phase sample (during previous periods), this process is slow with respect to any very fast event that may occur in the immediate preceding half-cycle. In these situations, sometimes the algorithm can produce an erroneous result. However, since the final average is obtained from the immediately previous samples, it works quite accurately when sudden events occur, lasting longer than a half-cycle. In this case, a swell voltage variation is detected perfectly.

2. This method works properly when a square disturbance is injected (Figure 11b). For square wave voltage superimposing tests (10 Hz and 1 kHz), the algorithm presents a good differentiation and detection (using a threshold level of 6%). If the phenomenon is persistent, the transient will gradually dissolve in the averaged waveform array (due to the filter effect).

3. For low-frequency sinusoidal oscillation (hundreds of Hz), the procedure has an equally good detection response because it compares the current signal samples with the previous averaged sine wave and without transients (Figure 11c). It could also detect the harmonic distortion (presence of the 4th harmonic).

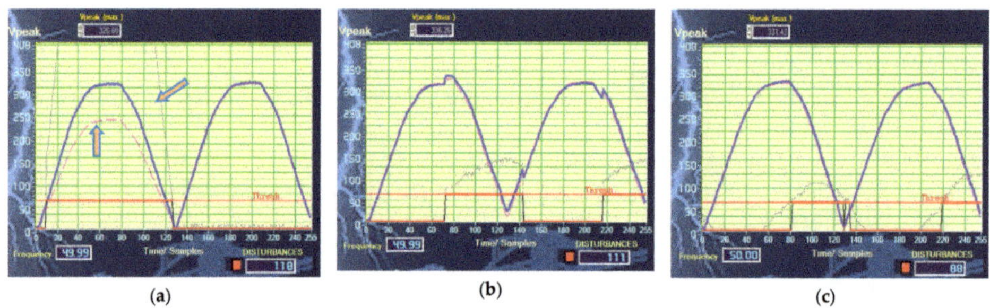

(a)   (b)   (c)

Figure 11. Integrator (LPF) and comparator method (Blue line: sampled input voltage; magenta: accumulative previous-integration (average); gray line: difference; *Red line*: error). (**a**) Swell: only detect half-cycle to full-cycle correcting cases. Detected disturbances: 118 D./cycle. (**b**) Square wave superimp., 10% of V_{peak}, 100 Hz, relative threshold 5%, 111 detections/cycle. (**c**) Oscillation, sine, 100 Hz, 10% of V_{peak}, and threshold level to 5%, up to 88 detections/cycle.

4.4. Performance in High Frequency

To verify the effectiveness of the ADC and sampling bypassing the antialiasing filter, we compare results of the three previous algorithms, considering an oscillation injected of 10 kHz, near to sampling frequency (Figure 12). The response was also good for the highest frequencies (up to 100 kHz).

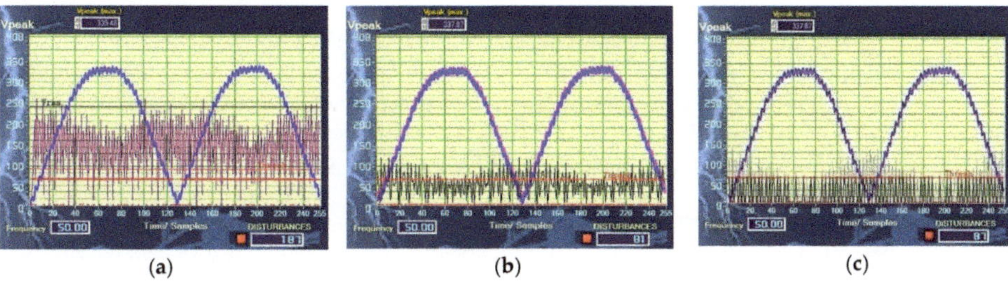

(a)   (b)   (c)

Figure 12. The response of each algorithm analyzed for a sinusoidal stimulus of 10% of V_{peak} and 10 kHz: (a) Helmholtz based method; (b) Kalman filter; (c) Successive Integrations (LPF) and comparison.

Table 2 summarizes the different results for each method discussed in this paper. Attending to the results, the Kalman algorithm has a better transient detections response. However, it represents a slight implementation complexity compared to other implemented algorithms.

Table 2. The response of the different algorithms analyzed.

Disturbance	Helmholtz Equation	Kalman Filter	Successive Integrations and Comparison
Sag/Swell	NO, False detections	Yes	Yes
Half-cycle off	NO, False detections	Yes	NO, slow for half-cycle
Pulse 100 Hz/10%	count any sudden pulses	Yes	Yes
Pulse 1 kHz/10%	Yes	Yes	Yes
Pulse 10 kHz/10%	Yes	Yes	Yes
Oscillation 200 Hz/10%	No sure detection (low freq.)	Yes	Yes
Oscillation 1 kHz/10%	Yes	Yes	Yes
Oscillation 10 kHz/10%	Yes, best if F > F sampling	Yes, work F > F sampling	Yes, work to F > F sampling

The integration (LPF) and comparison method requires several previous integrations cycles to a precise detection. For this reason, it works incorrectly in half-cycle disruption failures. For this case, it is easy to implement an additional algorithm based on peak value measurement.

Similarly, the Helmholtz-based method does not detect voltage variations cases. Any additional algorithm can be used to mathematically analyze the peak voltage variations.

5. Conclusions

This work presents a low-cost sensor-node supported on a microcontroller platform. It is aimed to be integrated into a cooperative wireless sensor network. In order to simplify, an attenuator-rectifier adapter was proposed to connect the analog-to-digital converter to the power line. The rectifier process improves the dynamic range of the ADC, but introduces a low-intensity spectral content in 200 Hz and with lower intensity, in 400 Hz.

Several simple processing algorithms were also analyzed, implemented and tested to assess the transient detection capability and specific oscillation failures (PQ). In order to accelerate the computational work, it was essential to execute integer mathematical algorithms. The search and selection of implementable algorithms and real-time achievement is an additional difficulty. Therefore, these small integrated systems could act correctly as very low-cost and low-power instruments to become sensor nodes. Frequency, RMS voltage and transient voltage measurements are feasible. The ADC (10 bits) resolution could comply with some standards as a lot of IEC-61557-12 specific paragraphs.

As the hall sensor was not included, it is not achievable at current or power measurements. However, the feature of simplicity makes them ideal parts of a PQ wireless sensor network.

In addition, a physical prototype based on an ATmega-328 microcontroller was built to verify the ideas and machine codes previously considered. It can also affirm that the use of embedded resources in simple microcontroller devices (ADC, interrupt vectors, counters, etc.) improves real-time execution performance. This prototype allows us to verify numerous detector algorithms and if necessary, we could replace the microcontroller architecture (e.g., replace ATmega by an ARM processor).

A software application (C++, LabWindows^{TM}) was developed and employed as an algorithms test-bench. Thus, it was possible to test the different detection methods described in this paper and to check the efficiency of each algorithm.

With this help, and as evidenced in Table 2, it can be considered that:

- In order to achieve a high penetration of sensor nodes, low-cost platforms are essential.
- For this reason, a special design is required, choosing low-energy and inexpensive hardware ad-hoc platforms and implementing specific algorithms to maximize the device performance.
- The 2nd-derivative Helmholtz method is quite sensitive to high-frequency interferences (brief oscillations) or pulses, while it is very insensitive to slow changes (harmonic distortion, sag or swell voltage).
- The Kalman filter option is the most effective and is able to show all kinds of anomalies that have been tested. The disadvantage is that it demands more memory and requires much more code implementation complexity.
- The 'integration (LPF) and comparison' method detects correctly for slow voltage variation or high-frequency disturbances (swell, sag, oscillations or short pulses, etc.). However, it is not capable of detecting short voltage variations with a shorter duration of a half cycle due to the integration time.

Future works may be conducted to complete the sensor network, increasing the number of nodes, to exploit the individual measurements acquired. It would also be interesting to design a breaker-node (remotely controlled) that disconnects a local section when an excessive phenomenon occurs and is detected by a remote sensor-node. In this way, this node could disconnect the power when another sensor node detects an anomaly (via radio 433 MHz).

Moreover, specific pattern recognition techniques may be considered for increasing nodes measurement capability and prediction. Thus, the node could allow predicting some anomalies.

Author Contributions: Formal analysis, Investigation, Software, Writing—original draft, J.-M.G.-R.; Investigation, Supervision, Methodology, C.C.-S.; Investigation, Methodology, Supervision, Writing—review and editing, J.-J.G.-d.-l.-R.; Investigation, Resources, D.S.-L.

Funding: This research was funded by the University of Cádiz and by the Spanish Ministry of Economy and Competitiveness through the project TEC2016-77632-C3-3-R.

Acknowledgments: This research is supported by the Spanish Ministry of Economy, Industry and Competitiveness and the EU (AEI/FEDER/UE) in the workflow of the State Plan of Excellency and Challenges for Research, via the project TEC2016-77632-C3-3-R-Control and Management of Isolable NanoGrids: Smart Instruments for Solar forecasting and Energy Monitoring (COMING-SISEM), which involves the development of new measurement techniques applied to monitoring the PQ in micro-grids.

Conflicts of Interest: The authors declare no conflict of interest.

References

1. IEEE Std. *1159-2009, IEEE Recommended Practice for Monitoring Electric Power Quality*; IEEE: Piscataway, NJ, USA, 2009.

2. IEC 61000-4-30. *Electromagnetic Compatibility (EMC), Part 4: Testing and Measurement Techniques. Section 30: Power Quality Measurement Methods*; International Electrotechnical Commission, IEC Central Office: Geneva, Switzerland, 2008.

3.   Neumann, R. The importance of IEC6100-4-30 Class A for the Coordination of Power Quality Levels. In Proceedings of the 9th International Conference Electrical Power Quality and Utilization, Barcelona, Spain, 9–11 October 2007. [CrossRef]

4.   IEC 61557-12. *Electrical Safety in Low Voltage Distribution Systems up to 1000 V a.c. and 1500 V d.c. - Equipment for Testing. Measuring or Monitoring of Protective Measures - Part 12: Performance Measuring and Monitoring Devices (PMD)*, 2nd ed.; International Electrotechnical Commission, IEC Webstore: Geneve, Switzerland, 2018; Available online: https://webstore.iec.ch/publication/64047 (accessed on 22 April 2019).

5.   González de la Rosa, J.J.; Sierra-Fernández, J.M.; Palomares Salas, J.C.; Aguera Pérez, A.; Jiménez Montero, A. An application of spectral kurtosis to separate hybrid power quality events. *Energies* **2015**. [CrossRef]

6.   González de la Rosa, J.J.; Aguera Pérez, A.; Palomares Salas, J.C.; Florencia-Oliveiros, O.; Sierra-Fernández, J.M. A dual Monitoring Technique to detect Power Quality Transients Based on the Fourth-Order Spectrogram. *Energies* **2018**, *11*, 503. [CrossRef]

7.   Palomares Salas, J.C.; González de la Rosa, J.J.; Sierra-Fernández, J.M.; Aguera Pérez, A. HOS Network-based classification of power quality events via regression algorithms. *EURASIP J. Adv. Signal Process.* **2015**. [CrossRef]

8.   Valtierra-Rodriguez, M.; De Romero-Troncoso, J.; Osornio-Rios, R.A.; Garcia-Perez, A. Detection and classification of single and combined power quality disturbances using neural networks. *IEEE Trans. Ind. Electron.* **2014**, *61*. [CrossRef]

9.   Borges, F.A.S.; Fernandes, R.A.S.; Silva, I.N.; Silva, C.B.S. Feature Extraction and Power Quality Disturbances Classification Using Smart Meters Signals. *IEEE Trans. Ind. Inform.* **2016**, *12*, 824–833. [CrossRef]

10.   Zygarlicki, J.; Zygarlicka, M.; Mroczka, J.; Latawiec, K.J. A Reduced Prony's Method in Power-Quality Analysis Parameters Selection. *IEEE Trans. Power Deliv.* **2010**, *25*. [CrossRef]

11.   Liu, Z.; Cui, Y.; Li, W. Combined Power Quality Disturbances Recognition Using Wavelet Packet Entropies and S-Transform. *Entropy* **2015**. [CrossRef]

12.   He, S.; Li, K.; Zhang, M. A Real-Time Power Quality Disturbances Classification Using Hybrid Methods Based on S-Transform and Dynamics. *IEEE Trans. Instrum. Meas.* **2013**, *62*. [CrossRef]

13.   Dash, P.K.; Chilukuri, M.V. Hybrid S-Transform and Kalman Filtering Approach for Detection and Measurement of Short Duration Disturbances in Power Networks. *IEEE Trans. Instrum. Meas.* **2004**, *53*. [CrossRef]

14.   Granados, D.; Romero, R.J.; Osornio, R.A.; Garcia, A. Techniques and methodologies for power quality analysis and disturbances classification in power systems: A review. *IET Gener. Transm. Distrib.* **2011**, *5*. [CrossRef]

15.   Barros, J.; Diego, R.I. Review of signal processing techniques for detection of transient disturbances in voltage supply systems. In Proceedings of the IEEE Instrumentation and Measurement Technology Conference, Minneapolis, MN, USA, 6–9 May 2013. [CrossRef]

16.   Artioli, M.; Pasini, G.; Peretto, L.; Sasdelli, R. Low-cost DSP-based equipment for the real-time detection of transients in power systems. *IEEE Trans. Instrum. Meas.* **2004**, *53*. [CrossRef]

17.   Gallo, D.; Landi, C.; Luiso, M.; Bucci, G.; Fiorucci, E. Low Cost Smart Power Metering. In Proceedings of the 2013 IEEE International Instrumentation and Measurement Technology Conference (I2MTC), Minneapolis, MN, USA, 6–9 May 2013; pp. 763–776. [CrossRef]

18.   Granados, D.; Valtierra, M.; Morales, L.A.; Romero, R.J.; Osornio, R.A. A Hilbert Transform-Based Smart Sensor for detection, Classification and Quantification of Power Quality Disturbances. *Sensors* **2013**. [CrossRef]

19.   Yingkayun, K.; Premrudeepreechacharn, S.; Watson, N.R.; Higuchi, K. Power Quality monitoring system based on embedded system with network monitoring. *Sci. Res. Essays* **2012**, *7*. [CrossRef]

20.   Quiros-Olozabal, A.; Gonzalez-de-la-Rosa, J.J.; Cifredo-Chacon, M.A.; Sierra-Fernandez, J.M. A novel FPGA-based system for real-time calculation of the Spectral Kurtosis: A prospective application to harmonic detection. *Measurement* **2016**, *86*. [CrossRef]

21.   Asha Kiranmai, S.; JayaLaxmi, A. Hardware for classification of power quality problems in three phase system using Microcontroller. *Cogent Eng.* **2017**, *4*. [CrossRef]

22.   Alam, K.; Chakraborty, T.; Pramanik, S.; Sardda, D.; Mal, S. Measurement of Power Frequency with Higher Accuracy Using PIC Microcontroller. *Procedia Technol.* **2013**, *10*. [CrossRef]

23. Malekpour, M.; Pouramin, A.; Malekpour, A.; Phung, T.; Ambikairajah, E. Monitoring and measurement of high-frequency oscillatory transient recovery voltage of circuit breakers. *IET Sci. Meas. Technol.* **2018**, *12*. [CrossRef]

24. Gajjar, S.; Choksi, N.; Sarkar, M.; Dasgupta, K. Comparative analysis of Wireless Sensor Network Motes. In Proceedings of the 2014 International Conference on Signal Processing and Integrated Networks (SPIN), Noida, India, 20–21 February 2014; ISBN 978-1-4799-2865-1. [CrossRef]

25. Maurya, M.; Shukla, S.R.N. Current Wireless Sensor Nodes (MOTES): Performance metrics and Constraints. *Int. J. Adv. Res. Electron. Commun. Eng.* **2013**, *2*. Available online: http://ijarece.org/wp-content/uploads/2013/08/IJARECE-VOL-2-ISSUE-1-45-48.pdf) (accessed on 19 February 2019).

26. Zhong, D.; Li, H.; Han, J.; Wei, Q. A Practical Combining Wireless Sensor Networks and Internet of Things: Safety Management System for Tower Crane Groups. *Sensors* **2014**. [CrossRef]

27. Alonso-Rosa, M.; Gil-de-Castro, A.; Medina-Gracia, R.; Moreno-Munoz, A.; Cañete-Carmona, E. Novel Internet of Things Platform for In-Building Power Quality Submetering. *Appl. Sci.* **2018**, *8*, 1320. [CrossRef]

28. Viciana, E.; Alcayde, A.; Montoya, F.G.; Baños, R.; Arrabal-Campos, F.M.; Manzano-Agugliaro, F. An Open Hardware Design for Internet of Things Power Quality and Energy Saving Solutions. *Sensors* **2019**, *19*, 627. [CrossRef] [PubMed]

29. Microchip—Atmel. Internet of Things Page Application. Available online: https://www.microchip.com/design-centers/internet-of-things/google-cloud-iot/avr-iot (accessed on 22 April 2019).

30. Kalman, R.E. A new approach to linear filtering and prediction problems. *Trans. Am. Soc. Mech. Eng. J. Basic Eng.* **1960**, *82*, 35–45. [CrossRef]

MDPI

St. Alban-Anlage 66

4052 Basel

Switzerland

Tel. +41 61 683 77 34

Fax +41 61 302 89 18

www.mdpi.com

Energies Editorial Office

E-mail: energies@mdpi.com

www.mdpi.com/journal/energies